The Future of the Teaching and Learning of Algebra
The 12th ICMI Study

New ICMI Study Series

VOLUME 8

Published under the auspices of the International Commission on Mathematical Instruction under the general editorship of

Hyman Bass, President   Bernard R. Hodgson, Secretary-General

The titles published in this series are listed at the end of this volume.

The Future of the Teaching and Learning of Algebra
The 12th ICMI Study

edited by

Kaye Stacey

Helen Chick

Margaret Kendal

The University of Melbourne, Australia

KLUWER ACADEMIC PUBLISHERS
Boston / Dordrecht / New York / London

Distributors for North, Central and South America:
Kluwer Academic Publishers
101 Philip Drive
Assinippi Park
Norwell, Massachusetts 02061 USA
Telephone (781) 871-6600
Fax (781) 681-9045
E-Mail: kluwer@wkap.com

Distributors for all other countries:
Kluwer Academic Publishers Group
Post Office Box 322
3300 AH Dordrecht, THE NETHERLANDS
Telephone 31 786 576 000
Fax 31 786 576 254
E-Mail: services@wkap.nl

Electronic Services <http://www.wkap.nl>

Library of Congress Cataloging-in-Publication Data

Title: The Future of the Teaching and Learning of Algebra: The 12th ICMI Study
Editors: Kaye Stacey, Helen Chick, Margaret Kendal
ISBN HB: 1-4020-8130-8
ISBN E-book: 1-4020-8131-6

A C.I.P. Catalogue record for this book is available
from the Library of Congress.

Printed on acid-free paper.

Printed in the United States of America.

Contents

Preface

Kaye Stacey, Helen Chick, and Margaret Kendal
The University of Melbourne, Australia

Abstract: This section reports on the organisation, procedures, and publications of the 12th ICMI Study, *The Future of the Teaching and Learning of Algebra.*

Key words: Study Conference, organisation, procedures, publications

The International Commission on Mathematical Instruction (ICMI) has, since the 1980s, conducted a series of studies into topics of particular significance to the theory and practice of contemporary mathematics education. Each ICMI Study involves an international seminar, the "Study Conference", and culminates in a published volume intended to promote and assist discussion and action at the international, national, regional, and institutional levels.

The 12th ICMI Study running from 2000 to 2004 was on *The Future of the Teaching and Learning of Algebra*, and its Study Conference was held at The University of Melbourne, Australia from December 9th to 14th, 2001. It was the first study held in the Southern Hemisphere. There are several reasons why the future of the teaching and learning of algebra was a timely focus at the beginning of the twenty first century. The strong research base developed over recent decades enabled us to take stock of what has been achieved and also to look forward to what should be done and what might be achieved in the future. In addition, trends evident over recent years have intensified. Those particularly affecting school mathematics are the "massification" of education—continuing in some countries whilst beginning in others—and the advance of technology. Algebra is centrally affected by both of these trends. For mass education, algebra teaching highlights questions of equity and relevance. For progression to higher mathematics, students need algebra but its abstraction makes it hard to learn and hard for beginners to see a reason for learning. Simultaneously, advancing technology provides rich prospects for improving teaching. However, it also provides a challenge to the existing curriculum because so many of the routines that have been the standard diet (and where students have been most successful) are now available "at the press of a button". The result is that an algebra curriculum that serves its students well in the coming century may well look very different from an ideal curriculum from some years ago. These ideas are further developed in the 2001 Discussion Document, reprinted in the conference proceedings (Chick, Stacey, Vincent, & Vincent, 2001).

Figure 1. Conference Organisers and Conference Book Editors from the University of Melbourne (L to R): Jill Vincent, John Vincent, Helen Chick, Kaye Stacey.

The International Program Committee (IPC) first met in January 2000 to draft the Discussion Document that was then disseminated throughout the international mathematics education community. In response to the call for contributions, over 150 papers were submitted from nearly 200 authors. This strong response indicated the wide concern with the teaching of algebra, the strong research base upon which we can go forward, and an optimism that attention to this issue can result in real gains for students around the world. The members of the IPC reviewed the submitted papers with assistance from additional readers. Taking into account paper submissions, areas of interest and expertise, the goals and objectives of the Study Conference, and geographic representation, potential participants were invited to the Study Conference. Finally, there were 110 participants from North and South America, Eastern and Western Europe, China, Indonesia, Israel, Japan, Malaysia, New Zealand, Singapore, and Tonga, as well as a strong contingent of Australians. The size of the Study Conference was kept small, consistent with ICMI aims, to maximise the opportunity for interaction between participants. Overall, only two-thirds of the submitted papers were accepted and in some instances, not even all the co-authors of accepted papers could be offered places. The papers were accepted on the basis they would stimulate discussion at the Conference and did not, in most cases, undergo revision prior to their publication in the Proceedings. Almost all participants gave individual presentations of their papers in parallel sessions. Since the Study Conference, many of these papers have been developed further appearing in journals, and where possible references to the updated papers are used in the relevant chapters of this book. The Conference Proceedings were published by the Department of Science and Mathematics Education, University of Melbourne in three volumes under the title *The Future of the Teaching and Learning of Algebra*

(Proceedings of the 12th ICMI Study Conference) and edited by Helen Chick, Kaye Stacey, Jill Vincent, and John Vincent. For most of the papers, authors were able to give permission for their paper to be openly distributed. These are in Volumes 1 and 2. Other papers were reprinted only for use at the Study Conference and were published in Volume 3.

Plenary Sessions

The Study Conference Program began with a welcome to Melbourne by Jeff Tye, a young indigenous Australian, who also gave a stunning performance on the didgeridoo. The academic program began with Carolyn Kieran's Plenary Address, *The core of algebra: Reflections on its main activities*, which is published in this book (Chapter 2). In addition, there were two Plenary Panels. The first, *Algebra around the world*, was organised by Romulo Lins to provide participants with illustrations of the international differences in curriculum and teaching for algebra. The speakers, Rosamund Sutherland (UK), Toshiakira Fujii (Japan), and Jarmila Novotná (Czech Republic) provided vivid descriptions of how algebra is taught in their own countries. Helen Chick (Australia) chaired the panel. This theme is expanded in Chapter 13 of this book.

Kaye Stacey organised the second Plenary Panel that was chaired by Desmond Fearnley-Sander (Australia). It addressed the theme *Why algebra, what algebra?* with speakers Luciana Bazzini (Italy), Gard Brekke (Norway), M. Kathleen Heid (USA), Jack Abramsky (UK), and Kaye Stacey (Australia). This theme was of fundamental importance to the goals of the whole conference. This session further exemplified some new practices in algebra, described in most chapters of this book, whilst identifying key issues for applying research to inform practice and curriculum priorities.

On the final day of the Conference, each Working Group presented a summary of their findings, which have formed the basis of many chapters of this book. Dr Sri Wayuni from Indonesia then spoke about what she had gained from the conference and expressed the hope that increasing links between her country and international researchers might assist in the teaching of mathematics in South East Asia and the Pacific. To the applause of all, she expressed her thanks in song. Finally, Helen Chick and Hugh Burkhardt, representing new and experienced researchers, provided personal reactions to the 12th Study Conference as a whole. A further developed version of Hugh's presentation with lessons beyond research in mathematics education has since been published in the Educational Researcher (Burkhardt, H. & Schoenfeld, A. H. (2003). Improving Educational Research: Toward a more useful, more influential and better funded enterprise. *Educational Researcher*, *32*(9), 3-14.)

During the conference dinner, we listened spellbound to the algebraic wisdom of Professor Alphonso Fratelini and Albie, his doctoral student and younger brother

(BSc, MSc, MSG, & MCG). In addition, the Fratelini brothers inspired us to sing, with great collegial spirit, in a range of languages.

Figure 2. The Fratelini brothers entertained the participants at the conference dinner: Front row (L): Albie Fratelini. Back row (R): Professor Alphonso Fratelini.

Figure 3. The Italian connection (L to R): Professor Alphonso Fratelini, Ferdinando Arzarello, Luciana Bazzini (Plenary Panel presenter), Albie Fratelini.

Working Groups

The main work of the Study Conference was achieved in Working Groups, whose brief titles are listed below. These Working Groups examined the central theme of the future of the teaching and learning of algebra from nine different viewpoints, as had been foreshadowed in the initial discussion document:

- Early Algebra
- Approaches to Algebra
- Goals of Algebra
- Technological Environments
- Computer Algebra Systems
- History of Algebra
- Symbols and Language
- Teachers' Knowledge and Practice
- Tertiary Algebra

As mentioned above, the results of these discussions form the basis of most of the chapters of this book. These chapters are preceded by two-page descriptions of the contributions of the members of the associated Working Group. The deliberations of each Working Group kept firmly in mind the goal of providing practical and imaginative advice on future directions that was soundly based in scientific research. One or more group members undertook the authorship of the chapter, aiming to consolidate as much as possible the collective viewpoint of their Working Group.

Our Thanks

Finally, we wish to offer our thanks to the many people who assisted us to make the 12th ICMI Study Conference, and this book, a success. Members of the International Program Committee, listed at the end of the Preface, have been involved in every aspect of this Study, offering sound advice based on wide experience and perspectives from around the world and assisting with refereeing submitted papers. The Conference was held under the auspices of the Australian Sub-committee of ICMI, chaired by Desmond Fearnley-Sander. Jill and John Vincent undertook the demanding role of Conference Secretariat, ably assisted by a group of volunteers from the Department of Science and Mathematics Education, University of Melbourne. Mollie MacGregor is specially thanked for proof-reading this book and assisting with refereeing. We gratefully acknowledge the sponsorship of the Australian Government's Department of Education, Training and Youth Affairs (Benchmarking, Assessment, and Numeracy Policy Sector); Shriro, the Australian distributors of Casio calculators; and Texas Instruments (Australia). Support from the Rotary Club of Melbourne enabled participation of delegates from Indonesia and Tonga, two countries in the Asia-Pacific region that have not previously been represented in international meetings on mathematics education. The leaders of the Working Groups are thanked for their effective organisation, and the authors of the chapters deserve special recognition for their conscientious efforts to showcase the ideas of the members of their Working Group.

Professor Michèle Artigue (France), a Vice President of ICMI, attended the Study Conference, and as a *free bird* went from group to group, contributing to discussion, monitoring the progress of the groups, and providing advice to the working group leaders and conference organisers. Professor Bernard Hodgson, Secretary of ICMI, provided strong support and valuable advice to the Study at all stages.

Conference proceedings

Helen Chick, Kaye Stacey, Jill Vincent, & John Vincent (Eds.). (2001). *The Future of the Teaching and Learning of Algebra* (Proceedings of the 12th ICMI Study Conference). University of Melbourne, Australia: Department of Science and Mathematics Education.

Volumes 1 and 2 are available from The Department Manager, Department of Science and Mathematics Education, University of Melbourne, Victoria 3010, Australia (or contact Kaye Stacey: k.stacey@unimelb.edu.au) These volumes contain the discussion document, papers in alphabetical order of the first author, and contact details of participants. Volume 3 contained papers that were available only to conference participants.

Kaye Stacey (Study Volume Editor and Program Chair)
Helen Chick (Study Volume Editor and Conference Secretary)
Margaret Kendal (Study Volume Editor and Study Participant)
February 2004

Figure 4. Conference participants (L to R): Michèle Artigue (Vice-President ICMI & Working Group "free bird"), Hugh Burkhardt (Final Presentation), Bernard Hodgson (Secretary ICMI & "free bird"), Kathleen Heid (Plenary Panel), Margaret Kendal (Editor).

Figure 5. International Program Committee for the 12th ICMI Study (L to R): Kaye Stacey, Teresa Rojano, Dave Carlson, Luis Puig, Bernard Hodgson, Toshiakira Fujii, Rom Lins, Carolyn Kieran, Barry Kissane, Desmond Fearnley-Sander, Jean-Philippe Drouhard.

Figure 6. Conference participants enjoying dinner. Front (L to R): Sylvia Johnson, Elizabeth Belfort, Sheryl Stump, David Driver. Back (L to R): Helen Doerr, Mary Enderson, Ivan Cnop.

Figure 7. The French connection (L to R): Elisabeth Delozanne, Jean-baptiste Lagrange, Brigitte Grugeon, Michèle Artigue, Bernard Hodgson.

About ICMI

http://www.mathunion.org/ICMI/

The International Commission on Mathematical Instruction, ICMI, is a commission of the International Mathematical Union (IMU), a non-governmental and non-profit-making scientific organisation to promote international cooperation. The official organ of ICMI is the journal *L'Enseignement Mathématique* and the ICMI Bulletin is accessible on the internet. ICMI was established at the International Congress of Mathematicians held in Rome in 1908. The first President was Felix Klein. The reports of earlier ICMI Studies are listed at http://www.wkap.nl/series.htm/NISS.

The 12th ICMI Study
International Program Committee

Program Chair: Kaye Stacey (Australia)

Dave Carlson (USA)
Jean-Philippe Drouhard (France)
Desmond Fearnley-Sander (Australia)
Toshiakira Fujii (Japan)
Carolyn Kieran (Canada)
Barry Kissane (Australia)
Romulo Lins (Brazil)
Luis Puig (Spain)
Teresa Rojano (Mexico)
Rosamund Sutherland (UK)
Bernard Hodgson (ex-officio, ICMI)
Conference Secretary: Helen Chick (Australia)

Australian Sub-Committee of ICMI (as at January 2000)

Chair: Desmond Fearnley-Sander, University of Tasmania
Michael Bulmer, University of Queensland
Peter Jones, Swinburne University of Technology
Barry Kissane, Murdoch University
Gilah Leder, La Trobe University
Bob Perry, University of Western Sydney
Derek Robinson, Australian National University
Kaye Stacey, The University of Melbourne
Steve Thornton, Australian Mathematics Trust, University of Canberra

Chapter 1

Solving the Problem with Algebra

Kaye Stacey and Helen Chick
The University of Melbourne, Australia

Abstract: This chapter draws together the major themes emerging from the 12th ICMI Study on *The Future of the Teaching and Learning of Algebra* and serves as an introduction to this book. The chapter begins with a short description of the major challenges that the teaching of algebra presents to researchers, curriculum writers and teachers. There follows a brief introduction to each chapter which surveys some key ideas presented, after which the significant suggestions for future algebra teaching and learning from that chapter are highlighted. The chapter finishes by drawing together the major themes that offer guidelines for making the future of the teaching and learning of algebra brighter than the past.

Key words: Algebra teaching, algebra learning, algebraic activity, early algebra, technological environments, computer algebra systems, history of algebra, teachers' knowledge, tertiary algebra, goals for algebra, algebra curricula

1.1 What is the Problem with Algebra?

The teaching of algebra has undergone a great deal of change in recent decades. The 12th ICMI study was an opportunity to reflect on the changes and to consider directions for the future. This chapter gives an overview of the directions for the future that are presented later in this book, drawing them together to present a composite view. We start by describing some of the features that have made algebra a problematic area of teaching mathematics; in other words to describe in general terms the problem that faced the 12th ICMI Study.

Several factors have stimulated changes in teaching algebra in many countries around the world. One of the most important factors is the growth of universal secondary education, occurring in developed countries during the twentieth century and continuing now around the world. Teaching algebra to a broad spectrum of the population raises issues of relevance and of equity. If algebra is interpreted just as symbolic manipulation, then it has little relevance to everyday life, in developed or undeveloped countries. Indeed, it can be a source of alienation of students from learning mathematics. The challenge, therefore, has been to reconceptualise algebra as a subject that does have relevance to students and to do this in a way that the students themselves can perceive the relevance.

Closely associated with relevance is the need to make the objects and processes of algebra meaningful to students and for teachers to have a clearer idea of what algebra might be beyond symbolic manipulation. This has been the stimulus for many experimental teaching approaches and curriculum ideas, and several chapters of this book report on what has been found and how these changes can be developed further and improved upon.

Algebra is often described as a gateway to higher mathematics, not least because it provides the language in which mathematics is taught. Consequently, it is important that all students be given a genuine opportunity to learn algebra. Without this, they are cut off from many occupations, either because algebra is really used there or because it is specified as a preliminary qualification. Equity in learning algebra is, however, hard to achieve in large school systems. It requires imaginative and well-founded teaching approaches that tap into the diverse strengths of students' intelligences. It also requires school structures that maximise the outcomes for learners of different abilities and interests.

Designing instruction to maximise learning opportunities also requires an in-depth understanding of the cognitive difficulties of learning algebra. These difficulties have many sources, including the level of abstraction, the surprisingly problematic and multi-faceted relationship with prior arithmetic learning that is touched upon in many chapters in this book, the demands of becoming fluent in symbolic manipulation, and the need to make transitions from procedural to structural thinking. In these areas, research in recent decades has made considerable progress, and this provides a substantial resource for improving the teaching of algebra. The problem is to make sure that the insights from this research are well known and used to good advantage, including and especially by teachers.

Algebra teaching is also being changed by advancing technology, which is providing new mathematical problem solving methods, so that skills that were rightly valued in previous decades may not be valuable now. Technologies such as spreadsheets have provided more accessible, numerically-based methods for workplace problem solving, and at a more advanced level, computer algebra systems can carry out complex symbolic manipulations quickly and reliably. Algebra curricula have to respond to these changes. At the same time, advances in technology also offer great potential for improving the learning of an unchanged curriculum, and so another challenge for algebra educators is to make sure that these new opportunities are harnessed to improve outcomes for students.

The paragraphs above have briefly summarised the major dimensions of the challenge to improve the teaching and learning of algebra. The next section introduces the various chapters of the book that give components of the solution to this problem.

1.2 Insights and Clues to a Solution

This book begins with a discussion of the main components of algebraic activity, and then looks at beginning algebra, firstly examining the possibility of starting early and then considering the options for students' first experiences, whether in elementary or secondary school. Later chapters examine the teaching of tertiary algebra and teacher education. Two chapters consider the impact of new technology. Special interests in research on algebra are represented in the chapters on the history of algebra and the nature of algebraic language. One chapter casts a critical eye over the claims of the benefits of algebra and reports on suitable goals for "algebra for all". The last chapter provides a necessarily selective guide to some of the ways in which algebra education varies around the world.

Nine chapters of this book, Chapters 4 to 12, arise from the deliberations of the Working Groups at the Study Conference, addressing issues identified in the Discussion Document (Program Committee, 2000a & 2000b). The authors of these chapters have not simply reported on the discussions that were held, but instead present synthesised accounts of the outcomes of the discussions and the resources provided in the conference papers of group members (Chick, Stacey, Vincent, & Vincent, 2001) and specially prepared research briefs. As such, these chapters are necessarily personal accounts, albeit strongly informed by the collective knowledge and experiences of members of the Working Groups. Drawing together the insights of any group with an eye to the future is a challenging task.

The sections below briefly introduce a few of the main ideas in each chapter and then highlight some of the directions for the future of the teaching and learning of algebra that arise from the chapter. In this way, a picture of the future of the teaching and learning of algebra is composed from the many separate viewpoints.

1.2.1 The core of algebra

Chapter 2, following this introductory chapter, presents Carolyn Kieran's Plenary Address, *The core of algebra: Reflections on its main activities,* which offers a helpful framework for organising school-level algebraic activity. The core activities of algebra are seen as generational activities (where situations, properties, patterns, and relationships are represented algebraically or interpreted), transformational activities (algebraic manipulation), and global/meta-level activities which are not algebra-specific but relate to the purpose and context for using algebra (e.g., to prove, or solve problems, or to notice structure etc.). This analysis posits a view of algebra as being much more than symbolic manipulation. In Chapter 3, Laurinda Brown offers, in reaction, vivid reports of two classroom examples that highlight the interaction of these core activities and the central role of the global/meta-level activity. Then Jean-Philippe Drouhard presents a thoughtful reaction to the Plenary Lecture, which examines the role of algebraic signs and symbols in the core activities.

The need for a framework for organising and defining algebraic activity becomes clear as one reads into the book and sees the great variety of mathematical work that is offered around the world under this one label of algebra. A working definition of what counts as algebra is also important as educators explore below the surface image of algebra as rule-based symbol manipulation, to find a livelier, more engaging, and more teachable subject for schools in the age of mass education. Many consequences of this work are evident throughout the book.

Kieran uses the framework to present an overview of recent history of teaching algebra. She characterises school algebra until the mid-1960s as focusing primarily on transformational work (i.e., on symbol manipulation). An important strength of algebra is that problems can be solved by following transformational rules without constant reference to the meaning of expressions. This is a very deep observation. However, at that stage of algebra education history, there was scant attention to meaning making. In response to dissatisfaction with outcomes and in a quest to make algebra meaningful to students, researchers and teachers experimented with new approaches to generational activity, often supported by the new technological tools. By the early 1990s in some countries, generational activity (e.g., describing numerical patterns algebraically) had all but replaced transformational activity. The hope was that the technical work would take care of itself, if students had strong algebraic understanding, but Kieran observes that this did not happen.

Kieran therefore sees the challenge for the future in giving meaning not only to the objects of algebra, but also to its manipulative processes. She sees that technology will be an invaluable aid to do this, and expands on this in Chapter 6. Equally important is taking up the more sophisticated view of conceptual understanding and manipulative skills that is being developed by a large team in France headed by Michèle Artigue (see, for example, Trouche, 2000). This view emphasises that the techniques of algebra are not only essential for solving problems (their pragmatic role) but also make an important contribution to the understanding of the algebraic objects involved. In other words, instead of thinking of concepts and skills as separate (i.e., generational and transformational activity as separate), a more productive view is to see that techniques involve both concepts and skills together. Learning about either part should involve learning about the other.

The surprising thing is that this deeper understanding of the role of technique has come about through investigations of the use of computer algebra systems in school mathematics. A naïve view may be that computer algebra systems would take away the need for transformational activity, so that teaching needs concentrate only on conceptual work. However, research and experience has shown that instead, the manipulative process is also a conceptual object. Kieran regards this new insight as a key to improving the future of teaching and learning algebra.

1.2.2 The early development of algebraic reasoning

In the agenda-setting Chapter 4, *The early development of algebraic reasoning*, Romulo Lins and James Kaput argue that it is possible to introduce very young students to the *culture of algebra* by encouraging them to think with a particular kind of generality early in their elementary school mathematics education. It rejects the traditional "arithmetic before algebra" curriculum sequence which has been boosted by theories that have led to algebra being regarded as cognitively more complex than arithmetic. Instead, the intention is to create an algebra that is accessible in parallel with arithmetic. This is the agenda of the "early algebra" movement.

The chapter does not specify exactly at what ages or stages algebra is to be introduced. Some of the proponents of early algebra believe that children can profitably start in the very early elementary years, whereas for others the intention is only to start earlier than is currently the case. Whenever it is, there is agreement that early algebra is *not* meant to be the same school algebra taught to younger children. Instead, Lins and Kaput envisage a new approach that focuses on teaching children about algebraic thinking, primarily but not exclusively in the context of number work. In this context, algebraic thinking has two characteristics: it involves deliberate generalisation with tools to express the generalities and it involves reasoning based on the forms of these generalisations (i.e., guided by syntax, not only by meaning). These two characteristics relate to aspects of Kieran's generational and transformational activities. The early Soviet work by Davydov (1962) is an inspiration for this approach.

Lins and Kaput begin the chapter with an overview of research on algebra and on students' understandings that they characterise as "sad stories". These studies focus on the difficulties that secondary students have and the obstacles that they face in learning algebra, either theoretically through theories of learning or of cognitive development or of mathematics, or empirically by reporting achievement levels of secondary school students learning algebra. They contrast these studies with "happy stories" which support the view that children (and especially young children) can do more than has previously been expected of them in school if given the opportunity.

Lins and Kaput explain that the possibility of early algebra results from the broadening of understandings of what algebra is—a theme of many chapters of this book. Early algebra is also seen as a possibility because old assumptions about what children can do have been reconsidered and because new technologies provide more opportunities for learning. However, the main evidence that early algebra may be a successful approach comes from a growing set of small experiments that have explored children's reactions to potential new curriculum offerings. These research studies are beginning to identify some of the major features of early algebra. In particular, they are beginning to make concrete what it means to introduce students to a culture of algebra, where students are familiar with the tools for working with generalities, but where manipulative skills do not dominate. The research questions at the end of the chapter describe how such work can be extended usefully.

Adopting early algebra as an approach is a major step for an educational system. It is not yet clear exactly what form this early algebra would take, and it is likely to take rather different forms in different countries. However, it is clear that there are substantial implications for teacher education—in Blanton and Kaput's (2001) words "building teachers' algebra eyes and ears"—as well as implications for the teaching of "normal algebra" which would need to adapt to capitalise upon students having an early start. Translating the "happy stories" from research into "happy stories for all" is the way forward, but it is a big undertaking.

1.2.3 A toolkit for analysing approaches to algebra

Chapter 5, *A toolkit for analysing approaches to algebra*, was coordinated by Rosamund Sutherland, from the deliberations of a large and diverse group of participants who were concerned with finding excellent ways of conveying the key ideas of algebra to beginning students (usually of secondary school age). This is an area where there has been a great deal of curriculum development and experimentation in recent decades, as countries around the world have struggled to make students' initial experiences with algebra meaningful and productive. It is also an area where there has been considerable research, not reviewed in the chapter but forming a background to the discussions. The Working Group began by exploring the options for beginning algebra using the scheme of Bednarz, Kieran, and Lee (1996) who categorised approaches to algebra as being from generalisation, problem solving, modelling, and functions. The key idea behind the categorisation is that algebra, even in the beginning stages, is a multi-faceted activity and different approaches expose the fundamental ideas to different extents and from different viewpoints. As a consequence, teachers and curriculum writers need to make choices.

The chapter engagingly describes how the Working Group came to understand that the Bednarz, Kieran, and Lee approach emphasises the problem domains that have been selected, yet other dimensions of the teaching and learning situation need to be considered in order to explore effective teaching approaches in depth. The chapter therefore reports on a four-dimensional toolkit for researchers and teachers to discuss various approaches to the teaching of introductory algebra. As well as the problem domain, there may be variations in the teaching approach. Even teachers using the same problems might nevertheless stress and ignore different underlying ideas. The theoretical perspectives also need to be made explicit. This dimension includes formal and informal theories of learning (such as how students can best be engaged, whether students should infer generalisations from specific cases) as well as perspectives on what is of greatest educational importance (e.g., links to the real world of the students or the development of ideas of structure and proof). The fourth dimension to consider is the community of students, noting that there are wide differences to be considered and that even with similar students, there are different ideas about how a teacher should work with students' perspectives.

Three different teaching approaches are then presented to illustrate these four dimensions. Monica Wijers and Martin van Reeuwijk report on the approach that is part of the Dutch Realistic Mathematics Education for early secondary school students. Laurinda Brown gives a lesson transcript that shows a teaching approach based on the philosophy of a community of inquiry. This teaching highly values an understanding of the process of mathematical conjecture and proof and portrays algebra as an increasingly useful language to communicate convincing ideas. Finally, Barbara Dougherty describes an elaborated Davydov approach that is being used with very young students. It is founded in students working with quantities rather than numbers, and making explicit students' thinking about informal relationships between quantities (such as the transitivity of the "greater than" relation).

How does this chapter inform thinking on the future of the teaching and learning of algebra? For curriculum design, it demonstrates the wide range of teaching possibilities, even for the beginning stages of teaching algebra, as well as the practical necessity of making choices between them. For teaching, it demonstrates how even operating within one problem domain, teachers can highlight different aspects of early algebra and therefore make the experience of students richer or poorer. For researchers, the chapter points to the futility of deciding which is the "best" approach to algebra because it is not possible to define an approach by one easily described characteristic, such as the problem domain. Instead, research needs to look for the strengths and weaknesses of approaches to algebra considering all of their dimensions of difference.

1.2.4 Research on technological environments

Research on the role of technological environments in algebra learning and teaching (Chapter 6), written by Carolyn Kieran and Michal Yerushalmy, presents an incisive and comprehensive survey of research on algebra learning and teaching that has been carried out with computers, graphics calculators, and other technological environments. The focus is on teaching beginning algebra, especially at the early secondary school level. This is similar to the focus of Chapter 5 but, for organisational convenience, approaches that use new technology were not considered there. Similarly, the special issues that arise with computer algebra systems are treated separately in Chapter 7.

The main themes of using technology to support algebra learning are its capacity to support multiple representations, the possibility of dynamic control over variables, and the possibility of providing a structured environment to support learning symbolic manipulation. An underlying aspiration is that these technological environments ought to provide a bridge to understanding algebraic symbolism and concepts and to support the learning of skills. The technological environment is also intended to enrich conceptual understanding of algebraic objects and processes.

The chapter begins, however, with a review of how the first use of technology to support algebra learning (and mathematics learning more generally) was with computer programming, sometimes with special languages for learners such as Logo. The use of programming and later technological environments raises questions of how to treat process/object duality, one of the main themes underlying current thought on algebra learning and touched upon in many of the chapters of this book. Process/object duality refers to the distinction between procedural (or operational) and structural views of algebraic objects and processes, and the way that, in many instances, students begin with a procedural view and progress to a structural view. With computer programming, students could first think of algebraic expressions, for example, as procedures that the computer would carry out. Equivalent expressions might then represent instructions that always produce the same output. This is regarded as a preliminary step towards understanding expressions as objects in their own right. Later these procedures become cognitively "encapsulated" as objects, and are then available for use within other procedures. In this way, the cognitive structures of mathematics are built up. As computers and graphics calculators with accessible programming features become more prevalent in classrooms, one possibility for the future is that programming may again find a similar place in algebra teaching, but current trends do not indicate that this will be so. Other possibilities of purpose-built software beckon.

This chapter is characterised by careful analysis of how a wide set of technological environments, both commercial and experimental, can contribute to learning algebra. These contributions are various: developing a rich and elaborated concept of function through multiple representations, acting as a bridge to the meaning of algebraic symbolism, supporting the development of manipulative skills, providing dynamic control of parameters and variables so that learning can tap into kinaesthetic senses, et cetera. This analysis will inform the future of the teaching of algebra. The chapter explains in detail how different technological environments support the learning of certain concepts and algebraic activity and not others; future curriculum choices can be guided by such knowledge.

The chapter leaves the reader with a strong sense of optimism that well-designed, imaginative technological environments will be able to improve the experiences of many students in the future. It also leaves the reader with a challenging list of questions for future research, because operating within environments other than pen and paper may be more different than we think.

1.2.5 Computer algebra systems and algebra

Chapter 7, *Computer algebra systems and algebra: Curriculum, assessment, teaching, and learning,* was co-authored by Mike Thomas, John Monaghan, and Robyn Pierce. Teaching (and hence research) involving Computer Algebra Systems (CAS) is a relatively new phenomenon. Consequently, all parts of this chapter indicate directions for the future or questions for the future. This chapter provides an extensive and informative account of this new literature, mostly relating to students in late secondary school or early university.

The first section examines how the availability of CAS can affect the algebra curriculum, by providing the option of extensions and additions, as well as supporting an emphasis on concept development, generalisation, and mathematical modelling. These affordances of CAS provide hope for a future where more students can do more algebra more competently, supported by this new tool. There are, though, already clear cautions from early experimental work that achieving enhanced results is certainly not an automatic consequence of having technology.

CAS is a very powerful tool and as such can have a strong effect on student learning. The chapter authors identify the key factors for algebra as the use of multiple representations, the development of ideas of function and parameters, and the opportunity for experimentation and generalisation.

Whilst there seems to be general agreement that CAS can be used to strengthen concepts, a central issue is the effect of CAS use on by-hand skills, which some feel may wither through lack of practice. This happens in some studies and in others, it does not: clearly the teaching is important. Along with Kieran in Chapter 2, Thomas, Monaghan and Pierce see value in the view of the analysis of Artigue and colleagues, which argues that viewing skills and concepts together as part of mathematical techniques leads to a more productive analysis of the roles of by-hand work in a CAS environment. The role of by-hand algebraic computation and the degree of efficiency of skill that students should attain will certainly be a prominent discussion point into the medium future.

The chapter also surveys several other areas that we can confidently predict will be important for future practice and future research: how assessment can or must change, how students can learn how to use such a powerful tool expertly, and the education for teachers upon which good learning with CAS will depend. Debate about the role of computer algebra systems and the place of symbol manipulation can confidently be predicted to be a feature of research and practice for the near future. Various chapters of the book make comments on the potential use of CAS, not all in favour, and there are underlying divisions between those who want to use technology only to support the development of traditional skills and those who are expecting change.

1.2.6 The history of algebra in mathematics education

The history of algebra in mathematics education (Chapter 8), co-authored by Luis Puig and Teresa Rojano, focuses on the teaching and learning of introductory secondary school algebra in terms of the lessons that can be learned from a historical perspective. Puig and Rojano say this is a two-way process. Twists and turns in the history of algebra can highlight aspects of mathematics that have presented challenges in the past and hence are likely to be challenging for students. On the other hand, knowledge of points of difficulty for present day students can pinpoint areas of history that are worth studying in depth.

In this carefully researched and detailed account, Puig and Rojano describe two medieval algebraic symbolisms and the corresponding ways in which problems are to be solved. They observe that the history of symbolism in algebra is the invention of a system that makes it possible to solve problems by manipulation of symbols according to rules, and without recourse to what the symbols mean. This is the legacy of Descartes and others. Once a problem in a real situation has been set up in terms of an equation or equations to be solved, the solution can proceed just in terms of the rules of manipulation, by hand or by machine. It is not required to know what the various algebraic expressions that arise as intermediate steps mean in terms of the original problem. This is an important part of the transition that modern day students need to make from arithmetic to algebraic thinking. Teaching which manages the consequent break between syntax and semantics, between the rules of algebra and meaning, is a recurring theme in the book. The chapter explains how Descartes came to develop this "Cartesian" method of problem solving and how it involves working with the unknown as if it were known. The chapter links this with modern experimental work with students experiencing the same transition.

Another theme recurring in the book and put in context by this chapter is the need to move from an arithmetic to an algebraic understanding of equality. Puis and Rojano explain that the "canonical form" of polynomials was not first our familiar

$$a_n x^n + a_{n-1}x^{n-1} + a_{n-2}x^{n-2} + \ldots + a_2x^2 + a_1x + a_0 = 0$$

but instead

$$x^n = a_{n-1}x^{n-1} \pm a_{n-2}x^{n-2} \pm \ldots \pm a_2x^2 \pm a_1x \pm a_0$$

which shows how the highest degree is calculated from the lower. Beginning students often restrict meaning of equality as an instruction of how to calculate one side from the other, rather than as a statement that two expressions are equivalent. By building an appreciation of these and other difficulties into curriculum sequences, teaching in the future may improve.

The chapter also shows how the development of algebra was accompanied by a different understanding of number and unknowns (e.g., al-Khwârizmî, who had different species of numbers and Bhâskara who had unknowns of different colours). A very simple example of this change in the nature of number that accompanies the development of algebraic processes is the familiar process of collecting like terms and simplifying an expression such as $12p-3-p$. Teachers may tell students to change the order of the terms to get $12p-p-3$, but what is the basis of this simple move? Subtraction is not commutative! Since we see subtraction as the inverse of addition and subtracting a negative number as the same as adding its inverse, teachers see $12p-3-p$ as the same as $12p+(-3)+(-p)$ and the order of the terms can be swapped because addition is commutative. Processes such as this were understood very differently in earlier times, when subtraction and addition were seen as two separate operations, when letters always stood for positive quantities and when negative numbers were not numbers at all. Puig and Rojano discuss this and much more with faithfulness to the history.

How does this chapter inform thinking on the future of the teaching and learning of algebra? For us, struggling to absorb the description of the mathematics of earlier times provided a vivid demonstration of how difficult it is to learn about mathematical ideas. This is so even where the ultimate object is familiar, such as solving equations using an apparently familiar method. Mathematics has an underlying universality, but the symbolisms and concepts are a human construction and change over time. Secondly, the chapter demonstrates that many of the difficulties that students are likely to have when learning present day algebra, have historical roots. Curricula in the future can benefit from these insights.

1.2.7 Symbols and language

It is often said that algebra is a language. This chapter arose from this observation, which, interestingly, the chapter rejects. Does treating algebra as a language, or at least a system of signs, lead to insights into teaching and learning? Are there useful parallels to be made with how languages are used and how they develop analogous processes in algebra? Are there parallels between learning one's mother tongue as a young child and learning algebra? Is the gradual development of symbolisation processes in early algebra learning parallel to other symbolisation? Is it fruitful to apply theories of learning languages to learning algebra?

Jean-Philippe Drouhard and Anne Teppo, in writing Chapter 9, *Symbols and language*, investigate algebraic symbols from both a linguistic and semiotic perspective. Theoretical frameworks are presented, especially Frege's notions of *sense* and *denotation* and Peirce's and Deacon's hierarchical assignation of meaning to symbols. In addition, factors that affect how students read collections of symbols and assign meaning to symbols are discussed and recommendations for areas for further research are presented.

The chapter begins by clarifying underlying questions about what is loosely called the language of algebra. Mathematical writings are separated into three components: the natural language component (such as "The number of matches is given by ..."), algebraic symbolic writings (such as $2x + 6 = 100$), and compound representations (such as a table or diagram). Both the natural language component and the algebraic symbolic writings component are languages, and hence their study can be in terms of linguistic concepts such as syntax and semantics, pragmatics, sense and denotation, ambiguity et cetera. According to this theory, such ideas can be helpful in analysing students' difficulties with understanding expressions, for example. One observation is to draw attention to the linguistic distinction between sense and denotation. The two expressions $2x + 6$ and $2(x + 3)$, for example have the same denotation, but the sense is different. On the other hand, the chapter proposes that linguistic ideas are not helpful in understanding the compound representations of algebra, where ideas of semiotics (the theory of signs) are instead applicable. Here the ideas of Peirce and Deacon are used to explain the dimensions of difference in the ways that different individuals (such as teachers and students) can interpret the same sign. The chapter discusses these ideas in depth and makes links with research, theory, and practice from many aspects of algebra learning.

The chapter represents an important attempt to bring a variety of apparently disparate phenomena in algebra learning together under the twin umbrellas of theories of languages and theories of signs. The challenge for the future is to evaluate whether this theoretical perspective is sufficiently well suited to the learning of algebra (and other branches of mathematics) to provide convincing explanations and a sound source of insights for teaching.

The chapter ends with a brief hint of another possibility. New mathematical and semi-mathematical notations are now becoming commonplace, in response to the needs of communication with calculators and computers. Moreover, as information technology influences mathematics, different ways of thinking may become prevalent and need expression with a different range of symbols, professionally and at school. Already we see these influences of new notations becoming widely understood such as

$$y = x^2 + 4x + \{1, 2, 3\}$$

to graph three parabolas with constant terms 1, 2, and 3 and

$$x^2 + 4x + 3 \mid x = 7$$

to substitute the value of 7 into the expression. These changes are minor, but meet a need. Mathematical notation can be expected to change more, just as it has through history, and in these changes, there may be possibilities to adopt symbols that assist learning.

1.2.8 Teachers' knowledge and the teaching of algebra

On behalf of her Working Group, Helen Doerr authored *Teachers' knowledge and the teaching of algebra,* Chapter 10. With respect to the teaching of secondary school algebra, this chapter provides an analysis of recent research findings on teachers' knowledge and practice and its development, discussion of critical issues, and suggestions for further research. Although there are many areas of teachers' knowledge for teaching algebra that have not been researched, the chapter assembles an impressive array of studies in the area.

Studies in one strand demonstrate that many teachers do not have a sufficiently strong understanding of fundamental concepts such as function and slope to make strong connections between lessons and to understand and explain students' reasoning. Moreover, some studies are able to show how expert teachers use their deeper knowledge of content and students' thinking in their teaching. This will be an important thread in future research: the research view of students' conceptions and misconceptions of algebra needs more frequently to be linked with teachers' practical and implicit knowledge of the same and then further linked to how teachers use this knowledge in the classroom. Similar observations apply to a second strand of research that has examined teachers' conceptions of algebra (e.g., the place of functions as a unifying theme in algebra, the identification of algebraic thinking other than in symbol manipulation, etc.), and considered how these relate to current curriculum thinking and how they affect classroom actions and decisions.

The chapter also reports several interesting studies of teacher education, both pre-service and in-service professional development, and makes clear indications of critical areas for future research. The chapter concludes with a summary of major questions for future research. Certainly, we need to know more about how teacher education can promote more effective teaching of algebra. This is important in training secondary teachers, who generally have a relatively strong background in mathematics including algebra. However, there are also very urgent questions for teacher education if the ideas espoused in Chapter 4 on the introduction of aspects of algebraic thinking into the early years of schooling are to be implemented successfully.

1.2.9 The teaching and learning of tertiary algebra

David Carlson wrote Chapter 11, *The teaching and learning of tertiary algebra*, drawing on briefs written by members of a very diverse Working Group. The chapter reports on a wide range of current educational issues related to the teaching and learning of tertiary algebra such as conceptual difficulties, issues of course design, students' motivation and its relation to abstraction and relevance, quality of teaching, and use of technologies. This chapter makes recommendations for improving practices for teaching tertiary algebra and proposes areas for further research. It also draws attention to the difficulties of disseminating educational findings and new curriculum ideas to teachers of tertiary mathematics and provides indications of some models that have proved effective in other reforms.

An interesting observation about tertiary algebra is that it is not the direct continuation of school algebra. Carlson and his Working Group decided to focus on courses in abstract algebra, linear algebra, number theory, and discrete mathematics as "tertiary algebra". In these areas of mathematics, the elements obey some but not all of the axioms obeyed by the real numbers, which are the main concern of school algebra. For example, linear algebra works with matrices, which do not have commutative multiplication and number theory works with congruence classes, only some of which have multiplicative inverses. The basic algebra of real numbers is left behind, to be used as a tool in "non-algebra" courses such as calculus and applied mathematics. Instead, algebra becomes the study of mathematical structures created from abstract sets of elements acted on by operations obeying specified axioms. Even the word "algebra" acquires a new specific meaning as just one of the axiomatically defined structures, such as groups, rings, and fields that are studied in this new part of mathematics.

This sea change from secondary to tertiary algebra means that there is a discontinuity in the challenges of teaching. Themes that lie below the surface, hardly noticed in the secondary school in many countries, now rise to grab the attention. Kieran's three core algebraic activities (generational, transformational, and global/meta-level activities) still apply, but it is aspects of the global/meta-level activity such as proving and noticing structure that now dominate. In the chapter, Carlson and the Working Group identify an interrelated set of problems with abstraction, proofs, symbolic logic, and the use of definitions as the major difficulties for students. In each of these areas, there is some research to guide future development, but not yet enough. One example of progress is in the research and development work of Dubinsky, who has constructed computer mini-worlds and programming tasks for tertiary students. These are based on a careful didactic analysis of the material to be learned, employing cycles of activities, classroom discussion, and exercises. Students meet the mathematical structure first as actions, then as processes, then objects, and culminating as schemas, drawing on the same observations about importance of the movement from process to object that are made in reference to beginning algebra.

There is a clear future agenda here for tertiary teachers, in seeking to help their students make the transition from school mathematics—which seems in many countries to be becoming increasingly informal—to the requirements of formal mathematical study.

1.2.10 Goals for the compulsory years

Mollie MacGregor authored Chapter 12, *Goals and content of an algebra curriculum for the compulsory years of schooling*. This chapter is concerned with the importance of algebra for *all* students during their compulsory years of schooling and the vexed question of what should constitute a basic algebra curriculum for students who have low interest or low achievement in mathematics. In writing this chapter, MacGregor has taken an outsider's sceptical view, trying to evaluate the arguments commonly given by enthusiasts for teaching algebra, although her own positive enjoyment of algebra peeks through. The chapter begins by reviewing the major pressures for change in teaching algebra. Firstly, the growth of universal education can lead to problems of disengaged learners who may be set goals that are too high. Secondly, decades of research and anecdotal information has revealed generally unsatisfactory outcomes for a high proportion of learners. Thirdly, new technology offers new opportunities for learning, as well as challenging the importance of some topics in the existing curriculum. The new prevalence of tools for fast numerical computation and dynamic visualisation is changing the balance of what all students who are unlikely to be users of advanced mathematics need to learn.

The chapter then addresses some of the reasons commonly advanced for having all students learn algebra, and notes slippage between the economic need for a highly mathematically trained population and the expectation that schools should teach algebra even to unwilling students. It rejects the notion that learning algebra is important for everyday life, beyond a basic but not traditional list of ideas. Schools have a difficult balance to draw between giving all students access to algebra as the gateway to higher mathematics (and hence to a range of other individual benefits from higher education and training) and providing students with an appropriate curriculum within their capabilities. Some of the different choices that educational jurisdictions make in this matter are outlined in Chapter 13.

Whilst endorsing many of the recommendations made in recent international reports and the many possibilities for improved teaching strategies offered elsewhere in this book, this chapter points to several other principles that can guide curriculum choices for low-achieving students. Firstly, algebra needs to be integrated with other mathematics topics, not just organisationally, but so that it is used as a language to express ideas on a regular basis. Secondly, especially for low achieving students, learning algebra must support and consolidate learning basic principles of arithmetic, rather as has been envisaged for younger students in Chapter 4. Calculators and other technology such as spreadsheets can be used for calculation and graphing where necessary, and will be used both as pedagogical aids and as important problem solving tools. This enables students to tackle problems with real data, enhancing their appreciation of the usefulness of what they are learning. The list of basic goals includes understanding the role of algebra as the language of science, even if the student himself or herself does not speak that language well. The basic goals include operating with graphs and tables and understanding the central properties of functions, especially linear and exponential, which have many everyday applications. A procedural view dominates the nature of the recommended curriculum, so that the orientation can be to substituting into formulas rather than working with equalities. These decisions are consistent with the research touched on elsewhere in the book, which identifies cognitive processes that are easier or more difficult to grasp. In this case, working with unknowns (required in transposing) is more difficult than working with knowns (required in substituting). Algebra for low-achieving students can be built with such principles in mind.

In his contribution to the Plenary Panel, Jack Abramsky from the UK Qualifications and Curriculum Authority, spoke about decisions that need to be made to deliver algebra education in a large school system. Abramsky presented the Study Conference with a wider list of macro-scale curriculum issues. Several of his issues were related to the rationale for algebra and the needs of low achieving students, as discussed in this Working Group. However, Abramsky also raised questions that had not been addressed systematically elsewhere about the needs of the most able students, the place of proof, the role of algebra in developing problem solving skills, what constitutes good assessment, and practical issues about teacher qualifications and supply.

1.2.11 Algebra, a world of difference

The final chapter, *Algebra: A world of difference* (Chapter 13), was compiled by Margaret Kendal and Kaye Stacey. It was inspired by the Plenary Panel *Algebra around the world* and draws in part, on information and cases presented in that session. It highlights through a range of examples, some of the dimensions of difference in the teaching of algebra around the world, which were at times barriers to communication and at other times greatly enriched the discussions during the 12th ICMI Study Conference. An important outcome of the Study Conference for participants was an enhanced awareness of the possibilities for content and approaches beyond their home approach.

The chapter reports on decisions around the world on two dimensions of school structures: whether classes are comprehensive or streamed and whether algebra is integrated into a wide mathematics curriculum, or taught on its own in the layer-cake curriculum. Whilst the teaching of all branches of mathematics is affected by these decisions, they are particularly likely to impact on the teaching of algebra since algebra is a relatively difficult and abstract area of mathematics and builds upon substantial prior knowledge.

The chapter also surveys some of the different ways in which the nature and purpose of algebraic activity is conceived in different educational jurisdictions. These theoretical perspectives are to view algebra: (i) as a way of expressing generality; (ii) as a study of symbol manipulation and equation solving; (iii) as a study of functions; (iv) as a way to solve certain classes of problems; (v) as a way to model real situations; (vi) or as a formal system involving set theory, logic, and operations on entities other than real numbers. Educational jurisdictions do not usually implement only one of these, but select from them within a given emphasis.

The chapter is also able to illustrate some of the differences between how various theoretical perspectives are implemented. For example, whilst it is the Western English-speaking countries which are most commonly associated with approaching algebra through generality and pattern (as discussed in Chapter 5), the Japanese curriculum extends this focus on pattern to a wider range of problem types.

Variations in expectations of the degree of expertise in symbol manipulation, the understanding of formal logic, the use of technology, the degree of formality expected in algebraic writing, and the way in which real world problems are used are also surveyed with examples from around the world. Cross-cultural research demonstrates that these differences are evident in what students' achievement. In terms of the toolkit in Chapter 5, this chapter illustrates differences in both the problem domains selected and the teaching approaches, as similar problems can be used to emphasise different aspects of algebra. We hope that this chapter will provide an accessible resource for the future development of algebra curriculum and teaching approaches, by drawing attention to the many rich possibilities for teaching that have been developed around the world.

1.2.12 Modelling to learn algebra or learning algebra to model?

In reviewing the above components of the solution to the problem with algebra, we see an important omission. The *Discussion document* (Program Committee, 2000a, 2000b) prepared by the International Program Committee invited contributions to the 12th ICMI Study on a wide range of topics. Only one of these topics did not receive sufficient submitted papers for a Working Group to be assembled. This was the topic on modelling the real world using algebra. Hugh Burkhardt commented on the absence of this orientation to using mathematics in his Plenary Address at the closing session:

> "The power of algebra as a tool for solving problems from outside mathematics received little attention. Applications were mainly illustrative, with the focus on understanding algebraic concepts and skills. There was little reference to learning the modelling skills needed to 'mathematise' practical problems with algebra, yet in the world for which we prepare students, most algebra is done for such purposes, not as a pure study. There was even a reference to "modelling as a bridge to algebra" —to me, a startling reversal of priorities".

In fact, the "modelling as a bridge to algebra" is a theme cutting across several of the book chapters. Many of the more recent approaches to generational activity, that are described in Chapters 2, 5, 6, and 13 use real or pseudo-real situations to stimulate algebraic symbolisation. Particular interest lies in the new possibilities for the use of real data generated in real time with data logging devices in technological environments. Luciana Bazzini, for example, in her address in the Plenary Panel described work reported in the paper by Arzarello and Robutti (2001), where students learn about graphing with the help of a motion detector linked to a logging and graphing device (e.g., a graphics calculator). As students run in the corridor, their motion can be graphed in real time and displayed for class discussion. Then, presented with a different graph of position or speed against time, students are challenged to match their motion with the graph. Bazzini analysed this situation in terms of the metaphorical qualities of the embodied cognition, and drew parallels with the metaphorical grounding evident in transcripts of other students' work on inequalities (Bazzini, Boero, & Garuti, 2001). In this work, it is hypothesised that the building of algebraic concepts is supported by students' (bodily) experiences in the physical world. Chapter 6, on technological environments, has other examples.

Why, at this point of time, is there an emphasis on "modelling as a bridge to algebra" rather than on instruction designed to help students use their algebra in modelling the world? It may simply reflect the dominance of interest in beginning rather than advanced algebra. It may reflect a feeling of stalemate in improving students' abilities to set up equations for solving word problems, whether presented traditionally or otherwise, and tackling traditional applications. Alternatively, it may reflect an intention to keep algebra and the real world closer together throughout instruction, so that special lessons to close the gap are irrelevant. In any case, it is important to recall the centrality of Kieran's global/meta-level activity. Proving, modelling, and applying are activities that give purpose to learning algebra.

1.3 Assembling the Solution

The sections above have looked at the future for teaching and research on algebra from eleven different points of view, and many disparate suggestions for curriculum, teaching, and research have emerged. We commend a close examination of these detailed suggestions to the reader. This section can do no more than assemble and review some major themes. In this we are mindful of Burkhardt's call to plan research that can make a positive impact on practice (Burkhardt & Schoenfeld, 2003).

First, it is clear that the attempts to define school algebra as much more than techniques for symbol manipulation have borne fruit, and that this work needs to continue. The impetus for further work here arises particularly from the need to cater better for low-achieving students and from the call for early algebra, which is based on the proposition that early mathematical learning would be enhanced if algebraic ideas were brought to the fore. It is especially important that teachers come to share the more sophisticated appreciation of what this algebra might be, rather than just identifying it with its surface features of dealing with symbols and graphs.

Second, research on the cognitive obstacles that students face when they learn algebra has reached some consensus positions that can be applied to designing tasks, learning sequences, and curricula. This book has drawn on this research, rather than discussing the findings in detail, but it contains many examples where teaching approaches, software, or materials have been analysed very usefully in terms of research on students' thinking and the likely pitfalls. A new challenge is to make this knowledge of student thinking more accessible to teachers. This is not just a call for wider dissemination of research results. Instead, reasonably complete bodies of research on student learning need to undergo a "didactic transposition" to create a body of knowledge that is learnable and useable by teachers. Research on students' difficulties in algebra is a good candidate for this.

Third, information and communication technologies offer great potential for better learning. Experimental work on designing better environments for learning algebra must continue and good products should be available for school use. New ways of linking real world experience with mathematics are evident. Research frameworks are helpful in analysing what aspects of algebra learning new environments support and what aspects they do not support. It is not known to what extent old understandings about learning algebra will apply with new environments.

Fourth, new technologies will continue to challenge what algebra needs to be learned, as they change how mathematics is done. For algebra education, one of the main issues arising is the desirable extent of by-hand skill in symbolic manipulation, for different categories of end-users. The contributions to this book have generally argued conservatively that symbolic manipulation skill levels should be maintained, but the arguments themselves are not (all) traditional. Instead, there are new theories about the contribution of symbolic manipulation to conceptual development to be tested and further developed. The importance of students' developing sufficient algebraic expectation to be able to guide themselves around in an algebraic environment is undisputed, but how it can be attained without traditional skill teaching is not clear.

Fifth, the many examples scattered throughout the book demonstrate clearly that school algebra is a multi-faceted construction. There are many options for the approaches, the problem domains, and the theoretical perspectives, and different decisions are taken in different places. As the world becomes smaller, we predict more interchange about options for teaching algebra, but not a convergence. There are too many choices available.

In summary, we commend the book to readers as a guide to assessing what algebra knowledge and skills will be of lasting value in the 21st century and to examining under what conditions these can best be learned.

1.4 References

Arzarello, F., & Robutti, O. (2001). From body motion to algebra through graphing. In H. Chick, K. Stacey, J. Vincent, & J. Vincent (Eds.), *The future of the teaching and learning of algebra* (Proceedings of the 12th ICMI Study Conference, pp. 33-40). Melbourne, Australia: The University of Melbourne.

Bazzini, L., Boero, P., & Garuti, R. (2001). Revealing and promoting the students' potential in algebra: A case study concerning inequalities. In H. Chick, K. Stacey, J. Vincent, & J. Vincent (Eds.), *The future of the teaching and learning of algebra* (Proceedings of the 12th ICMI Study Conference, pp. 53-60). Melbourne, Australia: The University of Melbourne.

Bednarz, N., Kieran, C., & Lee, L. (1996). *Approaches to algebra: Perspectives for research and teaching*. Dordrecht, The Netherlands: Kluwer Academic.

Blanton, M., & Kaput, J. (2001). Algebrafying the elementary mathematics experience. Part II: Transforming practice on a district-wide scale. In H. Chick, K. Stacey, J. Vincent, & J. Vincent (Eds.), *The future of the teaching and learning of algebra* (Proceedings of the 12th ICMI Study Conference, pp. 87-95). Melbourne, Australia: The University of Melbourne.

Burkhardt, H., & Schoenfeld, A. H. (2003). Improving educational research: Toward a more useful, more influential and better funded enterprise. *Educational Researcher, 32*(9), 3-14.

Chick, H., Stacey, K., Vincent, J., & Vincent, J. (Eds.) (2001). *The future of the teaching and learning of algebra. Proceedings of the 12th ICMI Study Conference.* Melbourne, Australia: The University of Melbourne.

Davydov, V. (1962). An experiment in introducing elements of algebra in elementary school. *Soviet Education, 8*, 27-37.

Program Committee. (2000a). Discussion Document for the 12th ICMI Study. The Future of the Teaching and Learning of Algebra. *Zentralblatt fur Didaktik der Mathematik*, 2000/4, 107-110.

Program Committee. (2000b). Discussion Document for the 12th ICMI Study: The Future of the Teaching and Learning of Algebra. *Educational Studies in Mathematics, 42*, 215-224.

Trouche, L. (2000). La parabole du gaucher et de la casserole à bec verseur: Étude des processus d'apprentissage dans un environnement de calculatrices symboliques. *Educational Studies in Mathematics, 41*, 239-264.

Chapter 2

The Core of Algebra: Reflections on its Main Activities

Carolyn Kieran
Université du Québec à Montréal, Canada

Abstract: This chapter is Carolyn Kieran's Plenary Lecture that was presented at the 12th ICMI Study Conference. It presents a model for conceptualising algebraic activity that is a synthesis of three principal activities of school algebra: generational activity, transformational activity, and global/meta-level activity. The model is used as a basis for reflecting on past research in algebra and on the changing perspectives possible, in both algebra classes and future research, in the presence of technology.

Key words: Algebra, research, generational activity, transformational activity, global/meta-level activity, technique, paper-and-pencil algebra, computer algebra, model of algebraic activity, technology

2.1 Introductory Remarks

There are several perspectives from which one may address the question of the core of algebra, that is, the essential ingredients of school algebra. For example, in the early 1990s, an international colloquium on algebra was held in Montreal, which focused on four approaches aimed at making algebra learning meaningful to students, approaches that could be said to encompass some of the basic ingredients of school algebra:

- Generalization of numerical and geometric patterns and of the laws governing numerical relationships,
- Problem solving,
- Functional situations, and
- Modelling of physical and mathematical phenomena (Bednarz, Kieran, & Lee, 1996).

Another point of view on what is important in school algebra was adopted by Lesley Lee during the mid-1990s within the context of a four-year study aimed at exploring algebraic understanding. She presented the question "What is algebra?" to a cohort of mathematicians, teachers, students, and mathematics education researchers. The seven themes that emerged from her interviews on the question of "What is algebra?" were:

- Algebra is a school subject,
- Algebra is generalized arithmetic,
- Algebra is a tool,
- Algebra is a language,
- Algebra is a culture,
- Algebra is a way of thinking,
- Algebra is an activity.

If there was one theme that tended to permeate all others among Lee's interviewees, it was *Algebra is an activity.* Lee (1997) wrote, "Algebra emerges as an activity, something you do, an area of action, in almost all of the interviews" (p. 187). For example, David Pimm, one of Lee's interviewees, is quoted as saying that *action* is the central feature of school algebra: "Algebra ... is so much more about doing, is actually about action on things, ... with attention being more on the *transformations* than on the objects themselves" (Lee, 1997, p. 187). On the other hand, Jim Kaput emphasised the alternate side of the coin when he pointed to the importance in algebraic activity of spending a great deal of time in the "building of algebraic objects" (Lee, 1997, p. 189). And Alan Bell expressed concern about the question of purpose, about students not really "having the experience of what algebra is for" (Lee, 1997, p. 196).

Taking up the theme of algebra as activity, I developed a model of algebraic activity, which was presented at the 1996 ICME-8 conference in Sevilla, and which attempted to deal with what I perceived to be the three principal activities of school algebra. The first part of my paper will synthesise the main features of this model. I will then use the model as a backdrop for reflecting on some of the past research in algebra and relate it to changing perspectives arising from the presence of technology in algebra classes.

2.2 A Model for Conceptualising Algebraic Activity

The activities of school algebra can be said to be of three types: generational, transformational, and global/meta-level (Kieran, 1996).

2.2.1 Generational activity

The *generational* activities of algebra involve the forming of the expressions and equations that are the objects of algebra. Typical examples include: (i) equations containing an unknown that represent quantitative problem situations, (ii) expressions of generality arising from geometric patterns or numerical sequences, and (iii) expressions of the rules governing numerical relationships. The underlying objects of expressions and equations are, of course, variables and unknowns.

Situations illustrating the three just-mentioned examples of generational activity are shown in Figure 2.1:

(i) *A quantitative problem leading to an algebraic equation*
There are 3 piles of stones; the first has 5 less than the third, and the second has 15 more than the third. There are 31 altogether. Find the number in each pile. (Bell, 1995, p. 61)

(ii) *A geometric pattern leading to an algebraic expression*
Say to yourself what you see in the picture sequence. Then state a rule for extending the sequence of pictures indefinitely. (Mason, 1996, p. 84)

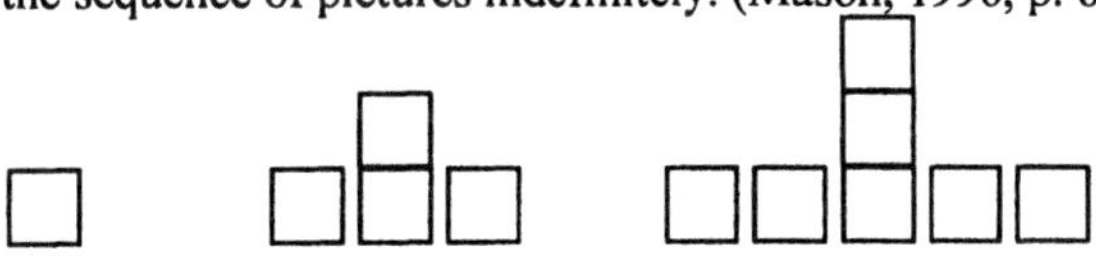

(iii) *A numerical relationship leading to an algebraic expression*
The sum of two consecutive numbers is always an odd number. Can you show why, using algebra? (Lee & Wheeler, 1987, p. 160)

Figure 2.1. Three examples of generational activity in algebra.

The focus of generational activities is the representation (and interpretation) of situations, properties, patterns, and relations. Much of the initial meaning-making of algebra is considered to be situated in this sphere of activity. With respect to meaning-making, there are two main frameworks that are used in a majority of algebra classes: One is the more recent "functions" framework and the other is the "generalised arithmetic" framework. These two frameworks each provide a unique transversal thread to the three categories of algebraic activity.

When a functions-based framework is adopted, the following aspects can be associated with it: (a) graphical and tabular representations can be used to develop meaning for the symbolic forms of expressions and equations; (b) unknowns tend to be considered a special case of variables, but it is variables that are usually the main focus; (c) first-degree equations in one variable can be interpreted as equalities of two functions, with the two functions often being represented as two linear graphs; (d) the solution of an equation can be viewed as the value of x for which both functions are equal; and so on. In much of the technology-supported research aimed at giving meaning to algebraic objects, it is the functional approach with its multiple representations that has been capitalised on.

When generalised arithmetic is the underlying framework for generating and interpreting algebraic objects, the unknown takes priority over the variable, and expressions and equations tend to be viewed as representations of numerical processes rather than functional relations.

2.2.2 Transformational activity

The second type of algebraic activity—the *transformational* (rule-based) activities—includes, for instance, collecting like terms, factoring, expanding, substituting, adding and multiplying polynomial expressions, exponentiation with polynomials, solving equations, simplifying expressions, working with equivalent expressions and equations, and so on. A great deal of this type of activity is concerned with changing the form of an expression or equation in order to maintain equivalence, yet it is only recently that significant research attention has shifted to the emergence and development of students' notions of equivalence (e.g., Cerulli & Mariotti, 2001; Lagrange 2000).

2.2.3 Global/meta-level activity

Lastly, there are the *global/meta-level* mathematical activities. These are the activities for which algebra is used as a tool but which are not exclusive to algebra. They include problem solving, modelling, noticing structure, studying change, generalising, analysing relationships, justifying, proving, and predicting—activities that could be engaged in without using any algebra at all. However, attempting to divorce these meta-level activities from algebra removes any context or need that one might have for using algebra. In fact, from the point of view of the curriculum, the global/meta-level activities cannot be separated from the other activities of algebra, in particular the generational activities, otherwise all sense of purpose is lost.

2.2.4 General remarks on these three types of activity

Algebra textbooks have traditionally emphasised the transformational aspects of algebraic activity, with more attention paid to the rules to be followed in manipulating symbolic expressions and equations than to conceptual notions that support these rules or to the structural underpinnings of the expressions or equations being manipulated. It is rather an exception that, as mentioned by Rosamund Sutherland (1997) in her Royal Society/JMC Working Group Report, the "French Brevet places emphasis on mathematical structure in that pupils are asked to present some answer in a particular form as opposed to being asked to complete the process of computation" (p. 14).

One of the strengths of algebra is that, for experts, a great deal of its transformational activity can be carried out in what appears to be a rather automated manner. Once one makes the transformation rules one's own, the algorithms of algebra can be executed, in a sense, without thinking. But not really; for, as Paolo Boero (1993) has emphasised, every algebraic manipulation contains an anticipatory element, a sense of the direction in which you want to be going and of what the desired expression will look like once you get there.

In much algebra teaching, conceptual understanding of the objects of algebra has tended to be segregated from the development of manipulative skill. Few have espoused the position that students' conceptual understanding grows as they engage in algebraic processes. Yet, Eric Love (1986) suggested the role that thinking about such processes might play when he argued:

> Algebra is now not merely "giving meaning to the symbols," but another level beyond that: concerning itself with those modes of thought that are essentially algebraic—for example, handling the as yet unknown, inverting and reversing operations, seeing the general in the particular. Being aware of these processes, and in control of them, is what it means to think algebraically. (p. 49)

I shall return to this issue of conceptual understanding of algebraic processes when I address some of the newer theoretical perspectives.

2.3 From Paper-and-Pencil Algebra to Electronic-Technology-Supported Algebra

Up until about the mid-1960s, algebra was a paper-and-pencil activity, focusing primarily on transformational work. A glimpse at the opening pages of an algebra text (Crawford, 1916) that was in use in some provinces of Canada prior to the mid-1960s (see Figure 2.2 for pages 1 and 2 of the Crawford text) shows a rather minimalist approach being taken to the question of creating meaning for the objects of algebra. To all intents and purposes, algebraic meaning-making took place in the first few pages of the text in the form of translating arithmetical statements into "signs and symbols" and doing a few numerical substitutions (e.g., "A boy has p marbles; he wins q marbles and then loses r marbles. How many has he now? How many if $p = 5$, $q = 11$, $r = 4$?" (Crawford, p. 4)). Pupils were then quickly launched into the transformational rules of "addition and subtraction of like terms" and the "use of brackets". Equation solving soon followed. There were few, if any, global/meta-level activities to provide a context or purpose.

In the ensuing years, the modern math movement and an importance given to problem solving created a few changes to the way in which algebraic activity came to be viewed and taught. For example, the 1970s series of *Modern Algebra* texts by Dolciani and Wooton introduced the language of variables and expressions; open sentences; and functions, relations, and graphs. However, the problems seen in these texts were not much different from the problems of several decades earlier, except that they now tended to be grouped as "motion problems", or "age problems", or "mixture problems", and so on.

HIGH SCHOOL ALGEBRA

CHAPTER I

ALGEBRAIC NOTATION

1. **Use of Arithmetical Signs.** In arithmetic, signs are used to abbreviate the work. In algebra the same signs are used, with the same meanings and for the same purpose.

EXERCISE 1

Write the following statements in the shortest way you can, using the signs and symbols with which you are familiar in arithmetic.

1. Two and two make four.
2. The sum of five, ten and twenty is thirty-five.
3. Six and four is the same as four and six.
4. Seven times eight is the same as eight times seven.
5. The difference between twelve and five is seven.
6. Ten exceeds six by four.
7. The excess of twenty over fifteen is five.
8. The defect of thirty from a hundred is seventy.
9. Thirty-six divided by four is nine.
10. Three score and ten is seventy.
11. One half of the sum of seven and five is six.
12. The sum or the product of three, five and seven is the same in whatever order they are written.
13. Three multiplied by four is twelve, therefore twelve divided by three is four.
14. The square of four is sixteen, therefore the square root of sixteen is four.

B

2 *ALGEBRA*

2. **Algebraic Symbols.** In the preceding exercise you have used symbols to represent the numbers stated and signs to show the operations performed on those numbers.

In algebra, symbols are used more extensively than in arithmetic.

If the length of this line be measured it will be found to be two inches. But without measuring it, we may say that the measure of its length is some definite number which might be represented by the letter a.

The measure of the length of another line might be represented by b. The cost of an article might be c cents, or the cost of a farm might be x dollars, or the weight of a stone might be m pounds.

Here a, b, c, x, m are **algebraic number-symbols**, or briefly algebraic numbers.

The symbols 1, 2, 3, etc., used to represent numbers in arithmetic are called **arithmetical number-symbols** or arithmetical numbers.

In algebra the number symbols of arithmetic are also used. For the present, when letters are used to represent numbers, it will be understood that each letter represents some integral or fractional number.

3. **Signs of Multiplication.** In this square the measure of the length of the side AB is a. What is the measure of the length of BC; of CD; of $AB+BC$; of $AB+BC+CD$?

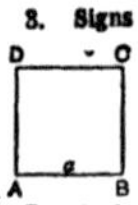

The measure of the perimeter (sum of all the sides) is $a+a+a+a$ or 4 times a or $4\times a$.

In algebra, $2\times a$ or $a\times 4$ is usually written $4a$, the sign of multiplication being understood. It is also written $4 . a$, the dot representing multiplication.

Thus, $4\times a=4 . a=4a$, and as in arithmetic, is a short way of writing $a+a+a+a$.

Thus, if $a=6$, the measure of the perimeter of the square is $6+6+6+6=4\times 6=24$.

Figure 2.2. Pages 1 and 2 of Crawford (1916).

But other factors began to intervene as well, factors that were to lead to more far-reaching changes in perspective on school algebra. One was the growing interest in algebra studies by the emerging community of mathematics education researchers; another was the arrival of electronic technology in an increasing number of mathematics classrooms from the mid-1980s onward.

The algebra research studies in the 1970s and early 1980s had yielded evidence of beginning algebra students' difficulties with both generational and transformational activity. Reasons for this included the fact that elementary school arithmetic is, in general, answer oriented; its focus is not the representation of processes or relations. Furthermore, when solving problems, students emerging from arithmetic would want to undo the operations of a given problem and found it difficult to first represent the situation. Thus, the research during these years, which was showing that the transition from arithmetic to algebraic thinking was not a straightforward matter, when combined with the large body of evidence on students' difficulties with the transformational activities of algebra, led researchers and educators to think that perhaps a great deal more time needed to be spent with creating meaning for the objects that were being manipulated.

Researchers turned to teaching experiments to try out new approaches related to generational activity. For example, in 1984, just as computing technology was beginning to make inroads in schools, Lesley Booth (1984) developed an approach for giving a "variable" meaning to letters (as opposed to viewing letters as "unknowns"), an approach that was analogous to programming a computer and that was designed around an idealised "mathematics machine" for which all instructions could be written using simply the language of mathematics. For instance, students worked at expressing processes such as "I want the machine to add 5 to any number I give it; how will I write the instructions?" ($x + 5$) and "I want the machine to add any two numbers I give it" ($x + y$). Students used letters to write rules to enable the virtual machine to solve whole classes of problems.

These modest beginnings with pseudo-technology were greatly extended in the studies that followed involving real computing technology (see Chapter 6 of this volume for more examples of the early use of computing technology in the learning and teaching of algebra). For instance, the research of Rosamund Sutherland and Teresa Rojano (1993) was to show that spreadsheet environments could not only sustain discussion of the role of letters as both variables and unknowns, but also afford meaningful experience with the creation of algebraic expressions, and allow a focus on the representation of problem relationships—at the same time as provide a tool with which to solve algebra problems.

However, very few of these "generational-activity" studies, where the aim was to create meaning for algebraic objects, were linking student work in this area to the transformational activities of school algebra. In fact, much of the algebra research with technology seemed to be minimising the importance of transformational activities. If students could solve algebra-type problems with spreadsheets and other such tools, there seemed little need to learn algebraic transformations. The view, *Algebra as a problem-solving tool*, appeared to be gaining in importance.

Indeed, in the UK for example, the search for meaning and the consequent suppression of symbolism led to a situation in the early 1990s where students were doing hardly any symbol manipulation (Sutherland, 1990). In various countries, problem solving, by whatever means, had all but replaced traditional algebra. The hope was that, in focusing on algebraic understanding (however this might be defined), the techniques would take care of themselves. The idea that universal access to the new technologies would "enable us to modify our skill-dominated conception of school algebra and rebalance it in favour of objectives related to understanding and problem solving" (Kieran & Wagner, 1989, p. 8), seemed an attractive one.

But it did not happen, as Michèle Artigue has argued (cited in Lee, 1997). The techniques did not take care of themselves.

2.4 Epistemic Role of Technique

So, in the mid-1990s, in France, when computer algebra systems (CAS) began to make their appearance in high school mathematics classes, research attention shifted to the question of technique, to the development of transformational abilities. A team headed by Michèle Artigue (Artigue, Defouad, Duperier, Juge, & Lagrange, 1998; Trouche, 2000) intensively observed (and questioned and interviewed students on) the use of DERIVE in French classrooms. The researchers found that the teachers were emphasising the conceptual dimensions while neglecting the role of the technical work in algebra learning. However, this emphasis on conceptual work was producing neither a clear lightening of the technical aspects of the work nor a definite enhancement of students' conceptual reflection (Lagrange, 1996). Jean-baptiste Lagrange has furthermore pointed out:

> Technical difficulties in the use of CAS replaced the usual difficulties that students encountered in paper-and-pencil calculations. Easier calculation did not automatically enhance students' reflections and understanding. (Lagrange, 2003, p. 271)

From their observations, the team came to think of techniques as a link between tasks and conceptual reflection, rather than as something that should be eliminated in the learning of mathematics. Lagrange has defined technique as follows:

> A technique is generally a mixture of routine and reflection. It plays a pragmatic role when the important thing is to complete the task or when the task is a routine part of another task. Technique plays an epistemic role by contributing to an understanding of the objects that it handles particularly during its elaboration. It also serves as an object for a conceptual reflection when compared with other techniques and when discussed with regard to consistency. Without CAS, paper-and-pencil techniques cannot be avoided, because of their pragmatic role in mathematics education. Therefore, teachers and researchers tend to consider only their practical role, neglecting the epistemic contribution. (Lagrange, 2003, p. 271)

In an example that Lagrange provides of a problem tackled by 10th graders (about 16 years of age), that of finding a general factorisation for $x^n - 1$, he argues that their poor ability to manage factors in DERIVE could not be separated from their (mis)understanding of the concept of factorisation. In those classrooms where the students had access to DERIVE on their classroom laptops and for personal work, and where the teachers were quite experienced (Mounier & Aldon, 1996), the students were successful. These teachers recognised the need for building techniques in using DERIVE and the epistemic role of these techniques in the understanding of algebra.

A deeper understanding of the complexity of the dialectic between technical and conceptual work can be gained from a reading of two recent doctoral theses in France, one by Luc Trouche (1997) and the other by Badr Defouad (2000). Even though the mathematical content of these two theses is quite advanced with respect to middle and beginning high school algebra (for pupils 11-15 years), the theoretical work related to instrumental genesis (see also Artigue, 2002) offers a tool for analysing transformational activity in technology-supported algebra classrooms at all levels.

Instrumental genesis combines two processes: (a) Instrumentation—by which the subject adapts to the tool; and (b) Instrumentalisation—by which the subject adapts the tool to him/herself. The dynamic interplay between these two processes involves three main components: technique, task, and theory (in the sense of mathematical conceptualisation). According to Lagrange (2000, pp. 16-17):

> Tasks are, first of all, problems. Techniques become elaborated relative to the tasks, and then arrange themselves hierarchically. Official techniques emerge and the tasks become routinised as they become means for perfecting these techniques. The theoretical environment constitutes itself to take into account the techniques, their functioning, and their limits. ... But one must not consider the techniques only under their routinised form. The work of constituting techniques in response to tasks, and of the theoretical elaboration of the problems posed by these techniques, remains fundamental to learning. ... The new instruments of mathematical work are of interest ... because they permit students to develop new techniques that constitute a bridge between tasks and theories. [Translated by the author].

This notion that mathematical theory can develop in the process of acquiring techniques resonates with the results of other recent studies, including those involving younger students. For example, Tenoch Cedillo (Cedillo, 2001; Cedillo & Kieran, 2003) carried out a study with grades 7, 8, and 9 students (13-15 years of age) on initiating them to algebraic activity with the TI-92 calculator. The tasks that the grade 8 students were initially presented with were of the following type:

Initiation into the use of algebraic language

I made the following table (see Table 2.1) using a program:

Table 2.1. *Table Prepared Using a Program*

Input Number	1.1	2.6	3	4.3	5
Output Number	3.2	6.2	7	9.6	11

i) What result will the calculator give me if I type the number 50 into my program? If I type the number 80? If I type the number 274?
ii) What operations did you carry out to obtain these results?
iii) Can you program your calculator to do the same as mine? Write your program below.
iv) Use the program you made to find the numbers missing from Table 2.2.

Table 2.2. *Table to Find Missing Numbers*

Input Number	17	35.02	89.73	107.06	299.1	307.09		
Output Number							511	613.03

v) What operations did you carry out to obtain the values associated with 511 and 613.03?

One of the principal findings of this long-term study was that students learned about algebraic code by using it. They were not given any definitions or rules prior to their working with the given tasks. They came to view the code as a "way of getting the calculator to do what they wanted it to do". As they perfected their technique of "writing programs" (e.g., for the above task, $2a + 1 \mid a = \{1, 3, 5, 7, 9\}$) that would produce the desired numerical sequences, their theory as to the meaning of letters and expressions deepened:

Student: The letter I use in a program serves to make the calculator recognise any number I input ... As many numbers as you want ... May I use the calculator? ... (types the program a*2+3)
If I input 5 the program gives 13, that is... five two times plus 3 ... See ... If I input 9 it does the same ... doubles 9 and adds on 3 ... And so on, it keeps doing the same operations I order the calculator to do, no matter which number you input.

Students' early development of the theory related to equivalent expressions had the same "instrumental" flavour:

Student: Two programs are equivalent if they produce the same values.

Findings from this growing body of research encourage us to think of techniques and conceptual understanding as an interrelation rather than in opposition to each other. Such a stance is quite far removed from the pragmatic, transformation-based approach to algebra that dominated up to the mid-1960s, as well as from its successor that emphasised almost exclusively meaning-building generational activity. We now find ourselves faced with evidence that the transformational activity of algebra can serve as a site for meaning making, that is, that techniques can have an epistemic dimension. It is ironic that, when consideration is given to technique in a technological environment, the epistemic factor can be greater than it ever was thought to be in paper-and-pencil environments. It is thus not only algebraic objects that can be made meaningful in algebra, so too can the processes that manipulate these objects. The manipulative process is, in fact, as much a conceptual object in algebra learning as are the typical algebraic objects—unknown, variable, expression, and equation—and one of the main manipulative processes is that which deals with equivalence of expressions and its conceptualisation.

2.5 Closing Remarks

In closing—and in keeping with the theme of this volume—I see the future of algebra teaching and learning as one geared toward giving meaning not only to the objects of algebra but also to its manipulative processes, and this with the help of technology. As has been emphasised in some recent research, a deeper conceptualisation of the manipulative processes of algebra can be obtained from stressing the mathematically relevant aspects of techniques of manipulation within technological environments. In particular, tasks aimed at creating a more thorough understanding of notions of equivalence have been argued as being especially pertinent.

However, a certain degree of caution is in order regarding the nature of the technologically-supported tasks that are used to give meaning to algebraic objects and processes. Not only the tasks, but certain tools as well, can make it quite easy to sidestep algebraic representation and algebraic transformations (c.f., Hershkowitz & Kieran, 2001). The algebra teacher has a crucial role to play both in bringing algebraic representations to the fore and in making their manipulation by students a venue for epistemic growth.

I began this paper with a description of a three-pronged model of algebraic activity, thus defining the core of algebra in terms of modes of activity rather than content. Even though a considerable portion of the paper has dealt with issues related to generational and transformational activity, this should not be construed as a devaluing of the importance of the global/meta-level activity of algebra. Noticing structure, justifying, and proving have been sorely neglected in school algebra. Nevertheless, my parting remarks do centre on the transformational activity of algebra. I return to Pimm who was quoted earlier as saying that "algebra ... is so much more about doing, is actually about action on things, ... with attention being more on the transformations than on the objects themselves" (Lee, 1997, p. 187). Because of the theoretical advances made recently in the area of transformational activity in CAS technological environments, we now have a lens—one that we have not had up to now—for researching students' emergent conceptualisations of algebraic transformations. The fact that conceptual understanding can come with technique will surely put the study of algebraic transformations among the fruitful and interesting areas of research to be carried out in algebra learning during the years to come.

2.6 References

Artigue, M. (2002). Learning mathematics in a CAS environment: The genesis of a reflection about instrumentation and the dialectics between technical and conceptual work. *International Journal of Computers for Mathematical Learning, 7*, 245-274.

Artigue, M., Defouad, B., Duperier, M., Juge, G., & Lagrange, J.-b. (1998). *Intégration de calculatrices complexes dans l'enseignement des mathématiques au lycée*]. Paris: Université Denis Didirot Paris 7, Équipe DIDIREM.

Bednarz, N., Kieran, C., & Lee. L. (Eds.). (1996). *Approaches to algebra: Perspectives for research and teaching*. Dordrecht, The Netherlands: Kluwer Academic.

Bell, A. (1995). Purpose in school algebra. In C. Kieran (Ed.), New perspectives on school algebra: Papers and discussions of the ICME-7 Algebra Working Group (special issue). *Journal of Mathematical Behavior, 14*, 41-73.

Boero, P. (1993). About the transformation function of the algebraic code. In R. Sutherland (Ed.), *Algebraic processes and the role of symbolism (working conference of the ESRC seminar group*, pp. 48-55). London: University of London, Institute of Education.

Booth, L. R. (1984). *Algebra: Children's strategies and errors.* Windsor, UK: NFER-Nelson.

Cedillo, T. E. (2001). Toward an algebra acquisition support system: A study based on using graphic calculators in the classroom. *Mathematical Thinking and Learning, 3*(4), 221-259.

Cedillo, T., & Kieran, C. (2003). Initiating students to algebra with symbol-manipulating calculators. In J. T. Fey (Ed.), *Computer algebra systems in school mathematics* (pp. 219-240). Reston, VA: National Council of Teachers of Mathematics.

Cerulli, M., & Mariotti, M. A. (2001). L'Algebrista: a microworld for symbolic manipulation. In H. Chick, K. Stacey, J. Vincent, & J. Vincent (Eds.), *The Future of the Teaching and Learning of Algebra* (pp. 179-186). Melbourne, Australia: The University of Melbourne.

Crawford, J. T. (1916). *High school algebra.* Toronto: Macmillan of Canada.

Defouad, B. (2000). *Étude de genèses instrumentales liées à l'utilisation d'une calculatrice symbolique en classe de première S.* Thèse de doctorat. Université Paris 7.

Hershkowitz, R., & Kieran, C. (2001). Algorithmic and meaningful ways of joining together representatives within the same mathematical activity: An experience with graphing calculators. In M. van den Heuvel-Panhuizen (Ed.), *Proceedings of the 25th annual conference of the International Group for the Psychology of Mathematics Education* (Vol. 1, pp. 96-107). Utrecht, The Netherlands: PME Program Committee.

Kieran, C. (1996). The changing face of school algebra. In C. Alsina, J. Alvarez, B. Hodgson, C. Laborde, & A. Pérez (Eds.), *8th International Congress on Mathematical Education: Selected lectures* (pp. 271-290). Sevilla, Spain: S.A.E.M. Thales.

Kieran, C., & Wagner, S. (1989). The Research Agenda Conference on algebra: Background and issues. In S. Wagner & C. Kieran (Eds.), *Research issues in the learning and teaching of algebra.* Reston, VA: NCTM; Hillsdale, NJ: Erlbaum.

Lagrange, J.-b. (1996). Analysing actual use of a computer algebra system in the teaching and learning of mathematics. *International DERIVE Journal, 3*, 91-108.

Lagrange, J.-b. (2000). L'intégration d'instruments informatiques dans l'enseignement: Une approche par les techniques. *Educational Studies in Mathematics, 43*, 1-30.

Lagrange, J.-b. (2003). Learning techniques and concepts using CAS: A practical and theoretical reflection. In J. T. Fey (Ed.), *Computer algebra systems in school mathematics* (pp. 269-284). Reston, VA: National Council of Teachers of Mathematics.

Lee, L. (1997). *Algebraic understanding: The search for a model in the mathematics education community.* Unpublished doctoral dissertation. Université du Québec à Montréal.

Lee, L., & Wheeler, D. (1987). *Algebraic thinking in high school students: Their conceptions of generalisation and justification* (research report). Montréal, QC: Concordia University, Mathematics Department.

Love, E. (1986). What *is* algebra? *Mathematics Teaching, 117*, 48-50.

Mason, J. (1996). Expressing generality and roots of algebra. In N. Bednarz, C. Kieran, & L. Lee (Eds.), *Approaches to algebra: Perspectives for research and teaching* (pp. 65-86). Dordrecht, The Netherlands: Kluwer.

Mounier, G., & Aldon, G. (1996). A problem story: Factorisations of $x^n - 1$. *International DERIVE Journal, 3*, 51-61.

Sutherland, R. (1990). The changing role of algebra in school mathematics: The potential of computer-based environments. In P. Dowling & R. Noss (Eds.), *Mathematics versus the National Curriculum* (pp. 154-175). London: Falmer Press.

Sutherland, R. (1997). *Teaching and learning algebra pre-19* (report of a Royal Society / JMC Working Group). London: The Royal Society.

Sutherland, R., & Rojano, T. (1993). A spreadsheet approach to solving algebra problems. *Journal of Mathematical Behavior, 12*, 353-383.

Trouche, L. (1997). *À propos de l'apprentissage des limites de fonctions dans un environnement calculatrice, étude des rapports entre processus de conceptualisation et processus d'instrumentation.* Thèse de doctorat. Université Montpellier II.

Trouche, L. (2000). La parabole du gaucher et de la casserole à bec verseur: Étude des processus d'apprentissage dans un environnement de calculatrices symboliques. *Educational Studies in Mathematics, 41*, 239-264.

Chapter 3

Responses to 'The Core of Algebra'

Laurinda Brown and Jean-Philippe Drouhard
University of Bristol, UK, and IUFM & IREM de Nice, UMR ADEF, France

3.1 Response 1: The Challenge of Learning Through Transformational Algebraic Activity

Laurinda Brown

3.1.1 Introduction

Thinking about students' need to use algebra, Kieran's (1996) model (see Chapter 2 in this volume) with its three types of algebraic activity—generational, transformational, and global/meta-level—prevents us placing too strong an emphasis on the transformational aspects of algebraic thinking. This raises the question of how algebra can be taught without starting by rehearsing the (seemingly meaningless to the students) transformations? Recognising that transformational activities *are* still important, however, means that we need to re-examine the role of transformation within algebraic thinking and learning. How can this be done without losing focus on mathematics that means something to students?

In my own work, the bridge between tasks and theories is the use of algebra in all of the mathematical activity engaged in by students. They always work on the global/meta-level tasks that permeate all mathematics such as proof (convincing and explaining) and structure. I feel that emphasising the meta-level reasoning aspects of algebraic work allows transformational skills to develop in a natural way since meaning supports the manipulations. Previous to reading Kieran's chapter, however, I had not considered that, similarly, transformational activity, if developed from contexts full of meaning, supports the development of new theory.

In this response I will explore these ideas within the context of work in secondary classrooms in the UK, using examples of practice with students aged 11-12 years old and their teacher Alf Coles, a co-researcher and co-teacher. This work is fully reported by Brown and Coles (1999) and Coles and Brown (2001). When we started to use Kieran's categories the global/meta-level activity seemed too all-encompassing, but it became the most important of the three activities because it pervaded the work and gave purpose to the activities. We provided students with contexts which encouraged them to seek reasons for why something worked; contexts that we recognised could be explored further with the tool of algebra. Alf demonstrated the use of algebraic thinking and manipulation, so that students could sense the power of this tool for explaining and justifying. The motivation was there to use transformational activity within all their mathematics, and justifying and proving became part of the culture of the classroom. Within this context and without explicitly teaching the students transformational techniques, algebraic activity became a natural part of the learning of mathematics, a mode of expression that was available to be used. Transformational activity was at the service of the meaning the students were creating for their mathematical work. It is within this context that I will look at one extended example and share my thoughts in relation to the issue of re-stressing transformational algebraic activity within the mathematics curriculum.

3.1.2 Using a functional approach

Before looking at a series of lessons in detail, consider the following problem, which might be presented to students as a *Guess My Rule* game. The game is played interactively. A number is written in the left-hand column and the person who knows the rule calculates the output that is needed in the right-hand column. The students start making their own suggestions for output values and are told by the rule-knower if their suggestions are correct or not (see Figure 3.1). There can be many reasonable and justifiable offers before they find the one that fits the rule-knower's rule. It is not necessary to get the right answer straight away but it is important that there are reasons for suggesting outputs.

2 ⟶ 15
1 ⟶ 8
4 ⟶ 29 ☹ 30 ☹ 35 ☺
3 ⟶

What's the rule?

Figure 3.1. Some results of the *Guess My Rule* game.

Students might generate the rule for the function in a number of different ways. For example the first two examples might suggest that the rule was multiply by 7 and add 1. Offering 29 as the number linked to 4, however, got a sad face, as it did not match the answer from the rule-knower's rule. Other students may identify different patterns, and so give alternative rules. Someone who spotted the pattern in the factors might give a factorised form (see (a) in Figure 3.2), or the pattern associated with the squares might give the squared numbers form (see (b) in Figure 3.2).

			(a)	(b)
0	→	3	1×3	4
1	→	8	2×4	9
2	→	15	3×5	16
3	→	24	4×6	25
4	→	35	5×7	36
x	→		$(x + 1) \times (x + 3)$	$(x + 2)^2 - 1$

Figure 3.2. Results of the *Guess My Rule* game expressed (a) in factorised form and (b) as a square pattern.

The natural question for students to ask given that the class is at this stage is: "Do both rules for this game work?" Other natural questions, given that both forms do seem to give the same output, are: "How or why are they both the same?" and "Can I get from one form to the other?" In more formal terms, the question becomes "Can I demonstrate the equivalence of the algebraic expressions?" This is a motivation for transformational activity. The students know that the expressions must be the same in some sense and so there is a self-checking mechanism for their explanations.

This is similar to the feedback that technology can provide. Using technology such as graphics calculators together with transformational activity, it then becomes possible to investigate the equivalence of the expressions. Students can work out on their graphics calculators that both formulae give rise to the same graphs but the natural question is why are the graphs the same when the expressions are different? This stimulates the need for transformation. Equivalence seems to be a powerful tool in the process of learning through the transformational activity.

3.1.3 Generalising arithmetic

What seems to be important is that students are asking their own questions within the task they are working on and wanting to try to answer them. What follows is an edited lesson write-up, illustrating what happened with Alf's class during the first lesson sequence of the school year with a new group of 11-year-old students entering secondary school. This activity could be said to be generalising arithmetic, but there are undercurrents of other important algebraic activity. The course of the lessons is influenced by the complex interactions of teacher (*Alf*) and the students.

Alf: I want you to write down in your books a three-figure number, where the first number is bigger than the last. I'm going to pick 472 for mine but yours should be different. (Alf writes this number on the board).

Okay, underneath your number will you write the same number the other way around. So I'm going to put 274.

All done that? Right. I'm going to ask you to subtract your second number from your first. It doesn't matter what method you use.

Okay, I'm going to do mine on the board. (Alf carries out the subtraction.) So, my answer is 198.

Now swap your answer around like you did last time, so I'm going to get 891.

This time I want you to add both your numbers. (See Figure 3.3 for the calculations shown on the board.)

		4	7	2
	−	2	7	4
		1	9	8
	+	8	9	1
	1	0	8	9

Figure 3.3. Alf's calculations written on the board.

Student: Sir, Nina's got the same as me.

Student: And Sophie's got the same as well.

Alf: *What did you all get?*

Student: 1089. (Many other students have this too.)

Alf: *Okay, we've talked about how asking questions is one of the key parts of becoming a mathematician, so what questions can we ask here?*

Student: Why does it work?

Alf: *There's a question before that.*

Student: How does it work?

Alf: *Yes, or even, does it always work? One challenge that I set you was: can you find me one that doesn't work?*

Student: I can't find one, sir.

Alf: *What do you mean?*

Student: They all come out 1089.

Alf: *Do you have any idea why?*

Student: No.

Alf: *Did you notice anything about the answers* [i.e., the intermediate results such as 198 above]?

Student: Yes, they all have a 9 in the middle.

By the fourth lesson (of seven on this activity) different students were working on different challenges (e.g., some were working on the same process with four digit numbers, sorting out how to subtract, trying to find a way of convincing themselves that three digit numbers always go to 1089).

By the sixth lesson a number of students were wondering why 9s appeared in various places in the calculation and were working on this question numerically. Everyone in the room was able to carry out the process of calculation but none had introduced algebraic notation and it seemed appropriate that Alf demonstrate the result for three digits algebraically. Alf describes how the following sequence took place:

> With twenty minutes to go I stopped everyone and went through the work shown here (see Figure 3.4), which I introduced as a way of proving what we found out for three digit numbers. Alongside the algebra I followed step by step with a numerical example and at each stage of both the numerical and algebraic example I elicited answers for what to write from the class. I then wiped the proof off the board and set the class the challenge of reproducing it and then extending it to prove things they had found out about the problem with different numbers of digits.

	~~a~~$^{a-1}$	~~b~~$^{b-1+10}$	c^{+10}		~~7~~^{6}	$\overset{1}{\cancel{5}}^{4}$	$^{1}2$
−	c	b	a	−	2	5	7
	a − c − 1	9	c + 10 − a		4	9	5
+	c + 10 − a	9	a − c − 1	+	5	9	4
1	$_{1}0$	8	9	1	$_{1}0$	8	9

Figure 3.4. Alf's algebraic proof with numerical example.

We judged that the students were comfortable enough with the process of the calculations to be able to look at that process as an object and to recognise that the algebraic transformational activity was capturing that process.

There was a moment of recognition amongst the students during the demonstration, when the 9 arrived in the central place of the subtraction. One student articulated "that's where the 9 comes from". The possibility of using algebra to know why things worked was now part of the culture of the classroom. We were not concerned at this time with how many students could perform the demonstration autonomously. Some students went back to their own train of thought, others tried to reproduce the process, and still others attempted to see what happens with two-digit numbers directly.

We work with a sense of algebra as an evolving language that can emerge from situations and contexts that are already laden with meaning. Algebra can be used to express and offer insights into those situations. It is in this emergent expression and consequent empowerment that students can discover a need for algebra. In the illustrative lesson the students are also asking questions and being introduced to tools for answering them without this being seen as prescriptive.

3.1.4 Conclusion

The crucial point is the power of algebra to gain insight into something that you could not do without it:

> Algebraic symbolism ... introduced from the very beginning in situations in which students can appreciate how empowering symbols can be in expressing generalities and justifications of arithmetical phenomena ... in tasks of this nature manipulations are at the service of structure and meanings. (Arcavi, 1994, p. 33)

Arcavi's point applies not only with arithmetical phenomena, but with any phenomena of objects and relationships between them. The different frameworks proposed by various authors to define the core of what algebra is, can each be used to give perspectives on algebraic activity. Sometimes these activities will be generational, sometimes transformational, sometimes functional, and sometimes involving generalised arithmetic, but all are encompassed by the global/meta-level activities that give purpose for algebra.

3.1.5 Acknowledgements

Thanks are extended to Alf Coles and his students who allowed me to participate in their mathematics lessons, taking observation notes which Alf augmented for the lesson write up.

3.1.6 References

Arcavi, A. (1994). Symbol sense: Informal sense-making in formal mathematics. *For the Learning of Mathematics*, 14 (3), 24-35.

Brown, L., & Coles, A. (1999). Needing to use algebra - A case study. In O. Zaslavsky (Ed.), *Proceedings of the 23rd annual conference of the International Group for the Psychology of Mathematics Education* (Vol. 2, pp. 153-160). Haifa, Israel: Program Committee.

Coles, A. & Brown, L. (2001). Needing to use algebra. In C. Morgan and K. Jones (Eds.), *Research in Mathematics Education* (Vol. 3, pp. 23-36). Hampshire: British Society for Research into Learning Mathematics.

Kieran, C. (1996). The changing face of school algebra. In C. Alsina, J. Alvarez, B. Hodgson, C. Laborde, & A. Pérez (Eds.), *8th International Congress on Mathematical Education: Selected lectures* (pp. 271-290). Sevilla, Spain: S.A.E.M. Thales.

3.2 Response 2: The Semiotics that Lie in the Core of Algebra

Jean-Philippe Drouhard

3.2.1 Introduction

As a reaction to Carolyn Kieran's chapter I will ask, and propose an answer to, four questions: Why is the question of the core of algebra so difficult? What role does semiotics plays in algebraic activities? Then, Is algebra a language? And finally, Why is the question of the core of algebra so important for the future?

3.2.2 Why is the question of the core of algebra so difficult?

The question of "What are the essential ingredients of school algebra?" is a quite thorny problem. As it appears in Kieran's quotations of Lee (1997), every answer may be at the origin of large discussions, since focusing on one essential feature lets in the shadow of the other—also essential—features. I attempt to explain why the core of algebra is difficult to define. There are two main ways to address the question of the nature of algebra: one is based on the nature of the *problems* you can solve with algebra ("What is algebra for?" says Alan Bell quoted in Lee (1997)), and the other is based on the specific "algebraic" *way* to solve the problems. Taking one path or another, you give priority to one type of characterisation (for instance, emphasising the notion of generalisation, or the role of the unknown) or another (for instance, emphasising the notion of meaning of the role of the variable). But what makes the question so uneasy is that these two types of characterisation are just *a bit* different. Different enough to make difficult the finding of a unique set of features to define the core of algebra; but similar enough to give the impression that everybody is speaking about the same thing—algebra.

Why it is so difficult to separate the *problems* approach from the *way of doing* approach? Well, the answer is to be found in the history and the evolution of algebra itself. Actually there is a *co-evolution* between the class of problems that can be solved algebraically and the way to solve them. Theoreticians of evolution call the parallel evolution of strongly related species co-evolution—for example, the parallel evolution of a flower and a butterfly who gathers nectar from this flower (see Gould, 1980). Neither the class of problems nor the way of doing can be properly defined with static, absolute terms, without referring to their historical (co-)evolution. There is no strict term-to-term correspondence between the problems and the way of solving them, either. New problems may be at the origin of new *procepts*. Gray and Tall (1994) coined the term *procept*. It describes symbols that represent both a process and the object of that process. The symbolic description of new mathematical processes can produce new mathematical objects. For example, finding the solutions of cubic and fourth degree equations much later led to negative and complex numbers, and the problem of looking for solutions of fifth degree equations later led to group theory. On the other hand, the introduction of a new *procept* may open a new field of problems. For example, the invention of letters, by François Viète in 1591 (see Dahan-Dalmedico & Peiffer, 1986), to represent (what we now call) parameters may be considered as the origin of an actual change of paradigm (using the word in the sense introduced by Thomas S. Kuhn (1970)). Clearly, doing algebra using letters as parameters is very different from doing algebra using particular numbers as coefficients—as occurred before Viète—or even later (the last of mathematicians who used the *old numerical fashion* was Pascal in some of his demonstrations, including his famous *triangle*).

Kieran's elegant solution to solve this difficulty was to shift slightly the problem and to focus on the question of *algebraic activity*. It is elegant because the notion of activity encompasses both aspects: algebraic activity is specific to algebraic problems and to an algebraic way of doing.

3.2.3 What role does semiotics play in algebraic activities?

Kieran's model of algebraic activities is based on three types of activities. What do these three types (generational, transformational, and global/meta) have in common? This question involves another one, more deeply related to the content of the model. An activity is *acting on something*, which can be called the *object* of the activity (actual or virtual, abstract or concrete, symbolic or not, that does not matter). So, what is, or, to put it in a better way, what *are* the objects of the algebraic activity? The answer to this question is that doing algebra is acting *on signs*. Signs are the objects of the algebraic activity. Letters, expressions, equations, graphs, written calculations, schemes, proofs, models, and so on, are signs and are made of signs.

But, a sceptical reader (Mr Sceptical) could object, "Everything in mathematics is expressed by ways of signs!" Carolyn Kieran herself presented, in a previous paper (1991), an essential distinction which can lead to an answer to our sceptical reader, calling *structural* the very specific way to act *on* signs, opposed to acting *with* signs (she called it *procedural* activity).

Procedural activity refers basically to arithmetic operations carried out on numbers to yield numbers. For instance, if we take the algebraic expression, $3x + y$, and replace x and y by 4 and 5, the result is 17. Another example involves the solving of $2x + 5 = 11$ by substituting various values for x until finding the correct one. In these two examples, which look ostensibly algebraic, the objects that are operated on are not the algebraic expressions but their numerical instantiations. Furthermore, the operations that are carried out on these numbers are computational—they yield a numerical result. Thus, both of these examples illustrate a procedural perspective in algebra.

The term *structural* refers, on the other hand, to a different set of operations that are carried out, not on numbers, but on algebraic expressions. For example, if we take the algebraic expression, $3x + y + 8x$; this can be simplified to yield $11x + y$ or divided by z to yield

$$\frac{3x + y + 8x}{z}$$

Equations such as $5x + 5 = 2x - 4$ can be solved by subtracting $2x$ from both sides to yield $5x - 2x + 5 = 2x - 2x - 4$ which can be subsequently simplified to $3x + 5 = -4$. In both of these examples, the objects that are operated on are the algebraic expressions, not some numerical instantiations. The operations that are carried out are not computational. Furthermore, the results are algebraic expressions.

So, we could reply to our sceptical reader that he is partially right: signs are everywhere in mathematics, but they are not used in the same way in every part. Roughly said, before algebra (and also in classical geometry) we act *with* signs, while in algebra and after, we work both *with* and *on* signs. This idea is what we want to express when speaking of *semiotics that lies in the core of algebra* in the title of this reaction to Kieran's chapter.

3.2.4 Then, is algebra a language?

Let us resume our discussion (the author, *Author*) with our imaginary reader, Mr Sceptical.

Mr Sceptical: So, you say that doing mathematics is working both *with* and *on* signs?"

Author: Well, actually this is doing algebra.

Mr Sceptical: Yes. Anyway, if you believe that doing algebra is *just* working with and on signs, then you assume that algebra is just a matter of *signs*; it is *just* a language. You revealed your true nature, you are a formalist, maybe worse, you are an old Bourbakist dinosaur!

Author: No. Algebra is not a language: algebra has a language, which is totally different.

We could explain to our fictitious reader that he made a logical mistake: we claimed that semiotics lies in the core of algebra, not that all algebra is semiotics.

We assume that language aspects cannot be removed from algebraic activities; this in no way means that we would like to reduce algebra to a mere language (or to a *language game*). However, it is not possible to carry on the discussion about the relationship between algebraic activities and language much further, since to do it properly, we would need a good (and common) definition of *language* (or *signs*). An extended discussion about the notions of language and semiology can be found in Chapter 9 of this book.

3.2.5 Why is the question of the core of algebra so important?

Mathematics educators are sometimes suspected of wasting their time in too general or abstract questions like the one concerning the core of algebra, and are urged to address questions more related to everyday classroom practice. The underlying prejudice is that theories have little to do with practices. We researchers in mathematics education know that this is not true, for more than one reason. Maybe the first reason is that theoretical ideas are not just a matter for theoreticians. Everybody who teaches or learns mathematics has his/her own answers to questions as theoretical as "What is algebra?"—this is clearly demonstrated by Lesley Lee's (1997) study quoted in Carolyn Kieran's lecture. Empirical evidence can easily be found in the literature on teachers' beliefs and practices. For example, Drouhard and Panizza (2003) present a model of knowledge where this kind of belief (on what algebra is) is classified as "third order knowledge" and plays an important role on learning and teaching mathematics. (See also Chapter 10 of this book that has a focus on teachers' knowledge.) The point is that the way you teach algebra depends dramatically on what you believe algebra is; therefore this question is worth addressing, particularly in a book on the *future* of teaching and learning algebra.

3.2.6 References

Dahan-Dalmedico, A., & Peiffer, J. (1986). *Une histoire des mathématiques, routes et dédales.* Coll. Points Sciences, Paris: Le Seuil.

Drouhard, J-Ph., & Panizza, M. (2003). What do the students need to know, in order to be able to actually do algebra? The three orders of knowledge. Paper presented to the 3rd European Conference on Research on Mathematics Education (CERME3), Bellaria, Italy. To appear in the *Proceedings of CERME3.*

Gray, E., & Tall, D. (1994). Duality, ambiguity, and flexibility: A proceptual view of simple arithmetic. *Journal for Research in Mathematics Education, 25*(2), 116-140.

Gould, S. J. (1980). *The panda's thumb.* New York: W. W. Norton.

Kuhn, T. (1970). *The structure of scientific revolutions.* Chicago: The University of Chicago Press.

Kieran, C. (1991). A procedural-structural perspective on algebra research. In F. Furinghetti (Ed.), *Proceedings of the 15th annual conference of the International Group for the Psychology of Mathematics Education* (Vol. 2, pp. 245-253). Assisi, Italy: Program Committee.

Lee, L. (1997). *Algebraic understanding: The search for a model in the mathematics education community.* Unpublished doctoral dissertation. Université du Québec à Montréal.

The Working Group on Early Algebra

Leaders: Romulo Lins and James Kaput

Working Group Members:
Maria Blanton, Bárbara Brizuela, Thomas Carpenter, David Carraher, Deanna de'Liberto, Megan Franke, Toshiakira Fujii, Lesley Lee, Milosav Marjanovic, John Olive, Analúcia Schliemann, Elizabeth Warren, and Gaye Williams.

The Working Group on Early Algebra. Seated (L to R): Gaye Williams, Bárbara Brizuela, Analúcia Schliemann. Standing (L to R): Toshiakira Fujii, Thomas Carpenter, Maria Blanton, Lesley Lee, James Kaput, John Olive, Romulo Lins, David Carraher, Elizabeth Warren, Milosav Marjanovic. Absent: Deanna de'Liberto, Megan Franke.

The majority of members of this group came from the USA (Maria Blanton, Bárbara Brizuela, Thomas Carpenter, David Carraher, Deanna de'Liberto, Megan Franke, James Kaput, John Olive, Analúcia Schliemann). However other countries were represented, Australia (Elizabeth Warren, Gaye Williams), Brazil (Romulo Lins), Canada (Lesley Lee), Japan (Toshiakira Fujii), and Yugoslavia (Milosav Marjanovic).

Prior to the Conference, each member of the Working Group prepared a paper for the 12th ICMI Study Conference Proceedings. These papers reflected members' interests and prior experiences in teaching and researching aspects of early algebra. The individual authors can be contacted using their e-mail addresses listed at the back of this book. The authors (sometimes with co-authors) and the titles of their papers are listed:

- Maria Blanton & James Kaput: *Algebrafying the elementary mathematics experience Part II: Transforming practice on a district wide scale* (pp. 87-95).
- Bárbara Brizuela & S. Lara-Roth: *Additive relations and functions tables* (pp. 110-119).
- Thomas Carpenter & Megan Franke: *Developing algebraic reasoning in the elementary school: Generalization and proof* (pp. 155-162).
- David Carraher, Bárbara Brizuela, & D. Earnest: *The reification of additive differences in early algebra* (pp. 163-170).
- Toshiakira Fujii & M. Stephens: *Fostering an understanding of algebraic generalisation through numerical expressions: The role of quasi-variables* (pp. 258-264).
- James Kaput & Maria Blanton: *Algebrafying the elementary mathematics experience Part I: Transforming task structures* (pp. 344-352).
- M. Kinzel & Deanna de'Liberto: *A framework for assessing algebra readiness* (pp. 721-724).
- Lesley Lee: *Early algebra – But which algebra?* (pp. 392-399).
- Miloslav Marjanovic & D. Kadijevic: *Linking arithmetic to algebra.*
- Analúcia Schliemann, S. Lara-Roth, & A. Goodrow: *Tables as multiplicative function tables* (pp. 531-540).
- L. Steffe: *What is algebraic about children's numerical operating?* (pp. 556-563). John Olive presented this paper.
- Elizabeth Warren: *Algebraic understanding: The importance of learning in the early years* (pp. 633-640).
- Elizabeth Warren & T. Cooper: *Theory and practice: Developing an algebra syllabus for Years P – 7* (pp. 641-648).

The members of the group regularly worked together and, on occasions, in small groups. Group discussions were quite open ended and provided ample opportunity for each member to briefly present his/her own work and for others to comment on, without pre-set questions. The focus of the Working Group became the approaches to early algebra that the members espoused. The members also devoted time to preparing the final presentation and this chapter was based on its structure.

The thoughtful and co-operative work of all of the participants of the Working Group on Early Algebra is gratefully acknowledged. Their contributions to the Conference Proceedings and discussions provided important insights that assisted Romulo Lins and James Kaput during the writing process.

Thanks are extended to Romulo Lins and James Kaput for their joint authorship of this chapter and to Lesley Lee for her gracious and patient review of earlier versions of the chapter. Romulo and James are also congratulated and thanked for their leadership of the Working Group on Early Algebra.

Chapter 4

The Early Development of Algebraic Reasoning: The Current State of the Field

Romulo Lins and James Kaput
Mathematics Dept/PGEM, UNESP-Rio Claro, Brazil, and Department of Mathematics, UMass-Dartmouth, MA, USA

Abstract: The main aim of this chapter is to argue that an early start to algebra education is possible and of great relevance for mathematics education because it provides a special opportunity to foster a particular kind of generality in our students' thinking. To argue this, we map the various views on algebra education found historically, and trace how the perceptions that mathematics educators hold about children's thinking and learning have changed. Overall, a great realisation that children can do more in mathematics than was previously believed leads to the adoption of more ambitious objectives for the initial years of school, and to the development of new classroom approaches to algebra education in the early grades. That does not mean teaching the *same old school algebra* in the same usual way to younger children, but rather to introduce them to new algebraic ways of thinking and immersing them in the *culture of algebra*. The chapter ends with a research agenda to further developments in this particular sub-field of mathematics education.

Key words: Early algebra, algebraic thinking, generalised arithmetic, symbolic arithmetic

4.1 Introduction

In this chapter, we attempt to provide the basis for an understanding of the early development of algebraic reasoning and the larger views of algebra education in which this may occur. Our intent is to help *algebra educators* move forward in the task of creating new approaches to algebra education that incorporate both the practices of the past that proved fruitful and the new possibilities offered both by the available technology and by recent views of cognition and learning. To achieve this, we step back to analyse the expression *algebra education*, in order to move beyond a content-centred characterisation.

Defining *algebra* is fraught with difficulty, especially if one expects tight and closed epistemological definitions, because what one takes to be algebra depends on many cultural and other factors that vary widely across and even within communities. Some of this variation is reviewed in Chapter 13 of this book. Nonetheless, we have been able to agree provisionally on two key characteristics of *algebraic thinking*. First, it involves acts of deliberate generalisation and expression of generality. Second, it involves, usually as a separate endeavour, reasoning based on the *forms* of syntactically-structured generalisations, including syntactically and semantically guided actions. This is a characterisation of the broad kinds of algebraic reasoning that helps us then discuss forms of algebraic thinking appropriate for young children and the conditions that may promote them. Among such conditions, for instance, is a need for greater integration of different mathematical topics, in order to promote the development of algebraic forms of thinking, which would yield better problem-solving abilities in students. Another consideration is the recognition that algebraic thinking empowers students by providing tools that allow a great degree of certain types of generality, something that has, of course, been taken to be true for a long time, but this time considering the empowerment of much younger students than usual.

The theme of this chapter, *early algebra*, allows us to look both sideways (how to integrate algebra education with other topics at all levels of schooling) and ahead (the implications of what is done *early* for the following grades). In other words, instead of algebra education being restricted to a more narrowly defined grade band or narrow sequence of courses or learning environments, we can examine the possibility of creating a *new algebra world* from the beginning. Because of this, Section 4.3 (about the implications) and Section 4.4 (a research agenda), assume a quite important role in this chapter, as it is there that we argue how the suggestions and indications of the more specific discussion on early algebra could become part of the bigger picture.

Two understandings of what *early algebra* means now seem to be current. The first, and for many years the more ubiquitous, refers to the first time students meet algebra in school. For many different reasons—sometimes tradition, sometimes dominant theoretical positions, sometimes the impact of published studies—that first encounter was likely to happen when students were about 12-13 years old, in some cases even older. This first understanding of early algebra applies to most of the other work reported in this book, including most but not all of the discussion on approaches to algebra (see Chapter 5). The second understanding, which only slowly and more recently has been gaining ground in the mathematics education community, takes *early algebra* to refer to the introduction of students to algebraic reasoning at a much earlier age, sometimes as young as seven years old. The approach we take in this chapter is to focus on fostering the development of *algebraic thinking*, and not at the teaching and learning of specific bits of algebra content. Whatever content or activity is useful in helping the teacher to achieve that goal might become part of early algebra. It would be impossible, of course, in the body of this chapter, to present the kind of examples that show more clearly how this can be done, so we strongly suggest that the reader take the many references we point to as an important follow-up to this chapter.

We will argue that the increased acceptance of the second view is related to the fact that it is only more recently that the mathematics education community began to realise seriously that younger children could do much more than was previously supposed. The old supposition is a consequence of the already mentioned combination of tradition and dominant theoretical positions acting as constraints and blinders, though we think that other factors are at work sustaining the historically received view of algebra education.

Changes in the views on what is learning and how formal education should be organised to integrate those new views, led to a more enlightened view of the way mathematics educators saw children's work. In coming to be seen as a truly long-term process, algebra education began to incorporate the idea that getting accustomed to particular aspects of algebraic activity (e.g., formulae and literal notation as well as written expressions containing indicated operations) was as relevant as mastering the syntactical structures of traditional formalisms.

Below we will argue that an early start in algebra education is not only possible but is necessary, and will focus on the different forms such early starts might take and the key assumptions that they are based upon.

4.2 Algebra Education in the Past

To understand the significance of the approaches to early algebra proposed in this chapter, it is necessary to examine the developments that preceded their emergence. Those developments may be grouped into three periods. During the first period, tradition ruled unchecked, reigning only for the reason of being tradition, and without support other than experience. The second period saw research begin to investigate the processes underlying the approaches adopted by the traditions of the first period. Finally, during the third period, the view that arithmetic should precede algebra began to be examined.

In this section, our overall aim is to give an overview but not, to any extent, to present a thorough literature review. The papers mentioned were chosen only for their exemplary character with respect to the points being discussed.

4.2.1 Relationship between arithmetic and algebra in school traditions

When we look at school traditions in different countries, the relationship between algebra and arithmetic is almost always characterised as *algebra is generalised arithmetic*. Lee (1997) reminds us that, in 1929, the advice of the British Mathematical Association was that "Historically, algebra grew out of arithmetic and so it ought to grow afresh for each individual" (p. 219). She goes on to say that:

> Chevallard, who undertook the examination of school textbooks over the centuries, certainly confirms that this was the direction taken in the introduction of algebra in schools up until the *new math* reforms of the 1960s. The justification and motivation for algebra lay in the presentation of solutions to some traditional arithmetic problems using the tools of algebra. (p. 211)

It seems safe to say that, even today, the *arithmetic then algebra* tradition persists in most countries, perhaps with a new justification added, namely, that (school) algebra is more abstract (and so, more difficult) than arithmetic, which is more concrete (and so, easier). Although some researchers strongly deny that claim (for example, Davis (1975, 1984) pointed to the complexity of certain arithmetic operations compared to core algebra activities such as solving linear equations), it is indeed a dominant view, and the reason for this can be found in the strong dominance of Piagetian constructivism. As algebra would require formal thinking, while arithmetic would not, and as formal thinking would correspond to a later developmental stage, algebra should come later than arithmetic (see Petitto (1979) for an explicit analysis of this assertion in relation to a series of teaching experiments). This is a very simplified version of the argumentation, but it contains the essential elements.

The work of Dietmar Küchemann for the Concepts in Secondary Mathematics and Science (CSMS) project, in the late 1970s and early 1980s, combined those two views, the *algebra as generalised arithmetic* and the Piagetian developmental one. On the one hand, although the original book-report from the CSMS survey refers to Küchemann's study under the heading of *Algebra* (Hart, 1984), Küchemann himself (1978, 1984) refers to it as an investigation of children's understanding of generalised arithmetic. On the other hand, the most visible result of Küchemann's work is a reported link between different uses of letters in *generalised arithmetic* and Piaget's levels of intellectual development. The Booth (1984) follow-up study, however, showed that suggestions about student learning made by the CSMS report were unconfirmed. It also indicated that appropriate teaching could eliminate a number of the reported error patterns.

One could argue that the choice of *generalised arithmetic* corresponded to a willingness to separate school algebra from abstract algebra. This would be a reasonable point particularly after the changes in mathematics that happened since the second half of the 19th century. In traditional school algebra, letters always stand for numbers, but in abstract algebra letters can stand for elements from any set where appropriate combining operations have been defined, including permutations, matrices, geometric transformations, and entirely abstract elements. Quite recently Nicholas Balacheff (2001) proposes a distinction between *symbolic arithmetic* and *algebra* in the editors' postscript to the book *Perspectives on School Algebra.*

> The students' solving world will contain symbolic representations (we may call them algebraic) as well as means to manipulate them, but the control structures—all through the solving process—will still refer to the external world of reference attached to the situation by the problem statement. ... Algebra is not there, but instead we see the functioning of what I would call symbolic arithmetic ... the forms of algebraic expressions. (p. 255)

In any case, with the exception of the pioneering work of Davydov and his colleagues in the former Soviet Union (1962, 1975, 1982, 1983), and, to some extent, parts of the work of Dienes (1973), up to the early 1990s practically all the attention of algebra educators was focused either on producing systems of *stages* related to the learning of algebra (developmental or otherwise) or a compendium of difficulties and their sources. We now consider some of that work.

4.2.2 Research on algebra education up to the 1990s

We begin with some exemplary efforts to produce systems of *stages* that could be related to the learning of algebra (although not always exclusively).

Küchemann has already been mentioned as having attempted to link different uses of letters in *generalised arithmetic* to the stages of development in Piaget. Biggs and Collis (1982) proposed the SOLO Taxonomy in which the structure of the observed responses was to be characterised (as uni-structural, multi-structural, relational, and extended abstract responses), rather than characterising the subject or the expected responses. They used algebra items among their examples and the discussion of those examples resonated with the research in algebra education at the time.

Garcia and Piaget (1984) argued that the mechanisms of transition between historical periods are analogous to those found in the transition between psychogenetic stages. They describe one of those mechanisms as the process which produces a succession of three stages: *intra-objectal*, *inter-objectal,* and *trans-objectal* (Lins, 1992). They refer to a history of algebra which they claim to have begun only with Vieta, who lived in France from 1540 to 1603, using as reference their own interpretation of Jakob Klein's 1968 classic, *Greek Mathematical Thought and the Origins of Algebra.* Further details of Vieta's work are given in Chapter 8 and other work on early symbolism is presented in Chapter 9 of this book.

Harper (1987) stops well short of Garcia and Piaget. In a paper discussed widely upon its publication, he attempted to correlate different notational presentations of solutions to problems (rhetorical, syncopated, and symbolic) to cognitive development (Lins, 1992). The notational categorisation is directly borrowed from Nesselmann, who presented it in 1842 with the sole purpose to distinguish the *presentation* of solutions, not their *production* (Heath, 1964).

Sfard (1989) proposed a model for concept formation based on the distinction between two ways in which a mathematical expression can be perceived: as a process (the *operational* aspect) or as a product (the *structural* aspect). Central to her model is the assumption that the operational aspect must *necessarily* precede the structural aspect because it is assumed to be less abstract. Later, at a presentation for the Algebra Working Group of ICME 1992, she returned to the same categories Harper had borrowed from Nesselmann, and used them to characterise the algebra taught at different school levels (cf. Lee, 1997).

These are examples—exemplary, though—of the kind of work done during the 1980s and 1990s, to produce normative systems of stages that could be used to inform algebra education. Although their thinking and methods varied, each of these studies contributed to the assumption that algebra was best left for later in school life.

A second broad group of research papers produced at that time was focused on producing a catalogue of students' difficulties with algebra and the sources of those difficulties. Those difficulties were frequently related to a particular set of proposed stages, in the sense of a misfit between stage of development and teaching, but also to issues related to notation (Becker, 1988; Filloy, 1987; Gallardo & Rojano, 1987; Herscovics, 1989; Kirshner, 1987, 1990; Pereira-Mendoza, 1987) and to difficulties caused by an insufficient understanding of arithmetic (Booth, 1989; Kieran, 1981). This group of papers, typical of the time, shows that the mood among the teachers, so to speak, was rather directed towards trying to improve algebra education by understanding what students were failing to do, either because teaching was out of synchrony with intellectual/cognitive development or because teaching failed take into account what particular students had previously failed to learn/understand in their prior school experience. As much as the effort to produce normative sets of stages, the perspectives taken here led, in almost all cases, to an interest in older students (12 years old and older).

Another set of consistently pessimistic studies examined error patterns in students' syntactical symbol-manipulation work, where it was often the case that this kind of activity was implicitly taken to be the essence of algebra (Lewis, 1981; Matz, 1980, 1982; McArthur, Stasz, & Zmunidzinas, 1990; Sleeman, 1986). This work, as well as work on interpretation of variables (e.g., Wagner, 1981) and reading of algebraic expressions (Wenger, 1987), repeatedly illustrated the fragility and superficial nature of student competence in operations on algebraic symbols and their interpretation.

To summarise: up to the early 1990s research in algebra education was focused on the *sad stories*, on what children could *not* do, rather than on ways to explore what they *could* do, and ways to tap the potential for development.

4.2.3 Research on algebra education opens the way for early algebra

If up to the early 1990s there were mainly sad stories, from then on things slowly began to change. In this sub-section we offer examples of *happy stories* reflecting a shift towards optimism in algebra education research in relation to what children could do.

There are three basic types of happy stories here. First, there is research that suggested directly that younger children could do more in mathematics than previously thought, particularly when provided with appropriate experiences and instruction. Second, there is research reporting changes in the perspectives on algebra education and algebraic thinking and third, research advancing the idea of using new technologies in algebra education. The first group is clearly related to early algebra as proposed in this chapter, in a sense that will soon become clearer. The second group refers to work that helped broaden the focus of research on algebra education, making it possible for subsequent work to have a more flexible view of it, and so making early algebra more acceptable. Quite frequently this research brought areas such as linguistics, history of mathematics, and epistemology closer to mathematics and developmental psychology, which dominated previous research on algebra education. On the third group we will comment later.

4.2.3.1 Children can do more if given the opportunity

A typical paper in this group is one by Carpenter, Ansell, Franke, Fennema, and Weisbeck (1993), which reported kindergarten children's problem-solving processes. The results suggest that "children can solve a wide range of problems, including problems involving multiplication and division situations, much earlier than generally has been presumed" (p. 439) and that "if specific multiplication and division schemata are required, these schemata are sufficiently well developed in many kindergarten children that they can solve multiplication and division problems by representing the action and relationships in the problems" (p. 440). The focus was on the problem-solving processes, but with an interest on "the potential for instruction to build upon and extend young children's problem solving processes" (p. 429).

Hativa and Cohen (1995) examined the feasibility of teaching certain negative number concepts and procedures to students of a much younger age than is presently done in schools and concluded with a positive answer. Working with low- and high-achievers, the study found that low-achievers gained at least as much as the high-achievers and suggested that teaching approaches based on students' pre-instructional intuitions can help students progress further than traditionally expected.

In a similar vein, Urbanska (1993) investigated the numerical competence of six-year-old children, concluding both that they have "a considerable degree of numerical competence" (p. 265) but that teachers in her study did not draw on this intuitive knowledge. Mulligan and Mitchelmore (1997, p. 328) argue, "the standard curriculum takes no advantage of the informal understanding that many students have developed well before grade 3."

Papers like these do not depart radically from the dominant theoretical views of the time. Rather, they expanded the boundaries of what could be said within those frameworks. This was a crucial contribution, enabling the idea that *children can do more* to be more readily accepted by the mathematics education community.

More directly to the point of algebra education, Mason (1991, 1996) reflects the optimistic point of view shared by most of the researchers in this group, that students come to school with natural powers of generalisation and abilities to express generality, and that the development of algebraic reasoning is, in large part, a matter of tapping into those naturally occurring capacities for didactic purposes. The pioneering work of Mason and his colleagues (Mason, 1989, 1991, 1996; Mason, Graham, Gower, & Pimm, 1985) provides a wide range of tasks and task-design principles that operationalise this fundamental observation. However, only more recently (with the notable exception of Davydov, discussed below), have there been empirical studies which explore learning and teaching implementing this approach. Much of this work has taken place in the USA and reflects the initiative of the National Council of Teachers of Mathematics (NCTM) in treating algebraic reasoning in a deliberately longitudinal way with roots in early mathematical development (NCTM 1989, 2000). This initiative is a response to a growing realisation of the failure of the approach to algebra in the USA, where it is introduced late, abruptly, and in relative isolation from other mathematics, with a focus on syntactic operation skill (see Kaput, 1998, 1999; Lacampagne, Blair, & Kaput, 1995; Moses, 1995; NCTM & Mathematical Sciences Education Board, National Research Council, 1998). Below, we will report further on the work in the USA spawned by this initiative and comparable efforts in other countries.

If those changes were mainly driven by curricula and failure considerations, work previously begun in the Soviet Union had been driven by theoretical considerations. Davydov's work precedes the more recent push towards building algebraic reasoning in elementary grades. A translation into English of *An experiment in introducing elements of algebra in elementary school* was published in 1962 in *Soviet Education* (Davydov, 1962) and in 1974 Hans Freudenthal published a paper on the Soviet work on the teaching of algebra at the lower grades of the elementary school (Freudenthal, 1974). However, it was only much later, in the 1990s that this work became better known in the West, and for this reason it will be considered among the happy stories.

The so-called Soviet School (Vygotsky, Luria, and Leontiev) proposed that learning precedes development, an assumption diametrically opposed to the

Piagetian idea that development precedes learning. Davydov and his colleagues aimed at producing an approach to primary mathematics that fostered generality in students' thinking by offering them thinking tools (socially, culturally, and historically developed) that could enable them to do things in mathematics that otherwise would take much longer. The case of the whole-part diagram and the use of literal notation are well known (Davydov, 1962). The title of his article, in English, is *An experiment in introducing elements of algebra in elementary school*, and its aim was to examine the level of generality those elements could bring to children's thinking. Davydov was quite naturally interested in primary school children, given his theoretical assumptions. So was Dienes (1973); the difference was that, while Davydov was centrally interested in fostering a mode of thinking, Dienes was interested in developing a concrete meaning for the rules of algebra, in a sense aiming primarily at content. Although Davydov's work showed that children could do more, unlike the papers mentioned above it actually bases that assumption on theoretical grounds. The key point is that children will, indeed, do more, if we offer them access to appropriate cultural tools—for instance, diagrams, and special notations.

This is the kind of research and development work that, so to speak, raised the banner *children can do more*, paving the way for early algebra as presented in this chapter. This was no small deal, given the dominant views at that time. While it is true today that most research in algebra education still falls outside early algebra (that is, it is still directed towards the education of older students), the work of these pioneering researchers and others opened the path for studying an early introduction to the ideas of algebra for mainstream students rather than merely for gifted students.

4.2.3.2 Opening the algebra education door even wider

As we said earlier, much of the research on algebra education up to the 1990s was dedicated either to producing systems of developmental stages or to producing catalogues of errors made by children. This work was oriented to the content of algebra and closely tied to the traditions of mathematics education, including historic relations between school arithmetic and school algebra.

For reasons not so easy to pin down, beginning in the late 1980s and continuing during the 1990s, this began to change, through the use of the history of mathematics as a source of insights into the difficulties students have with algebra (e.g., Gallardo, 1990; Radford, 1995; Sfard, 1995); through a more explicit discussion of the underlying epistemological aspects (e.g., Balacheff & Sutherland, 1994; Brousseau, 1983; Kaput, 1979; Lins, 1992, 1994, 2001; Radford, 1994; Sfard, 1991); and with the broadening of the discussion of linguistic aspects, beyond the usual syntax-semantics dichotomy (e.g., Arzarello, Bazzini, & Chiappini, 2001; Boero, 2001; Kirshner, 1990, 2001; Nemirovsky, 1996; Pimm, 1987; Chapter 9 in this book). Technology acted as a destabilising factor as well, and is discussed further below (see also Chapter 6 and Chapter 7 in this book).

Many of these issues emerged over a few years, in the sessions of the Algebra Working Group at conferences of the International Study Group on the Psychology of Mathematics Education (PME), supported by the diversity of backgrounds and interests of the researchers who took part in it. The book by Sutherland, Rojano, Bell, and Lins (2001) emerged from these discussions. This was most likely a real reflection of what was happening across the world in the algebra education research community.

This enriched perspective on algebra education allowed for more flexible views of what algebra education could or should mean. Levels of intellectual development were, more and more, being considered together with the effects of contexts. The use and learning of natural language was informing our understanding of the use of the language of algebra. We were getting insights on possible sources of difficulties from history. In addition, we observed that young students were facing difficulties that were previously faced by grown-up mathematicians. Had the human brain developed so quickly?

All this contributed to stimulate part of the mathematics education community to take, as we said, a more flexible view of algebra education, helping to open a door to early algebra.

4.2.3.3 New technologies: More challenges, more opportunities

From the late 1980s onwards, the increasing availability of computers and other technologies made it more and more appealing to consider algebra without necessarily associating it with the tradition of having manipulation of algebraic expressions as the core of algebra education. Certainly there was already a push in this direction (Fey, 1989; see also the review in Kaput, 1992), but the new technologies allowed students and teachers to integrate algebraic expressions into richer, more concrete and meaningful contexts, with much greater ease (Fey, 1984).

Probably the first widespread approach linking computers and algebra education was the use of programming languages (Camp & Machionini, 1984; Feurzieg, Lucas, Grant, & Faflick, 1969; Soloway, Lochhead, & Clement, 1982; Sutherland, 1989, 1993). This was quickly followed by the use of spreadsheets (Dettori, Garuti, & Lemut, 2001; Sutherland & Rojano, 1993) and specially designed software (Kieran, Boileau, & Garançon, 1996), as well as Computer Algebra Systems, some of which became embedded in hand-held devices, and used mainly with older students. Modelling and real data activities were also greatly stimulated (Kaput, 1994; Nemirovsky, 1996). A more extensive review here is unnecessary as the reader may consult Chapter 6 and Chapter 7 in this book that are specifically about these themes.

For the purposes of this chapter, what is important to emphasise is that the new technologies allow students to work with algebra in a variety of contexts, before they have mastered the by-hand manipulation of expressions. In addition, they greatly facilitate multi-representational activities (Kaput, 1986; also see the review in Kaput, 1992). The twofold implication is that, on the one hand, younger children can do these kinds of actions with the assistance of tools and, on the other hand, by doing them they will most likely develop an integrated perception of algebra and its applications, something found more difficult in the "first algebra, then applications" tradition, or even in the use of concrete settings to facilitate the transition or to help bridging the gap. The overall effect was further to call into question what is possible and appropriate with younger children, as well as raising questions of the nature of mathematics (Kaput, Noss, & Hoyles, 2001).

4.2.3.4 Conditions leading to change in algebra education

This combination of changes in our perception of children's thinking, changes in the scope and basis of research on children's thinking, and changes in the availability and roles of technologies, proved to be a powerful stimulus for algebra education. In a number of countries as mentioned above, the drive for reform of mathematics education in general, including mathematics teaching, has also provided adequate background for early algebra to grow, but we think that the underlying support came from the changes in foundational conditions mentioned above.

We suggest that a thorough study of this *transition* period could further enlighten our understanding of mathematics education today and towards the future. Given that algebra plays such a central role in school mathematics and in the thinking and planning of policy makers, curricula makers, teachers and pupils, and that it is associated, rightly or wrongly, with much of the failure in schools, there is a reasonable chance that such a study would be of interest to the community at large.

4.3 What Early Algebra Can Mean Today and For the Future

In this section, we will look at work representing that of people who participated in the Working Group on Early Algebra at the ICMI Study, reflecting, by and large, a shared perspective of *early algebra* as proposed in this chapter and amounting to an overview of the current early algebra landscape.

Research in and implementation of such a different perspective involves several considerations: for instance, integrating new instructional materials and/or teaching practices with existing ones, and perhaps replacing certain existing approaches with new ones. These considerations typically have multiple levels, ranging from detailed cognitive and classroom practice issues to larger scale questions associated with teacher professional development, assessment, and other systemic factors. Given the strong shift of perspective involved the discussion of early algebra naturally invites reflection at a theoretical level, an invitation that we accept.

A common assumption of the supporters of early algebra is that an *algebrafied* elementary mathematics would empower students, particularly by fostering a greater degree of generality in their thinking and an increased ability to communicate that generality. There are, however, two different views on what this algebrafication means and how it should proceed. One view is that we should build on what is already algebraic in young children's thinking, particularly with respect to their numerical, or arithmetical, reasoning. Another view is that changes in students' thinking are better promoted if we offer them tools such as notations and diagrams, which allow them to operate at a higher level of generality. In practice those views are not necessarily conflicting (Confrey, 1991).

Making this distinction highlights the need to examine the assumptions behind each of the views, leading to a clearer understanding of the possible roads for early algebra. Les Steffe (2001, p. 557) says, quite rightly, that school mathematics should be viewed "as a product of the functioning of children's intelligence". Steffe arrives at his statement from a Piagetian angle, and since there are different views on the functioning of children's intelligence, we are bound to find different specifications for early algebra. The suggestion remains, however, that a shift from content-centred planning would be of interest.

4.3.1 Arithmetic as a basis for early algebra

A group of papers focus on what is algebraic in arithmetic, that is, what can we find in arithmetic that may serve as the basis for developing students' algebraic understanding. Fujii (2003) and Fujii and Stephens (2001) propose the notion of *quasi-variables*, which are numbers within "a number sentence or group of number sentences that indicate an underlying mathematical relationship which remains true whatever the numbers used are" (p. 259). An example of such sentences is:

$$78 - 49 + 49 = 78$$

where both 78 and 49 can be considered as acting as quasi-variables, indicating the relationship that a number (e.g., 78) remains unchanged if something (e.g., 49) is subtracted and then added to it. They observe that the intention is not to introduce children to expressions like $a - b + b = a$, but rather to get them to understand that this sentence belongs to a *type* of number sentence which is true whatever number is taken away and added back. Carpenter and Levi (1999) provided a comparable analysis. More general task-design principles suggesting sequences of unexecuted number sentences have been offered by Blanton and Kaput (in press) and Thompson (1993). The key idea is that children can become acquainted with the important concept of variable either well before they are introduced to formal algebraic notation or as an intrinsic part of learning variation.

Brizuela and Lara-Roth (2001a, 2001b), as part of a larger team including Carraher, Schliemann, and others operating from the same general principle (see Carraher, Brizuela, & Earnest, 2001; Carraher, Schliemann, & Brizuela, in press), explicitly state their interest in bringing out the algebraic character of arithmetic. Working with additive relations and function tables, Brizuela and Lara-Roth's focus is on "uncovering the understandings already present by analysing the original self-designed tables constructed by young children" (2001a, p. 111). Similarly, Carraher, Brizuela, and Earnest (2001) worked with young children on the notion of difference, developing in the process what they termed *variable number lines*, number lines in which instead of specific numbers there are expressions like N – 3, N – 2, N – 1, N, and N + 1. Schliemann, Lara-Roth, and Goodrow (2001) explored multiplication tables as function tables. Marjanovic and Kadijevich (2001) offer five topics through which arithmetic can be linked to algebra: first steps in addition and subtraction, the invariant manner of expressing arithmetic procedures explicitly (an approach similar to Fujii and Stephens' quasi-variables), equations, inequalities, and discovering a rule.

Although also working in a numerical context, Carpenter and Franke (2001) focus on the processes of generalisation and proof, addressing the aspect of algebra that is generalised arithmetic: "We characterise the development of elementary school children's algebraic reasoning as reflected in their ability to generate and justify generalisations about fundamental properties of arithmetic" (p. 155). Warren (2001) examines children's understanding of the commutative law in the early years and just before they begin formal algebra studies, pointing to several implications and recommendations for teaching and learning algebra.

4.3.2 Algebrafying the elementary mathematics experience

Those papers share the idea that the study of arithmetic, both numbers and operations, already involves a degree of generalisation and thus a useful way into algebra is to exploit that generality by building and expressing new generalisations. This is a central idea in what Kaput and Blanton (2001) call "algebrafying the elementary mathematics experience" (p. 344). They propose that this process has three dimensions: (1) The process of building task-opportunity for generalisation and progressive formalisation of mathematical patterns and structure; (2) Building teachers' *algebra eyes and ears* so that they can recognise opportunities for such work in daily practice; and (3) Creating classroom practice and culture that support such work. They also argue that introducing algebra early would open curricular space needed at the secondary level and add a new level of coherence, depth, and power to elementary mathematics. Crucially, it is necessary for "democratising access to powerful ideas [...] thereby making opportunities for achievement more equitable" (p. 345).

Blanton and Kaput (2001) describe the implementation of early algebra by a teacher with her grade 3 students (8-9 years old) prior to its implementation on a district-wide scale, a task that involves deep changes in the practice and thinking of teachers. It also illustrates the point that the changes needed to implement early algebra involve many systemic educational factors beyond classroom teaching, curriculum, and learning. It should be noted, however, that many of these factors vary greatly from country to country (see Chapter 13 in this book).

Lee (2001), in discussing six views of what algebra means, proposes that *Algebra as a culture* makes it possible to pull together the other five views (as a language, as a way of thinking, as a kind of activity, as a tool, and as generalised arithmetic), "and weave them into a rich tapestry of what early algebra is or might become" (p. 397). From there, she tells an algebra story (first proposed by Kaput and Blanton (2001)) for elementary school, involving engagement in algebraic activities and communication in an algebraic language.

Taking *a culture of algebra* from a slightly different point of view, Lins (2001) proposes the notion of *legitimacy* as crucial in algebra education, arguing that an early introduction to the culture of algebra promotes, for instance, a natural legitimacy for calculating with letters. He also argues that students' difficulties documented by research may have a strong root in the fact that teachers too often fail to make clear to students the subtle shifts in the mode of meaning production. An example is shifting between *equation as scale balance* and *equation as a numerical sentence.* This suggestion is consistent with the finding that children can do more if given the opportunity and supports the value of viewing early algebra as a process of enculturation, allowing for the integration of *algebrafying from arithmetic*—as found in most papers discussed in this section—with *algebraic objects as tools for general thinking in problem-solving*—as found in the work of Davydov.

4.4 A Research Agenda

In view of what we have considered so far in this chapter, three broad areas seem to deserve attention from the algebra education community in the coming years.

The first is assessment and curriculum development from the point of view of *early algebra* seen as an *early start in algebra education.* The second is the study of the relationship among research, policies, and practice in this context, with special attention to teacher education, both for beginning and for experienced teachers. The third is a study of the implications for later grades of changes in earlier grades. This would naturally involve again the two areas identified above, now with respect to those older students. And of course, if we believe that younger students can do more then, perhaps, we have set the stage for also believing that older students can do more. Indeed, it should be recalled that most of the sad stories of Section 4.2 concerned older students, products of the current algebra education system, whereas most of the happy stories concerned younger students whose mathematical experiences vary significantly from the traditional norm.

In the sub-sections that follow, we do not mean to be exhaustive in any sense. Also, the specific suggestions presented are meant only as examples, to help to clarify the scope of those specific sub-areas. Research related to environments incorporating new technologies and to teacher education is quite relevant to all sub-sections but for these we refer the reader to Chapters 6, 7, and 10 in this volume.

4.4.1 Basic yet practical research

Basic yet practical research is needed on cognition, development, culture and change in mathematics education. We suggest that algebra education may be a venue for fundamental research while at the same time being of highly practical value. Algebra has historically been at the centre of the debates about what children can and cannot do at given ages or levels of development, and as we have seen, because of its use of powerful cultural tools, it raises deep issues regarding the relation between cultural tools and development. Analysis of these issues might lead to better theoretical syntheses than have been achieved to date. Furthermore, given the centrality of algebraic reasoning to mathematics itself and hence to school mathematics, understanding how it develops will serve mathematics education more broadly. Lastly, studying the many issues that arise when such fundamental change as the introduction of early algebra is attempted, can lead to much understanding with practical value, particularly if international differences are kept in mind so that these issues are understood in full generality and more robustly.

4.4.2 Research on forms of algebraic thinking

Another aspect of research should focus on the *teaching* side, in an effort to anticipate those aspects of algebraic thinking that could or should be presented, promoted and emphasised in the classroom. It would draw both from research indicated in Section 4.4.1 and other studies.

Several directions can be pursued. For instance, one might be interested in integration between algebra and other content areas, not only arithmetic, but geometry or the mathematics of data, for instance. Or one might be interested in which tools (diagrams, notations, graphs) can successfully lead students to develop more powerful, general algebraic ways of thinking. Or one might be interested in the enculturation aspect such as getting students to be familiar with the use of literal notation in different contexts, beginning working with *quasi-variables* for example, and getting students familiar with the idea of reasoning from the *forms* of expressions, and directly manipulating expressions (including numerical ones) to obtain new, and hopefully more useful, ones.

Of course, those same aspects could also be and have been of interest for someone focusing on older students' algebra education. The fact that we are here talking about much younger students, however, suggests that this area of research should be considered afresh.

4.4.3 Students failing in algebra

As we have pointed out in Section 4.2, much research has been conducted in the past concerning students' error patterns and misconceptions and concerning stages or levels of development. This effort seemed directed towards improving algebra education by anticipating the bad things that could happen in the classrooms and recognising putative developmental constraints on students' learning.

Few questions, however, seem to have been posed to the students themselves, especially those who are failing, regarding what sense they make of their condition. Research on students' beliefs has usually been directed to general beliefs about mathematics, not algebra (for a recent and similar example from science education, see Davis (2003)). We can ask, for instance, in their terms "What factors do they attribute their failure to? What things in algebra do and do not make sense to them and why?" Much more could be investigated here, including eliciting how students categorise the things that are being presented to them. It is difficult to anticipate what kind of insights we will get from research in this area, but we should at least expect to develop better ways of *reading* students' thinking. This may make it possible, perhaps, to promote *non-deficit reading* as an idea that is useful in the classrooms. This means looking at what students are *actually* thinking about and with, rather than at what they are failing to do and checking this against what they are *expected* to do. Franke and Carey (1997), while examining young children's perception of mathematics in problem-solving environments, provide an interesting example of the kind of research meant in this sub-section.

4.4.4 Curriculum development and intervention studies

In Section 4.4.1 we pointed to theoretical development, in Section 4.4.2 we pointed to the teachers' side, and in Section 4.4.3 we pointed to the students' side. In Section 4.4.4 these three areas come together to inform curriculum development based on long-term intervention studies. Different trails followed in the three previous sub-sections will probably lead to different approaches to this theme.

One question could be how to algebrafy the whole (or parts) of early mathematics. Another one might be how to combine different traditions to produce innovative and efficient approaches such as implementing the Davydov-Elkonin curriculum in a way compatible with Western traditions in school mathematics. A major effort along these lines, the Measure Up Project, is underway at the University of Hawaii led by Dougherty and colleagues (Dougherty & Zilliox, 2003; see also Chapter 5). Questions like "Does traditional arithmetic affect some children's abilities to reason algebraically?" and "How do mathematical learning and development evolve in early algebra environments?" are also good pointers to the kind of studies we are suggesting in this sub-section. Carraher, Schliemann, Brizuela, and colleagues are addressing these kinds of questions in a longitudinal project studying student's progress through the elementary grades with a consistent set of new curriculum materials (Carraher, Schliemann, & Brizuela, in press).

It seems clear, however, that a substantial part of this effort should explicitly take into consideration how strongly traditional views are present in schools, and that includes a permanent concern with the interface between the *new* and the *old.* Studies that look at how teachers adapt to change in early mathematics (towards early algebra) are particularly relevant. Three large scale efforts along these lines are currently underway, one by Blanton and Kaput (2002, in press), another by Carpenter and Franke (2001), and a third by Schifter, Bastable, and Monk that is extending the *Developing Mathematical Ideas* teacher professional development program to early algebra (see Schifter, 1999, for an illustration of the use of teachers' own writings and voices in the style of this effort).

4.4.5 Implications of early algebra for later grades

Changes in curricula for earlier grades naturally have implications for later grades. In this case, there are good reasons to believe that early algebra would have a significant impact on the curricula for those later grades, for two reasons. First, topics that the students would traditionally meet at later grades already will have been studied. That does not mean those topics will have been explored fully, rather that at later grades there may be further studies of those topics instead of an introduction to them. This may require major changes in curricula. Second, it is reasonable to suppose that the exposure of young students to algebra, even if only to some aspects of it, is bound to change their thinking about other topic areas in many ways. For instance, based on ongoing work at many sites, it is likely that their numerical thinking will generally be different from that of students who have not been exposed to early algebra, as will their thinking about such transition problem areas such as the equal sign. Finally, of course, there is the issue of what new kinds of ideas will be within the capacity of students whose introduction to algebraic thinking began in the early grades. Advocates of and researchers in early algebra should pay close attention to such issues, as these will certainly be part of the process of trying to implement the proposed changes for the long term.

4.4.6 Policy and practice in the context of change

Finally, there is scope for studies that tackle the complex relationship among all the issues presented in Sections 4.4.1 to 4.4.5, including the implementation processes, in which the matter of policy is central. Such complex studies are usually best conducted by larger groups working in collaboration, possibly involving multiple countries for the reasons mentioned earlier. Indeed, plans for such collaboration are underway as of this writing. This is especially important since the potential impact of the changes implied by early algebra could encompass the whole of mathematics education in schools. Such studies will benefit from what is already known from previous major curricular reforms and will inform future efforts. In a sense, that closes the circle: the issues in Section 4.4.1 can be said to be more theoretical, while in this sub-section we face the systemic and institutional issues which, while of a much more practical nature, are themselves subject to substantial theory development.

4.5 Final Remarks

It is fair to say that the proposition of an early start to algebra has various roots, and it is safe to say that its branches have broad and far-reaching implications. On the one hand, roots will be found in theoretical developments that led to curricular development, as in the case of the so-called Soviet School. On the other hand, roots will also be found in studies (markedly the studies in the spirit of the happy stories of Section 4.2.3) that led to the reconsideration of some theoretical assumptions. We reiterate our suggestion that algebra education may be an ideal place for the interplay of theory and practice.

In both cases there is a common consequence: the strengthening of the idea that young children can do more than we expected before. That, in itself, can answer the question "Why early algebra?" simply because our students deserve the chance to develop to the best of their potential.

Besides everything said so far in this chapter, we emerge from this process with a renewed awareness of the need to pay attention to what our students *are being*, rather than focusing on anticipating what they are, are not, or will be. That makes the suggested research of Section 4.4.3, which proposes research into students' perspectives, a rather intriguing area to be investigated.

Finally, if it is not yet sufficiently clear, early algebra as proposed here aims at promoting flexible, articulated, and powerful thinking (with emphasis on generality, a central aspect of what makes mathematical thinking), not at *making kids better in algebra manipulation.* Technique is only a part of the story and is certainly not the main target.

4.6 References

Arzarello, F., Bazzini, L., & Chiappini, G. (2001). A model for analysing algebraic processes of thinking. In R. Sutherland, T. Rojano, A. Bell, & R. Lins (Eds.), *Perspectives on school algebra* (pp. 61-82). Dordrecht, The Netherlands: Kluwer Academic.

Balacheff, N. (2001). Symbolic arithmetic vs algebra: The core of a didactical dilemma. In R. Sutherland, T. Rojano, A. Bell, & R. Lins (Eds.), *Perspectives on school algebra* (pp. 249-260). Dordrecht, The Netherlands: Kluwer Academic.

Balacheff. N., & Sutherland, R. (1994). Epistemological domain of validity of Microworlds: The case of LOGO and Cabri-géometre. In R. Lewis & P. Mendelson (Eds.), *Lessons from learning* (pp. 137-150). North Holland: Elsevier Science B.V.

Becker, G. (1988). A classification of students' errors in secondary level algebra. In A. Borbás (Ed.), *Proceedings of the 12th annual conference of the International Group for the Psychology of Mathematics Education* (Vol. 1, pp. 131-138). Veszprém, Hungary: Program Committee.

Biggs, J., & Collis, K. (1982). *Evaluating the quality of learning: The SOLO Taxonomy*. New York: Academic Press.

Blanton, M., & Kaput, J. (2001). Algebrafying the elementary mathematics experience. Part II: Transforming practice on a district-wide scale. In H. Chick, K. Stacey, J. Vincent, & J. Vincent (Eds.), *The future of the teaching and learning of algebra* (Proceedings of the 12th ICMI Study Conference, pp. 87-95). Melbourne, Australia: The University of Melbourne.

Blanton, M., & Kaput, J. (2002). Design principles for tasks that support algebraic thinking in elementary school classrooms. In A. D. Cockburn & E. Nardi (Eds.), *Proceedings of the 26th Conference of the International Group for the Psychology of Mathematics Education* (Vol. 2, 105–112). University of East Anglia: Norwich, UK.

Blanton, M., & Kaput, J. (in press). Design principles for instructional contexts that support students' transition from arithmetic to algebraic reasoning. Elements of task and culture. In R. Nemirovsky, B. Warren, A. Rosebery, & J. Soloman (Eds.), *Learning environments: The encounter of everyday and disciplinary experiences*. Mahwah, NJ: Lawrence Erlbaum.

Boero, P. (2001). Transformation and anticipation as key processes in algebraic problem solving. In R. Sutherland, T. Rojano, A. Bell, & R. Lins (Eds.), *Perspectives on school algebra* (pp. 99-120). Dordrecht, The Netherlands: Kluwer Academic.

Booth, L. (1984). *Algebra: Children's strategies and errors*. Windsor, UK: NFER-NELSON.

Booth, L. (1989). Grade 8 students' understanding of structural properties in mathematics. In G. Vergnaud, J. Rogalski, & M. Artigue (Eds.), *Proceedings of the 13th annual conference of the International Group for the Psychology of Mathematics Education* (Vol. 1, pp. 141-148). Paris, France: Program Committee.

Brizuela, B., & Lara-Roth, S. (2001a). Additive relations and function tables. In H. Chick, K. Stacey, J. Vincent, & J. Vincent (Eds.), *The future of the teaching and learning of algebra* (Proceedings of the 12th ICMI Study Conference, pp. 110-119). Melbourne, Australia: The University of Melbourne.

Brizuela, B. M., & Lara-Roth, S. (2001b). Additive relations and function tables. *Journal of Mathematical Behavior, 20*(3), 309-319.

Brousseau, G. (1983). Les Obstacles epistemologiques et les problemes en mathematiques. *Revista Reserches en Didactique des Mathématiques, 4*(2), 165-180.

Camp, J., & Marchionini, G. (1984). Programming and learning: Implications for mathematics education. In V. P. Hanson & M. J. Zweng (Eds.), *Computers in mathematics education, 1984 Yearbook* (pp. 118-126). Reston, VA: National Council of Teachers of Mathematics.

Carpenter, T., Ansell, E., Franke, M., Fennema, E., & Weisbeck, L. (1993). Models of problem-solving: A study of kindergarten children's problem-solving processes. *Journal for Research in Mathematics Education, 24*, 429-441.

Carpenter, T., & Franke, M. (2001). Developing algebraic reasoning in the elementary school: Generalization and proof. In H. Chick, K. Stacey, J. Vincent, & J. Vincent (Eds.), *The future of the teaching and learning of algebra* (Proceedings of the 12th ICMI Study Conference, pp. 155-162). Melbourne, Australia: The University of Melbourne.

Carpenter, T. P., & Levi, L. (1999). *Developing conceptions of algebraic reasoning in the primary grades.* Paper presented at the Annual Meeting of the American Educational Research Association, Montreal, Canada (Available from the first author at the University of Wisconsin-Madison).

Carraher, D., Brizuela, B., & Earnest, D. (2001). The reification of additive differences in early algebra: Viva La Différence! In H. Chick, K. Stacey, J. Vincent, & J. Vincent (Eds.), *The future of the teaching and learning of algebra* (Proceedings of the 12th ICMI Study Conference, pp. 163-170). Melbourne, Australia: The University of Melbourne.

Carraher, D., Schliemann, A.D., & Brizuela, B.M. (in press). Treating operations as functions. In D. W. Carraher, R. Nemirovsky, & C. DiMattia, C. (Eds.), *Media and Meaning*. CD-ROM issue of Monographs for the Journal of Research in Mathematics Education.

Confrey, J. (1991). Steering a course between Vygotsky and Piaget. *Educational Researcher, 20*(8), 28-32.

Davis, E. (2003). Untangling dimensions of middle school students' beliefs about scientific knowledge and science learning. *International Journal for Science Education, 25*(4), 439-468.

Davis, R. B. (1975). Cognitive processes involved in solving simple algebraic equations. *Journal of Children's Mathematical Behavior, 1*(3), 7-35.

Davis, R. B. (1984). ICME-5 report: Algebraic thinking in early grades. *The Journal of Mathematical Behavior, 4*, 195-208.

Davydov, V. (1962). An experiment in introducing elements of algebra in elementary school. *Soviet Education, 8*, 27-37.

Davydov, V. (1975). The psychological characteristics of the "prenumerical" period of mathematics instruction. In L. P. Steffe (Ed.), *Soviet studies in the psychology of learning and teaching mathematics* (Vol. 7, pp. 109-206). Chicago, IL: The University of Chicago Press.

Davydov, V. (1982). The psychological characteristics of the formation of elementary mathematical operations in children. In T. Carpenter, J. Noser, & T. Romberg (Eds.), *Addition and subtraction: A cognitive perspective* (pp. 224-238). Hillsdale, NJ: Lawrence Erlbaum Associates.

Davydov, V. (1983). A concept of educational activity for school children. *Soviet Psychology, 11*(2), 50-76.

Dettori, G., Garuti, R., & Lemut, E. (2001). From arithmetic to algebraic thinking by using a spreadsheet. In R. Sutherland, T. Rojano, A. Bell, & R. Lins (Eds.), *Perspectives on school algebra* (pp. 191-208). Dordrecht, The Netherlands: Kluwer Academic.

Dienes, Z. (1973). *The six stages in the process of learning mathematics* (P. L. Seaborne, Trans.). New York: Humanities Press.

Dougherty, B. J., & Zilliox, J. (2003). Voyaging from theory to practice in teaching and learning: A view from Hawai'i. In N. Pateman, B. Dougherty, & J. Zilliox (Eds.), *Proceedings of the 27th annual conference of the International Group for the Psychology of Mathematics Education* (Vol. 1, 13-31). Honolulu, Hawaii: Program Committee.

Feurzeig, W., Lucas, G., Grant, R., & Faflick, P. (1969). *Programming languages as a conceptual framework for teaching mathematics.* (Technical Report No. 1899). Cambridge, MA: Bolt, Beranek, & Neuman.

Fey, J. T. (1984). *Computing and mathematics: The impact on secondary school curricula.* Reston, VA: National Council of Teachers of Mathematics.

Fey, J. (1989). Technology and mathematics education: A survey of recent developments and important problems. *Educational Studies in Mathematics, 20*, 237-272.

Filloy, Y. (1987). Modelling and the teaching of algebra. In J. Bergeron, N. Herscovics, & C. Kieran (Eds.), *Proceedings of the 11th annual conference of the International Group for the Psychology of Mathematics Education* (Vol. I, pp. 295-300). Montreal, Canada: Program Committee.

Franke, M., & Carey, D. (1997). Young children's perceptions of mathematics in problem-solving environments. *Journal for Research in Mathematics Education, 28*(1), 8-25.

Freudenthal, H. (1974). Soviet research on teaching algebra at the lower grades of elementary school. *Educational Studies in Mathematics, 5*, 391-412.

Fujii, T. (2003) Probing students' understanding of variables through cognitive conflict problems: Is the concept of a variable so difficult for students to understand? In N. Pateman, B. Dougherty, & J. Zilliox (Eds.), *Proceedings of the 27th annual conference of the International Group for the Psychology of Mathematics Education* (Vol. 1, pp. 49-66). Honolulu, Hawaii: Program Committee.

Fujii, T., & Stephens, M. (2001). Fostering an understanding of algebraic generalisation through numerical expressions: The role of quasi-variables. In H. Chick, K. Stacey, J. Vincent, & J. Vincent (Eds.), *The future of the teaching and learning of algebra* (Proceedings of the 12th ICMI Study Conference, pp. 258-264). Melbourne, Australia: The University of Melbourne.

Gallardo, A. (1990). Avoidance and acknowledgement of negative numbers in the context of linear equations. In G. Booker, P. Cobb, & T. de Mendicuti (Eds.), *Proceedings of the 14th annual conference of the International Group for the Psychology of Mathematics Education* (Vol. 2, pp. 43-50). Oaxtepex, Mexico: Program Committee.

Gallardo, A., & Rojano, T. (1987). Common difficulties in the learning of algebra among children displaying low and medium pre-algebraic proficiency levels. In J. Bergeron, N. Herscovics, & C. Kieran (Eds.), *Proceedings of the 11th annual conference of the International Group for the Psychology of Mathematics Education* (Vol. I, pp. 301-307). Montreal, Canada: Program Committee.

Garcia, R., & Piaget, J. (1984). *Psicogénesis e História de la Ciencia.* Mexico City, Mexico: Siglo Veintuno Editores.

Harper, E. (1987). Ghosts of Diophantus. *Educational Studies in Mathematics, 18,* 75-90.

Hart, K. (1984). *Children's understanding of mathematics: 11-16.* London: John Murray..

Hativa, N., & Cohen, D. (1995). Self learning of negative number concepts by lower division elementary students through solving computer-provided numerical problems. *Educational Studies in Mathematics, 28,* 401-431.

Heath, T. (1964). *Diophantus of Alexandria: A study in the history of Greek algebra.* New York: Dover.

Herscovics, N. (1989). Cognitive obstacles encountered in the learning of algebra. In C. Kieran & S. Wagner (Eds.), *Research issues in the learning and teaching of algebra* (pp. 60-86). Reston, VA: National Council of Teachers of Mathematics, Hillsdale, NJ: Lawrence Erlbaum Associates.

Kaput, J. (1979). Mathematics and learning: Roots of epistemological status. In J. Clement & J. Lochhead (Eds.), *Cognitive process instruction* (pp. 289-303). Philadelphia, PA: Franklin Institute Press.

Kaput, J. (1986). Information technology and mathematics: Opening new representational windows. *The Journal of Mathematical Behavior, 5*(2), 187–207.

Kaput, J. (1992). Technology and mathematics education. In D. Grouws (Ed.), *Handbook on research in mathematics teaching and learning* (pp. 515-556). New York: Macmillan.

Kaput, J. (1994). The representational roles of technology in connecting mathematics with authentic experience. In R. Biehler, R. W. Scholz, R. Strasser, & B. Winkelman (Eds.), *Mathematics didactics as a scientific discipline* (pp. 379-397). Dordrecht, The Netherlands: Kluwer Academic.

Kaput, J. (1998). Transforming algebra from an engine of inequity to an engine of mathematical power by "algebrafying" the K–12 curriculum. In National Council of Teachers of Mathematics and Mathematical Sciences Education Board (Eds.), *The nature and role of algebra in the K–14 curriculum: Proceedings of a national symposium* (pp. 25-26). Washington, DC: National Research Council, National Academy Press.

Kaput, J. (1999). Teaching and learning a new algebra. In E. Fennema & T. Romberg (Eds.), *Mathematics classrooms that promote understanding* (pp. 133-155). Mahwah, NJ: Lawrence Erlbaum Associates.

Kaput, J., & Blanton M. (2001). Algebrafying the elementary mathematics experience. Part I: Transforming task structures. In H. Chick, K. Stacey, J. Vincent, & J. Vincent (Eds.), *The future of the teaching and learning of algebra* (Proceedings of the 12th ICMI Study Conference, pp. 344-351). Melbourne, Australia: The University of Melbourne.

Kaput, J., Noss, R., & Hoyles, C. (2001). Developing new notations for learnable mathematics in the computational era. In L. D. English (Ed.), *The handbook of international research in mathematics* (pp. 51-73). London: Kluwer Academic.

Kieran, C. (1981). Pre-algebraic notions among 12- and 13-year-olds. In Equipe de Recherche Pedagogique Laboratoire (Eds.), *Proceedings of the 5th annual conference of the International Group for the Psychology of Mathematics Education* (pp. 158-164). Grenoble, France: Program Committee.

Kieran, C., Boileau, A., & Garançon M. (1996). Introducing algebra by means of a technology-supported functional approach. In N. Bednarz, C. Kieran, & L. Lee (Eds.), *Approaches to algebra* (pp. 257-293). Dordrecht, The Netherlands: Kluwer Academic.

Kirshner, D. (1987). The myth about binary representation in algebra. In J. Bergeron, N. Herscovics, & C. Kieran (Eds.), *Proceedings of the 11th annual conference of the International Group for the Psychology of Mathematics Education* (Vol. I, pp. 308-315). Montreal, Canada: Program Committee.

Kirshner, D. (1990). Acquisition of algebraic grammar. In G. Booker, P. Cobb, & T. de Mendicuti (Eds.), *Proceedings of the 14th annual conference of the International Group for the Psychology of Mathematics Education* (Vol. 2, pp. 75-84). Oaxtepex, Mexico: Program Committee.

Kirshner, D. (2001). The structural algebra option revisited. In R. Sutherland, T. Rojano, A. Bell, & R. Lins (Eds.), *Perspectives on school algebra* (pp. 83-98). Dordrecht, The Netherlands: Kluwer Academic.

Klein, J. (1968). *Greek mathematical thought and the origins of algebra*. Cambridge, MA: The MIT Press.

Küchemann, D. (1978). Children's understanding of numerical variables. *Mathematics in School*, September, 23-26.

Küchemann, D. (1984). Stages in understanding algebra. *Journal of Structured Learning, 8*, 113-124.

Lacampagne, C., Blair, W., & Kaput, J. (Eds.). (1995). *The algebra initiative colloquium: Vols. 1 and 2*. Washington, DC: U.S. Department of Education.

Lee, L. (1997). *Algebraic understanding: The search for a model in the mathematics education community*. Unpublished doctoral thesis. Université du Québec à Montréal, Canada.

Lee, L. (2001). Early algebra - But which algebra? In H. Chick, K. Stacey, J. Vincent, & J. Vincent (Eds.), *The future of the teaching and learning of algebra* (Proceedings of the 12th ICMI Study Conference, pp. 392-399). Melbourne, Australia: University of Melbourne.

Lewis, C. (1981). Skill in algebra. In J. R. Anderson (Ed.), *Cognitive skills and their acquisition* (pp. 85-110). Hillsdale, NJ: Lawrence Erlbaum Associates.

Lins, R. (1992.) *A framework for understanding what algebraic thinking is*. Unpublished Ph.D. Thesis. Shell Centre for Mathematical Education/Nottingham University, UK.

Lins, R. (1994). Eliciting the meanings for algebra produced by students: Knowledge, justification and semantic fields. In J. P. da Ponte, & J. F. Matos (Eds.), *Proceedings of the 18th annual conference of the International Group for the Psychology of Mathematics Education* (Vol. III, pp. 184-191). Lisbon, Portugal: Program Committee.

Lins, R. (2001). The production of meaning for algebra: A perspective based on a theoretical model of semantic fields. In R. Sutherland, T. Rojano, A. Bell, & R. Lins (Eds.), *Perspectives on school algebra* (pp. 37-60). Dordrecht, The Netherlands: Kluwer.

Marjanovic, M., & Kadijevich, D. (2001). Linking arithmetic to algebra. In H. Chick, K. Stacey, J. Vincent, & J. Vincent (Eds.), *The future of the teaching and learning of algebra* (Proceedings of the 12th ICMI Study Conference, pp. 425-429). Melbourne, Australia: The University of Melbourne.

Mason, J. (1989). Mathematical abstraction as the result of a delicate shift of attention. *For the Learning of Mathematics, 9*(2), 2-8.

Mason, J. (1991). *Supporting primary mathematics: Algebra.* Milton Keynes, UK: Open University Press.

Mason, J. (1996). Expressing generality and roots of algebra. In N. Bednarz, C. Kieran, & L. Lee (Eds.), *Approaches to algebra: Perspectives for research and teaching* (pp. 65-86). Dordrecht, The Netherlands: Kluwer.

Mason, J., Graham, A., Gower, N., & Pimm, D. (1985). *Routes to / roots of algebra.* Milton Keynes, UK: Open University Press.

Matz, M. (1980). Towards a computational theory of algebraic competence. *Journal of Mathematical Behavior, 3*, 93-166.

Matz, M. (1982). Towards a process model for high school algebra errors. In D. H. Sleeman & J. S. Brown (Eds.), *Intelligent tutoring systems* (pp. 25-50). New York: Academic Press.

McArthur, D., Stasz, C., & Zmunidzinas, M. (1990). Tutoring techniques in algebra. *Cognition and Instruction, 7*, 197-244.

Moses, R. (1995). Algebra, the new civil right. In C. Lacampagne, W. Blair, & J. Kaput (Eds.), *The algebra colloquium: Vol. 2. Working group papers* (pp. 53-67). Washington, DC: U.S. Department of Education.

Mulligan, J., & Mitchelmore, M. (1997). Young children's intuitive models of multiplication and division. *Journal for Research in Mathematics Education, 28*, 309-330.

National Council of Teachers of Mathematics. (1989). *Curriculum and evaluation standards for school mathematics.* Reston, VA: Author.

National Council of Teachers of Mathematics. (2000). *Principles and standards for school mathematics.* Reston, VA: Author.

National Council of Teachers of Mathematics, & Mathematical Sciences Education Board, National Research Council (Eds.). (1998). *The nature and role of algebra in the K–14 curriculum: Proceedings of a national symposium.* Washington: National Academy Press.

Nemirovsky, R. (1996). A fundamental approach to algebra: Two issues that emerge. In N. Bednarz, C. Kieran, & L. Lee (Eds.), *Approaches to algebra* (pp. 295-313). Dordrecht, The Netherlands: Kluwer Academic.

Pereira-Mendoza, L. (1987). Error patterns and strategies in algebraic simplification. In J. Bergeron, N. Herscovics, & C. Kieran (Eds.), *Proceedings of the 11th annual conference of the International Group for the Psychology of Mathematics Education* (Vol. I, pp. 331-337). Montreal, Canada: Program Committee.

Petitto, A. (1979). The role of formal and non-formal thinking in doing algebra. *Journal of Children's Mathematical Behavior, 2*(2), 69-82.

Pimm, D. (1987). *Speaking mathematically.* London: Hodder & Stoughton.

Radford, L. (1994). Moving through systems of mathematical knowledge: From algebra with one unknown to algebra with two unknowns. In J. P. da Ponte & J. F. Matos (Eds.), *Proceedings of the 18th annual conference of the International Group for the Psychology of Mathematics Education* (Vol. IV, pp. 73-80). Lisbon, Portugal: Program Committee.

Radford, L. (1995). Before the other unknowns were invented: Didactic inquiries in the methods and problems of medieval Italian algebra. *For the Learning of Mathematics, 15*(3), 28-38.

Schifter, D. (Ed.). (1999). *Voicing the new pedagogy: Teachers interpret the rhetoric of mathematics education reform.* New York: Teachers College Press.

Schliemann, A., Lara-Roth, S., & Goodrow, A. (2001). Tables as multiplicative function tables. In H. Chick, K. Stacey, J. Vincent, & J. Vincent (Eds.), *The future of the teaching and learning of algebra* (Proceedings of the 12th ICMI Study Conference, pp. 531-540). Melbourne, Australia: The University of Melbourne.

Sfard, A. (1989). Transition from operational to structural conception: The notion of function revisited. In G. Vergnaud, J. Rogalski, & M. Artigue (Eds.), *Proceedings of the 13th annual conference of the International Group for the Psychology of Mathematics Education* (Vol. 3, pp. 151-158). Paris, France: Program Committee.

Sfard, A. (1991). On the dual nature of mathematical conceptions: Reflections on processes and objects as different sides of the same coin. *Educational Studies in Mathematics, 22*, 1 - 36.

Sfard, A. (1995). The development of algebra: Confronting historical and psychological perspectives. In *New perspectives on school algebra: Papers and discussions of the ICME-7 algebra working group. Journal of Mathematical Behavior (Special Issue), 14*, 15-39.

Sleeman, D. (1986). Introductory algebra: A case study of misconceptions. *Journal of Mathematical Behavior, 5*(1), 25-52.

Soloway, E., Lochhead, J., & Clement, J. (1982). Does computer programming enhance problem solving ability? Some positive evidence on algebra word problems. In R. J. Seidel, R. E. Anderson, & B. Hunter (Eds.), *Computer literacy* (pp. 171-185). New York: Academic Press.

Steffe, L. P. (2001). What is algebraic about children's numerical operating? In H. Chick, K. Stacey, J. Vincent, & J. Vincent (Eds.), *The future of the teaching and learning of algebra* (Proceedings of the 12th ICMI Study Conference, pp. 556-563). Melbourne, Australia: The University of Melbourne.

Sutherland, R. (1989). Providing a computer-based framework for algebraic thinking. *Educational Studies in Mathematics, 20*, 317-344.

Sutherland, R. (1993). Connecting theory and practice: Results from the teaching of LOGO. *Educational Studies in Mathematics, 24*, 95-113.

Sutherland, R., & Rojano, T. (1993). Spreadsheet approach to solving algebra problems. *The Journal of Mathematical Behavior, 12*(3), 353-383.

Sutherland, R., Rojano, T., Bell, A., & Lins, R. (2001). *Perspectives on school algebra.* Dordrecht, The Netherlands: Kluwer Academic

Thompson, P. W. (1993). Quantitative reasoning, complexity, and additive structures. *Educational Studies in Mathematics, 25*(3), 165–208.

Urbanska, A. (1993). On the numerical competence of six-year-old children. *Educational Studies in Mathematics, 24*, 265-275.

Wagner, S. (1981). Conservation of equation and function under transformations of variable. *Journal for Research in Mathematics Education, 12*, 107-118.

Warren, E. (2001). Algebraic understanding: The importance of learning in the early years. In H. Chick, K. Stacey, J. Vincent, & J. Vincent (Eds.), *The future of the teaching and learning of algebra* (Proceedings of the 12th ICMI Study Conference, pp. 633-640). Melbourne, Australia: The University of Melbourne.

Wenger, R. (1987). Some cognitive science perspectives concerning algebra and elementary functions. In A. Schoenfeld (Ed.), *Cognitive science and mathematics education* (pp. 217-252). Hillsdale, NJ: Lawrence Erlbaum Associates.

The Working Group on Approaches to Algebra

Leader: Rosamund Sutherland

Working Group Members:
James Aczel, Alifeleti Atiole, Luciana Bazzini, Murray Britt, Laurinda Brown, Tom Cooper, Barbara Dougherty, Gill Hatch, Valerie Henry, Marjorie Horne, John Mason, Swee Fong Ng, John Pegg, Martin van Reeuwijk, Monica Wijers, Anne Williams, and Rina Zazkis.

The Working Group on Approaches to Algebra. Front row (L to R): Martin van Reeuwijk, Valerie Henry, Anne Williams, John Pegg, Gill Hatch, James Aczel. Middle row (L to R): Marjorie Horne, Rosamund Sutherland, Barbara Dougherty, Laurinda Brown. Back row (L to R): John Mason, Murray Britt, Luciana Bazzini, Alifeleti Atiole, Rina Zazkis, Swee Fong Ng, Tom Cooper. Absent: Monica Wijers.

The Working Group on Approaches to Algebra (APPA) was a very diverse, international group of people. There were teachers, mathematicians, researchers, and curriculum developers who came from Australia (Tom Cooper, Marjorie Horne, John Pegg, Anne Williams), Canada (Rina Zazkis), England (James Aczel, Laurinda Brown, Gill Hatch, John Mason, Rosamund Sutherland), Italy (Luciana Bazzini), New Zealand (Murray Britt), Singapore (Swee Fong Ng), The Netherlands (Martin van Reeuwijk, Monica Wijers), Tonga (Alifeleti Atiole), and the USA (Barbara Dougherty, Vilma Mesa, Valerie Henry).

Prior to the Conference, members of the Working Group prepared papers for the 12th ICMI Study Conference Proceedings and the individual authors can be contacted using their e-mail addresses listed at the back of this book. The authors (sometimes with co-authors) and the titles of their papers are listed:

- James Aczel: *Towards a theoretical synthesis of research in the early learning of symbolic algebra* (pp. 13-20).
- Luciana Bazzini, P. Boero, & R. Garuti: *Research based instruction: Widening students' perspective when dealing with inequalities* (pp. 53-60).
- Luciana Bazzini & P. Tsamir: *Research based instructions: Widening students' perspective when dealing with inequalities* (pp. 61-68).
- Murray Britt: *Linear equations and introductory algebra* (pp. 103-109).
- Laurinda Brown & A. Coles: *Natural algebraic activity* (pp. 120-127).
- Barbara Dougherty: *Access to algebra: A process approach* (pp. 207-212).
- Gill Hatch: *Making algebra meaningful to pupils* (pp. 288-295).
- Valerie Henry: *An examination of educational practices and assumptions regarding algebra instruction in the United States* (pp. 296-304).
- Marjorie Horne & L. Lindberg: *A scaffolding for linear equations* (pp. 313-319).
- John Mason: *On the use and abuse of word problems in moving from arithmetic to algebra* (pp. 439-430).
- Swee Fong Ng: *Secondary school students' perceptions of the relationship between the model method and algebra* (pp. 468-474).
- T. Norton, Tom Cooper, & A. Baturo: *Same teacher, different teaching behaviours when students are using an ILS to learn algebra* (pp. 487-493).
- T. Rojano & Rosamund Sutherland: *Arithmetic world - Algebra world* (pp. 515-522).
- Rosamund Sutherland: *Algebra as an emergent language of expression* (pp. 570-576).
- Martin van Reeuwijk: *From informal to formal, progressive formalization: An example on "Solving Systems of Equations"* (pp. 613-620).
- Monica Wijers: *How to deal with algebra skills in realistic mathematics education* (pp. 649-654).
- Anne Williams & Tom Cooper: *Moving from arithmetic to algebra under the time pressures of real classrooms* (pp. 655-662).
- Rina Zazkis: *From arithmetic to algebra via big numbers* (pp. 676-681).

Thanks are extended to all of the members of the Working Group on Approaches to Algebra (APPA). They worked as a community of inquiry with everyone openly and creatively contributing, asking questions, and moving the ideas forward.

Rosamund Sutherland is particularly thanked for her leadership of the Working Group, for her introductory contribution to the chapter on behalf of the APPA Group, and for monitoring the development of the entire chapter. Special thanks are also extended to Laurinda Brown, Barbara Dougherty, and Martin van Reeuwijk with Monica Wijers who, on behalf of the APPA Group, undertook the authorship of individual sections of this chapter.

Chapter 5

A Toolkit for Analysing Approaches to Algebra

The APPA Group, led by Rosamund Sutherland
University of Bristol, Bristol, UK

Abstract: This chapter presents a toolkit for analysing approaches to algebra which was developed as a response to the diverse views of members of the Working Group on Approaches to Algebra (APPA) with respect to both what school algebra is and theories about teaching and learning. The interrelated dimensions of the toolkit incorporate a focus on the problem domain, the teaching approach, the theoretical perspective and the community of students. Within this context three approaches to algebra are analysed using the toolkit to illuminate differences which are grounded in classroom practice. The Working Group maintains that such differences often remain implicit within meta-level discussions of the mathematics education community.

Key words: Algebraic approach, toolkit, problem domain, teaching approach

5.1 Introduction

5.1.1 Background

What is an algebraic approach to solving problems in mathematics? The starting point for the deliberations of our Working Group was a consideration of some of the different ways in which the teaching of algebra can be approached, using categories which have been outlined in "Approaches to algebra: Perspectives for research and teaching" (Bednarz, Kieran, & Lee, 1996). These categories are: a generalisation approach, a problem-solving approach, a modelling approach, and a functional approach. [These are also mentioned in Chapters 2, 3, 10, and 13, together with discussion about the nature of algebraic activities.]

A generalisation approach centres around the idea that appreciating generality lies at the heart of mathematics and that all children can learn to appreciate invariance amidst change (Mason, Graham, Pimm, & Gowar, 1985). John Mason has written widely about this approach and his views are encapsulated in the quote "at the heart of teaching mathematics is the awakening of pupil sensitivity to the nature of mathematical generalisation" (Mason, 1996, p. 65). This approach to algebra has mainly influenced curricula in English-speaking countries, where it has tended to be confined to finding formulae for patterns, whether from a geometric source or in the structure of numbers (see, for example, Stacey & MacGregor, 2001). As Mason and Sutherland have recently pointed out (Mason & Sutherland, 2002) "Expressing generality is not a skill that is mastered and then transcended, but rather an ongoing process of increasing sophistication" (p. 25).

A problem-solving approach involves introducing pupils to algebra through traditional word problems, with a focus on solving equations and viewing letters as unknowns (see, for example, Filloy, Rojano, & Rubio, 2001). Researchers generally distinguish between two types of solutions in this context: (a) algebraic solutions, which involve "thinking with the unknown", and (b) arithmetic solutions, which involve thinking with the knowns and working from these to the unknowns. Some problems are amenable to both types of solutions; others *require* an algebraic approach (see, for example, Boero, 2001). More recently, modelling approaches to algebra have been incorporated into what has been called a problem-solving approach (see, for example, Nemirovsky, 1996). A modelling approach introduces algebra through the means of "mathematical narratives" which are constructed in describing events and situations (see Nemirovsky, 1996). These require a focus on the variables and invariants involved in modelling a situation.

A functional approach to algebra evolved from a Bourbakian perspective which views functions as fundamental mathematical objects. Functions can be thought of in a number of ways such as a table of values, a graph of a relationship, and as a rule, usually expressed in algebraic symbols; a functional approach to algebra focuses on developing an understanding of these representations and the idea of a variable. A functional approach is greatly enhanced by the availability of appropriate software such as graphical calculators and spreadsheets which enable the user to move between different representations of functions. This approach is being advocated in the USA by, for example, Confrey (1992), Yerushalmy and Schwartz (1993), and Kaput (1999).

5.1.2 A motivation for rethinking approaches

In the Working Group it soon became clear that there was no way in which we would be able to agree on what an approach to algebra is or should be. There were almost as many perspectives on this as there were members of the group (see for example, Aczel (2001); Bazzini in Bazzini, Boero, & Garuti (2001) and Bazzini & Tsamir (2001); Britt (2001); Brown in Brown & Coles (2001); Cooper in Norton, Cooper & Baturo (2001); Dougherty (2001); Hatch (2001); Henry (2001); Horne in

Horne & Lindberg (2001); Mason (2001); Ng (2001); Sutherland in Sutherland (2001) and Rojano & Sutherland (2001); van Reeuwijk (2001); Wijers (2001); Williams in Williams & Cooper (2001); and Zazkis (2001)). We also wanted to call an approach to algebra something which more explicitly focused on teaching and learning. The following quotes reflect our discussion:

> "Approach" includes a lot of things ... it isn't just 'hey I'm going to use a pattern approach' but it includes all sorts of aspects of the pedagogy, all sorts of aspects of what happens in the classroom as part of the approach.
>
> Every time we came up with a problem ... we found that everybody would look at it in different ways.
>
> Also we had a problem with separating algebra from mathematics ... for most of the things we came up with we had to say 'but this is true of all mathematics'.
>
> One aspect of algebra which is different is the use of letters ... and not just letters but other symbols ... and with this is the link with other sign systems and representations ... operating on them in some way ...
>
> A major point we made is that a particular problem which might be traditional in appearance can lead to many different aspects of algebra.

Some members (and particularly those who were not from English-speaking countries) found it difficult to use the categories discussed above (a generalisation approach, a problem-solving approach, a modelling approach, a functional approach) to make sense of algebra education in their own country. This is not surprising as there are quite considerable differences in emphasis in the algebra curricula around the world (Sutherland, 2000; see also Chapter 13). For example, in France and Hungary there is an emphasis on algebra as a study of a system of equations which develops into a relatively formal approach to functions and transformations of functions, but it is not clear that this could be called either a functional or a problem-solving approach to algebra. Even those who believed that an approach to algebra could be labelled as one of the four categories above felt that some aspects of "approaching" algebra were missing.

The idea of a generalisation approach to algebra, for example, has been re-interpreted over the years. John Mason, commented on a draft of this chapter:

> One of the basic misunderstandings which has still not been thoroughly clarified is that the 'generalising approach' is not simply about expressing generality concerning patterns of matchsticks or other similar objects forming diagrammatic patterns. Generality underlies mathematics, indeed some might say that without generality there is no mathematics.

So, where does this lead us? A generalisation approach has become linked to a class of problems, although for Mason and colleagues it was and still is about a whole way of working with mathematics. This approach is probably the same for the categories 'a problem-solving approach', 'a modelling approach', and 'a functional approach' suggesting that there is more to analysing the learning and teaching of algebra than just labelling the 'approach'. In particular, we all agreed when these 'approach' categories are used there tends to be too much of an emphasis on classes of problems and not enough emphasis on the way in which they are used in the classroom. As a mathematics education community we need to become more explicit about the often taken-for-granted aspects of teaching and learning which are linked to our uses of particular types of problems.

This became clearer to us when the group discussed the use of generalisation problems in the classroom. John Pegg suggested that the difficulty with the use of generalisation problems is that they carry a high cognitive load, and he illustrated this by talking about problems involving finding the number of matches in a sequence of matchstick shapes (see Pegg & Redden, 1990a, 1990b; Pegg & Tall, 2002). Laurinda Brown, on the other hand, argued that it was not possible to talk about 'cognitive load' without discussing how such problems were used in the classroom. Laurinda suggested that it is the 'jump' from seeing a pattern to *expressing* the observed generalisation which is what students find difficult (see also Stacey & MacGregor, 2001), and that it is possible to present such problems to students in such a way that this 'jump' is avoided. This discussion possibly reflected different theoretical perspectives and it soon became clear that the only way to get a sense of any difference in the way these problems might be presented to pupils was to actively work on a particular example.

5.1.3 Focusing on an example—a matchstick problem

Laurinda Brown offered a particular example for the members of the group to solve. Figure 5.1 was drawn and presented to the group to look at, without anyone having seen it being drawn. The invitation from Laurinda was for everyone to draw the diagram for themselves and to be organised about how they drew it. "Watch yourself drawing it in such a way that you could give instructions to others about how to draw any number of squares". (This problem is based on Mason, Graham, Pimm, & Gowar, 1985, p. 9, and Giles, 1972).

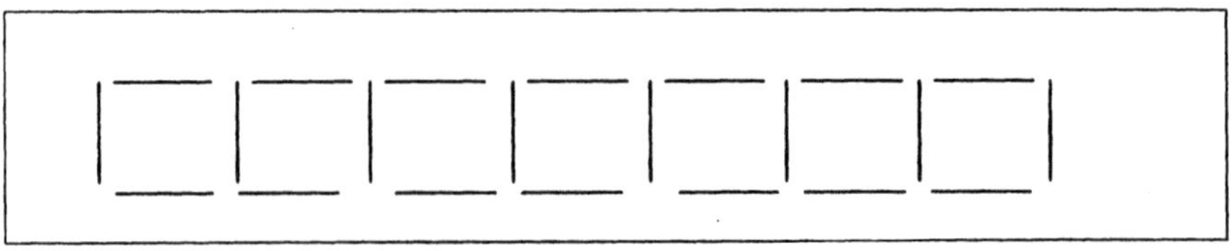

Figure 5.1. Diagram presented to Working Group members to copy in an organised way.

A range of suggestions was forthcoming and we all tried to draw what was said, with Laurinda doing this publicly (see Figures 5.2-5.4 that match Suggestions 1-3). We all began to see that the *way* that the drawing was produced made it possible to go straight to the general. Moreover, these descriptions lent themselves to being able to obtain an expression for the number of matches given the number of squares.

Suggestion 1: *Draw one vertical match and then the same number of 'reverse Cs' as there were in the original set of squares.* This suggestion is depicted diagrammatically in Figure 5.2.

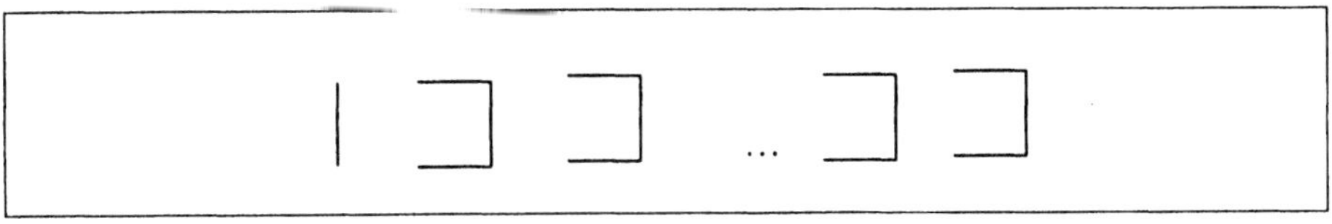

Figure 5.2. Diagrammatic interpretation of Suggestion 1

Rule: 1 + 3 × (number of squares) = 1 + 3n

Suggestion 2: *Draw a square with 4 matches and then 'reverse Cs'. Whatever the number of squares you want draw one less of the 'reverse Cs'.* This suggestion is depicted diagrammatically in Figure 5.3.

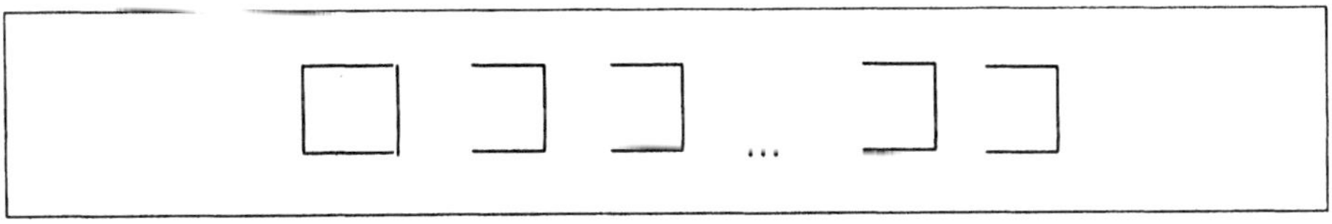

Figure 5.3. Diagrammatic interpretation of Suggestion 2

Rule: 4 + 3 × (number of squares take away 1) = 4 + 3(n – 1)

Suggestion 3. *Draw the top row of horizontal matches, then the bottom row, one each end and fill in the rest.* This suggestion is depicted in Figure 5.4.

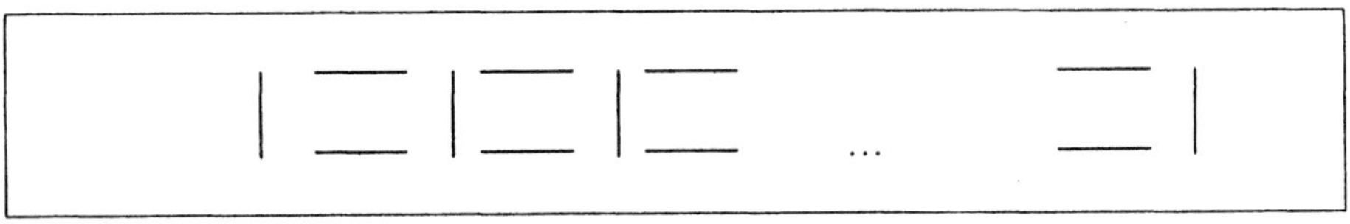

Figure 5.4. *Diagrammatic interpretation of Suggestion 3*

Rule: n + n + 1 + 1 + (n – 1)

As we worked together we could see that there are many different ways in which this 'matchstick pattern' can be drawn. As a consequence there was a naturally arising question about whether the algebraic expressions so formed were equivalent. (which additionally provides an opportunity to show that all the algebraic expressions produced are equivalent). In presenting the problem in this way certain aspects of algebra could be highlighted. To a different audience and with a different presentation it is likely that other aspects of algebra might be illustrated.

This example suggested that while the same class of problems can be specified in different curricula around the world they may be interpreted by teachers in very different ways.

5.1.4 A multi-dimensional toolkit

As we worked together the idea emerged that we should develop a multi-dimensional analytical toolkit for teachers, teacher educators, researchers, and students to analyse algebraic activity. This idea of an analytical toolkit developed as a direct response to the diverse and often conflicting views of members of the group, with respect to not only what school algebra is but also theories and beliefs about teaching and learning. Through our discussions we had agreed that any approach to algebra has to take into account the problem domain, the choice of pedagogic strategies, beliefs, theories about learning and teaching, and the knowledge and experience of the students themselves. Our proposed toolkit incorporates these facets, and allows those with an interest in algebra learning and teaching to examine approaches to algebra in a multi-dimensional way. As John Mason said in the Working Group:

> We propose what at first may appear to be more difficult for teachers and perhaps may even be thought to be more difficult for learners, but which may turn out to be richer and more effective for both, by maintaining all of the interwoven strands which contribute to the richness of the fabric we call algebra.

There are four interrelated dimensions to the toolkit, which will be discussed in the following sections.

- Problem domain,
- Teaching approach,
- Theoretical perspective,
- Community of students.

5.1.5 Problem domain

Here, the focus is on problem situations which may give access to various aspects of algebra such as operating-on-the-unknown, working with variables, generalising, and modelling. These are necessarily overlapping and the toolkit will enable the user to analyse the 'affordance' of a particular set of problems with respect to a range of aspects of algebra. This dimension relates most closely to what is often called 'content'. We also incorporated into the problem-domain dimension a lens which draws attention to the ways in which letters are used as symbols, including:

- Letter as yet unknown,
- Letter as generaliser,
- Letter as variable,
- Letter representing actions as objects.

The availability of particular problem-solving tools is incorporated into this dimension of the toolkit.

5.1.6 Teaching approach

The second dimension of the toolkit focuses on teaching strategies as they relate to the choice of problem domain and the other dimensions of the toolkit. Here we agree with Davis, Sumara, and Luce-Kepler (2000) that:

> Learning to teach and transforming one's teaching practices, then, are not simply matters of deliberately selecting and enacting particular pedagogical strategies. They are, rather, complex matters of embodying different habits of perception, of speaking, of theorising, and of acting. (p. 23)

In particular it deals with the way in which the teacher focuses the students' attention when they are working within a particular problem domain. For example, the teacher can use the idea of the lens (described above) to ask questions which focus on different aspects of the use of 'letter'. It is noted that this teaching approach dimension is likely to be influenced by other dimensions in the toolkit, including theoretical perspective and the community of students.

5.1.7 Theoretical perspective

This dimension is concerned with the theories and beliefs about teaching and learning which might influence both the choice of teaching strategies and the choice of problem domain. Examples of such theories are constructivism, social constructivism, socio-cultural theory, interactionism, and enactivism. For a further discussion of how theories of learning influence teaching see Davis, Sumara, and Luce-Kepler (2000) and Cobb and Bauersfeld (1995). Here it is important to recognise that all teachers have their own 'informal' theories about teaching and learning which are influenced by their experience of being a learner and their experience of teaching. This dimension includes both formal and informal theories.

5.1.8 Community of students

This dimension emphasises the importance of the 'community of learners' within any learning and teaching situation and takes into account aspects such as purpose of learning algebra for a particular community, stage of schooling, and students' previous experiences. Nowadays an awareness that students bring their own perspectives to a learning situation is a strong aspect of most theories of learning. However, as we shall see in the three examples which are presented later in the chapter, there is often disagreement about how a teacher should work with such students' perspectives.

Within the following sections of the chapter, different members of the Working Group present three rather different approaches to algebra which have been developed around the world. The multi-dimensional toolkit is used to analyse each approach and to probe the similarities and differences between approaches. What this toolkit allows us to see is that there are likely to be many differences in the realisation of these approaches in the classroom.

5.2 Using the Toolkit: RME and A Realistic Approach to Algebra

Monica Wijers and Martin van Reeuwijk

We provide an example of an approach towards the learning and teaching of algebra based on the philosophy, or theory, of Realistic Mathematics Education (RME), an approach which had its genesis in the Netherlands. We will elaborate on what is actually 'real', when, and for whom.

5.2.1 The theoretical perspective - Realistic Mathematics Education

In the theory of Realistic Mathematics Education (RME) (Gravemeijer, 1994; Treffers, 1987), realistic is to be interpreted as 'real to the students'. This does not imply that *all* problems should be embedded in real daily life contexts. If an 'artificial' situation is real to the students then we would call this a real situation. In terms of the multi-dimensional toolkit that we described above, a realistic approach to algebra can be seen as an integrated approach in which all aspects of the learning and teaching process play a role. In general RME is characterised by several features:

1. The use of realistic (as in 'real to the learner') contexts as sources from which to develop mathematics and as situations in which the problems to be solved are presented.
2. The use of models to develop mathematical concepts, thinking and skills. Models evolve from 'models of' to 'models for' (Gravemeijer, 1994), where the student moves from the concrete informal level of doing mathematics to the

more abstract and formal level. This process is known as progressive formalisation.

3. The use of guided reinvention. Students are encouraged to reinvent the mathematics, guided by the teacher and the instructional materials.
4. The use of students' own productions and own constructions to demonstrate understanding and to reflect on the learning process. This means that students are asked to create problems themselves.
5. The use of various instructional modes (individual, group work, pairs, with and without technology), together with interaction (e.g., through discussion) is the key to explicate the learning and teaching.
6. The use of intertwined learning strands. Mathematics is seen as one subject. At primary and secondary school there are no separate courses on algebra, geometry, and calculus, but the topics are integrated in one course named mathematics.

RME is an overall theory about learning and teaching mathematics, so it also deals with assessment, teacher training (in-service and pre-service), and other issues related to learning and teaching. In the following section we zoom in to one particular example in order to illustrate what an RME approach could be with respect to the learning and teaching of equations in the early secondary school.

5.2.2 The problem domain - Fruit salad algebra, Realistic algebra

The problem illustrated in Figure 5.5 is used in *Comparing Quantities* in the *Mathematics in Context* curriculum (Wisconsin Center for Educational Research and Freudenthal Institute, 1998) as an introduction to the algebra of systems of equations. It is a well-known kind of problem.

In terms of the multi-dimensional toolkit, the problem domain here emphasises 'systems of equations', 'unknowns', 'variables', 'generalising' (since solving this problem may lead to strategies that can be applied to other systems of equations), and 'modelling'. In other words, the problem could be used to introduce various aspects of algebra. The problem as presented in the *Mathematics in Context* curriculum is intended for students of about 12 years old and is at the beginning of a trajectory in which students will move from informal strategies (for example guessing and checking) through pre-formal strategies (for example the combination chart, which is an informal chart to represent linear combinations) to formal strategies (an algorithm to solve systems of equations). The major contexts used in the instructional sequence were 'shopping problems' where the value of the individual items is unknown. Since it is an introductory problem, students are expected to operate at a concrete and informal level.

Spending Money at the Game

1. How much does a T-shirt cost?
   And how much is a soda?
   Explain how you got your answers.

Figure 5.5. T-shirts and Sodas Problem

5.2.3 Teaching approach and the community of students

The teaching strategies that are used with this problem are based on the RME approach. The idea is that this problem will provoke students to (re-)invent strategies to deal with the problem situation, building on their intuitive knowledge. The RME belief (theory) about learning and teaching is that we do not want students to start by translating the information into symbols and solve the problem using a 'standard algorithm' which might be the case in other teaching approaches.

The *T-shirts and Sodas* problem can be described in terms of equations, using T and S as abbreviations for (the prices of) a T-shirt and a Soda:

$$2T + 2S = 44$$
$$1T + 3S = 30$$

This is sometimes referred to as 'fruit salad algebra'. Teachers often comment on this type of problem with questions like: "How can you add two different items (like apples and bananas in 2A + 3B = 70)?" The teacher is viewing the letters as objects rather than as variables, and this perspective could lead to students developing difficulties later on in the learning sequence with the use of variables.

Some questions which arise are 'How *real* is the problem to the students?' and 'What do students make of this problem?' Does the presentation confuse students and does a problem like *T-shirt and Sodas* contribute to misconceptions instead of to an understanding of the underlying mathematics?

The problem was presented in pictures (not in formal equations with letters), and students were free to come up with their own notations like shortcuts and abbreviations. Some students kept on using drawings, while other students came up with abbreviations like letters. When asked what their drawings or letters meant, they could explain that they stood for the objects. However, when they were confronted with the fact that a T-shirt is not the same as dollars, they said that it actually is *the price of the T-shirt*. Students' reactions were often somewhat irritated: "... isn't that clear, of course a T-shirt is not the same as 18 dollars, the price of a T-shirt is 18 dollars ...".

From a mathematical point of view, T and S stand for the unknown values (prices) of the T-shirt and soda, and they do not stand for the objects themselves. It is important to be very accurate with defining what the T and S stand for. The students do not worry about these issues. They are looking for strategies to solve the problem, and they come up with strategies and notations that are meaningful to them, and meaningful in terms of the context in which the problem is presented. In the process they talk about T and S and manipulate the letters in ways that, from a formal mathematical point of view, are not sound, but make sense to the students in terms of the context. To the students this use of symbols is realistic.

Of course, the notations students come up with are discussed in class. Not all students will (re)invent letters, but the use of letters is a topic of discussion. Some students may need more guidance in the reinvention process. In the discussions the meaning of the letters is explicated and students become aware of the need to be clear about the meaning of letters.

The *T-Shirt and Sodas* problem was used in numerous classes. As part of a research project, we investigated in detail what happened with this problem and the associated sequence of instructional materials. By the end of the unit all students understood formal notation and equations. Average students seemed to benefit most: they developed a conceptual understanding of solving systems of equations in a sound way. Because of the variety of models that were offered (visual, dynamical, and other representations), students could select those models that made most sense to them and thus could attach meaning to the mathematics of solving equations (for an alternative approach to the use of models for teaching algebra see Ng, 2001). More capable students broadened their (formal) understanding, and learned to reflect on the meaning of formal mathematics. Less capable students could solve problems with informal and pre-formal strategies.

This teaching approach inevitably involves focusing on the 'community of students'. As discussed above, the intention is that students learn mathematics in a way that makes sense to them and evolves from their 'own' pre-formal knowledge. A consequence is that this may result in students using (temporarily) mathematical actions which are not fully formed.

5.3 Using the Toolkit: Algebraic Activity in a Community of Inquiry

Laurinda Brown

As the Working Group discussed issues over a number of sessions we were able to start to explore what each of us really meant in action by what was said. I was struck by how comfortable I felt as the Dutch were describing the Realistic Mathematics Education (RME) in theory and that I only experienced a first jolt when they presented some materials that students worked on. At first the problem domain had seemed comfortable and familiar; then, as we looked more closely at the details of particular practices in order to compare with our own approaches, it suddenly felt very different. This I would now argue is because I was understanding their words through images from my own research, and teaching and learning contexts. I had not asked the fundamental questions of how these ideas were turned into practice and, of course, there are many ways of doing that. In applying the multi-dimensional toolkit I was able to start looking at my own work as if from the outside and to explore differences between our differing situations. This was not to decide that one or other of the approaches were better or worse—we are all bound by our cultural contexts—but to highlight what we might not have considered in our own contexts.

5.3.1 Background

I co-teach and co-research with Alf Coles who is currently a head of secondary mathematics at a comprehensive school (11-18 year old students) in the UK. The focus since 1995 has been on his development as a mathematics teacher, as we research a single extended case study of his practice (see, for example, Brown (with Coles), 1997; Brown & Coles, 1997; Brown & Coles, 2000; Brown & Coles, 2001). The work is therefore grounded in what has worked for him and, by extension, with the teachers in the department that he now manages. I visit the school at least once a week and observe lessons. There is joint planning and discussion before the visits and some reflection after the visits. The observation notes are written ones and I record more or less everything that Alf says and as much as possible of what the students say in those parts of the lessons when the whole class works together. I know in detail the developing classroom culture of the classes that I observe.

Here is an extract from a lesson that I will use to talk about teaching approaches, the problem domain, and the community of students. This extract is from the most recent lesson that I observed with Alf's Year 7 class at the time this section was written. All pupils are within the age range 11-12 years old.

Alf Coles teaching 28 mixed ability 11-12 year olds, January 2003

Key: *A* – Alf (Teacher), S – Student

A: *I'd like someone to begin by saying what they were working on last lesson. What were the questions? Ideas?*
S: We had a sheet of shapes, picked one, worked out the area.
A: *What are these?*
S: Quad …
A: *Quadrilaterals (smiling). What does it mean to work on a shape?*
S: Make a conjecture to make it easier to find the area. More accurate.
A: *Anyone got a comment?*
S: Algebra rules.
S: To make it quicker to find the area.
S: Think more mathematically.
A: *What do you mean by that?*
S: Instead of counting squares to use sort of knowledge.
A: *Yes, they're things mathematicians do like getting easier, faster …*
S: Using mathematical language.
A: *Such as? What are you trying to use?*
S: Conjecture, theorem, quadrilateral …
A: *What's the difference between a conjecture or theorem?*
S: It works for all cases …

[Lengthy discussion involving many students contributing and talking to each other. Students look at the statements on a 'common board', see Figure 5.6.]

A: *Look at this list, are they theorems or conjectures?*
S: The first two are theorems … it works for all the ones I've done.

Area of a square = length × width
Area of a rectangle = length × width
Area of a parallelogram = length × height
Area of a kite = ½ height × length

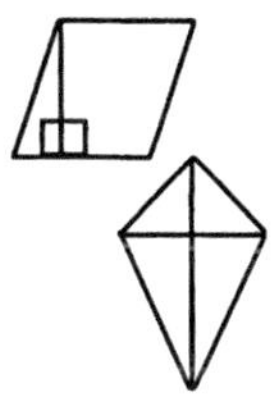

Figure 5.6. Statements on a 'common board' referred to as 'this list' in the transcript.

S: If it works on one square it should work on them all.
S: All that changes is the area, say 4 by 4, times them together and that's the number of squares. It has to work … because …
S: You got 4 there, 4, 8, 12, 16.

[Discussion amongst the students about the merits of the approaches.]

S: It's an easier way and fits in …
S: I'm going to find a conjecture for the arrowhead.
A: *I'd be interested in someone writing out a proof.*
S: Test out conjectures first and then tell the class and they can work on it and try to turn it into a theorem. Experiment.
S: Just find out something that's the same about two things.

5.3.2 Problem domain

At this stage it is worth saying that it is difficult in the work I do to separate the four domains of the toolkit because the work is essentially 'action research', starting from the focus of how to work with students so that they develop a 'need for algebra' (Brown & Coles, 1999). We do not therefore look for contexts that might be real for the students, but expect that students will develop that need for algebra through asking and answering their own questions, as illustrated in the example above.

Students construct problems for themselves because they are asking and answering their own questions in relation to the problem under discussion. They reflect on the learning process at the end of each sequence of lessons, through writing about what they have learnt both about mathematics and about being a mathematician. In this sense there is a focus on students' own constructions. In this approach we are teaching mathematics without separate courses in, say, algebra or geometry and so we are interleaving content strands.

What are the similarities and differences of Alf's approach in relation to RME? The structured, almost pre-planned, development is not here and there are few instructional materials through which to progress. There are, in fact, only 8 starter problems for Alf's class for the whole year 7 (11-12 year olds) and the lesson illustrated above is part of a sequence that started with 'Pick's Theorem', developed a formula (proved) for the area of a triangle, and finally led into investigating the area of quadrilaterals. Thus the problem domain here in Alf's class and that of RME differ, but the toolkit gives us a framework for discussing that difference.

5.3.3 Theoretical perspective, teaching approach, and community of students

The students in Alf's current year 7 class talk about using the knowledge they have gained to do problems more efficiently. Algebra is part of everything that they do because using structure and proof is part of what they think doing mathematics is. *Progressive formalisation* is a good way of describing this process. Student language becomes progressively more aligned with the standard forms of describing the world mathematically, but emerges out of their natural descriptions. As can be seen above, ideas of 'proof' are still under construction. Ownership is where the students start from, in that they develop a need for algebra through expressing and exploring their own ideas in relation to a problem.

The teaching approaches (or what we call teaching strategies) that are now part of Alf's practice in all his classes include the use of 'common boards' to collect conjectures or questions or examples of objects for later classification, discussing homework in whole-class discussions when students have been working on their own questions, and taking comments in discussions on the previous point raised before movement to a new point. The developing discourse throughout the students' work on different problems is about 'doing what mathematicians do' but not to the teacher's structure. Discussions are between student and student, within the whole class (including the teacher) and, through writing about their mathematics as they do it, constructing an internal dialogue with themselves. In this sense *interaction and discussion* are an important part of both the teaching approach and the community of students. We 'do' mathematics but beliefs and practices are crucial here. We work as a 'community of inquirers' (after Schoenfeld, 1996), where questions such as 'Why?' and 'How?' are seen as primarily important in the classroom. We are working to become mathematicians or to think mathematically and these behaviours are stressed and commented upon when Alf observes students working. These comments about actions do not all become part of the culture of the classroom. What becomes the culture of the classroom is how the students themselves talk about what they are doing and this reflection is an important part of their practices.

Within the RME approach students are guided by the teacher and there is a focus on skills from time to time through the use of instructional materials. I now see that there is a strong divergence in the teaching approach between RME theory and the approach that Alf Coles and I are developing. The main point, however, is that the 'teaching approach', 'theoretical perspective', and 'community of students' dimensions of the toolkit thus make important contributions to a framework that allows us to discuss an individual algebra teaching situation, and also identify characteristics of different approaches.

5.4 Using the Toolkit: An Elaborated Davydov Approach

Barbara Dougherty

The toolkit provides a way of examining the features of the approach used within a project called "Measure Up" (MU) in which algebra is being introduced to pupils at the beginning of primary school in Hawai'i (Dougherty & Zilliox, 2003). This approach is based on a Russian framework created by the melding of multiple theories posited from groups of psychologists, mathematicians, and educators (e.g., Davydov, 1966, Vygotsky, 1978, 1986). Motivated by children not doing well in middle and high school mathematics, the MU group began with a complete rethinking of what constitutes the fundamentals or basic structure of mathematics for young children, by considering how children should begin their formal mathematical learning.

Most contemporary elementary mathematics programmes begin with counting and computational work with single digit numbers. Children are expected to make their own generalisations based on seeing multiple examples of number problems. These generalisations would then be used to build an understanding that could be used in algebraic settings. But, there is a key question: Does number provide a basic and fundamental understanding so that students can succeed when formal algebra, with all of its skills, is introduced in later grades? Perhaps algebraic thinking and manipulations may be more readily attained if experiences with abstract thinking are part of early mathematics for children. Yet, most educators would not be inclined to give young children such tasks because abstract reasoning is not thought to be possible in early grades.

5.4.1 Theoretical perspective and problem domain

So what is the elaborated Davydov approach? MU has interpreted it to be a presentation of mathematical concepts in the context of measurement, using algebra—and particularly algebraic symbolisation—as the means by which students convey their thinking, describe relationships, and generalise their conjectures (see Chapter 4 for further discussion of the elaborated Davydov approach). For the MU team, this approach would begin with six-year-old students. McClain, Cobb, Gravemeijer, and Estes (1999) suggested a similar focus on algebra as "developing understanding rather than the correct use of tools" (p. 94). They noted that when children learned to reason quantitatively through measurement, they were able to "reconceptualize their understanding of what it means to know and do mathematics" (p. 105). Kaput (1995) has also argued that algebra as generalised quantitative reasoning is superior to the generalised arithmetic approach because it links physical experiences with the mathematics. Thompson (1993) defined this focus as considering the measurable aspects of a quantity related to a situation or context. These measurable aspects could include length, mass, area, volume, and so on.

Before working with numbers, children begin in first grade to compare attributes of both two- and three-dimensional objects. They talk about the attributes that can be compared and then make oral statements or drawings about the comparison using language (or representative pictures) such as shorter, longer, more than, less than, and equal to. They quickly realise that it is difficult to communicate these comparisons to someone. By students' own suggestion, a 'shorthand' method of writing or archiving their comparisons is developed. When asked how we could write statements, children suggest that we could use letters for the measurements. For example, if they found the mass of a bag of rice to be less than the mass of another bag of rice, they would write $L < C$, where L designates the mass of the first bag and C designates the mass of the second bag. Students are quite comfortable writing the statements and reading them, carefully noting that *mass* L is less than *mass* C. Interestingly, students spontaneously comment that if we know that mass L is less than mass C, then we also know that mass C is greater than mass L. The same is true when the relationship between the quantities is one of equality—students note that $F = S$ is the same as $S = F$. With little effort, students develop a working knowledge of the symmetric property.

Using the MU approach, our team found that before long students are creating a means by which two unequal amounts can be made equal by subtracting an amount from the larger quantity, or adding an amount to the smaller quantity. Through experimentation with a variety of materials comparing area, length, volume, or mass, students decide that the amount added and the amount subtracted are the same, and represents the difference. These six-year-olds also determine that if two quantities are equal, you have to add (or subtract) the same amount from each in order to keep the quantities equal. Or, if two quantities are unequal, you can add or subtract the same quantity from either one, and still maintain the inequality.

Continuing without numbers, these young students can determine relationships among three quantities without directly comparing or measuring them. For example, students wrote the following after comparing areas (using regions whose areas were readily comparable).

$G > K$

$K = P$

They were asked to write a statement about that comparison and orally explain how area G compares to area P, without directly comparing it. Students explained that area G must be greater than area P because 'we know that area G is greater than area K and area K is the same as area P so the area G has to be bigger than area P' (Alicia, 6 years old).

These experiences provide a generalised approach to relationships and operations. Students become comfortable operating in a symbolic notation system, and in approaching mathematical ideas from a perspective that uses non-specified quantities. Understanding that equality can be maintained by adding or subtracting the same amount from two equal quantities or, in a similar vein, make two unequal amounts equal by adding (or subtracting) the difference to one quantity is a precursor to developing the properties of real numbers that are used in a more formal way in middle and secondary school algebra.

What about the issue of applying what is learned in a generalised form to the specific cases of real numbers? First graders now have to think about the question, "what if we wanted to know exactly how volume J compared to volume Q? What would we have to do?" Through some experimentation, students decide that we have to have a way to quantify the amounts. A smaller container could act as a unit and the 'exact' amount of volume could be determined. With this volume-unit, students explore the concept of unit before it is used to find a specific number.

Imagine first graders able to compare units by merely inspecting statements written about comparisons. For example, one group of students may have been measuring length W using length-units E and Y. They wrote:

$$\frac{W}{E} = 5 \text{ and } \frac{W}{Y} = 8$$

The notation represented here is not indicative of formal work with fractions or division. Rather, it is a way students are taught to communicate their informal measures of a quantity, in this case length. The numerator W represents the total quantity; E and Y represent the length of the units used to measure the quantity. The resulting numbers indicate the number of times the unit was used to measure the quantity. Being able to record the measures in multiple ways helps students draw conclusions about the relationships of quantities and units. When another group of students was asked what conclusion they could make about the relationship between length-units E and Y, they responded that length-unit Y must be smaller than E because it took more of the Y units to measure length W. Thus they wrote the statement $E > Y$.

From this beginning, children can approach numbers differently. Instead of thinking of the discreteness of numbers, they can think about decomposition in multiple ways and focus on part-whole relationships.

Even though measurement of continuous quantities is the basis for developing mathematical concepts, the principles that students use to solve and represent measurement contexts can be applied to more discrete number problems. Consider the following problem which might be used with second graders:

Jared had 73 pencils. Margot gave him some more so that he had a total of 108. How many pencils did Margot give him?

Students know multiple ways of approaching this problem. One of these is Reed's method. Figure 5.7 shows what Reed wrote.

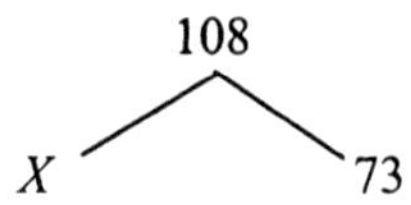

$$108 = X + 73$$
$$108 = 73 + X$$
$$108 - 73 = X$$
$$108 - X = 73$$

Figure 5.7. Reed's method

He noted that 108 represents the whole and that X and 73 represent the parts. From that part-whole relationship, four equations can be written. In MU this is called a *fact team*. Note that Reed is able to write equations where the equals sign appears in different places; in particular, note that he does not limit his notation use so that the "answer" follows the equals sign. Reed said, "I can use the equation $108 - 73 = X$ to solve it because 108 is the whole and 73 and X are the parts." He performed the subtraction and reported that Margot gave 35 pencils to Jared.

In this approach, algebra is not treated as a separate topic. It is integrated within the development of all topics so that it becomes a natural part of any concept and skill acquisition. Children become comfortable with expressing generalisations early in first grade; it is anticipated that this will extend throughout the remainder of the grades. This natural inclusion of symbolism, as well as the focus on relationships between and among quantities, leads to a greater access to algebraic language in much earlier grades than students have traditionally experienced.

MU's preliminary findings would suggest that if the context of the problem is relevant to children, they can attempt and solve much more sophisticated mathematical questions. However, context does need some clarification. For MU students, the context is one in which children operate naturally. At the age at which they enter school, they are very concerned with who has more, who has less, and who has the same.

5.4.2 Teaching approach and community of students

As discussed above, beginning from 'where children are' dramatically changes the mathematics they are able to handle. This may be where MU differs substantially from RME and relates to both the teaching approach and the problem domain. MU is based on beginning with generalisations rather than specific instances. In this way, children can see the concepts in action rather than trying to build the bigger picture from multiple specific examples, as is the case when a curriculum begins with natural numbers. Vygotsky (1978) makes a distinction between spontaneous and scientific concepts. Spontaneous or empirical concepts are developed when children can abstract properties from concrete experiences or instances. Scientific concepts, on the other hand, develop from formal experiences with properties themselves, to identifying those properties in concrete instances. As an example, spontaneous concepts progress from natural numbers to whole, rational, irrational, and finally real numbers, in a very specific progression. Topics are taught within each number system, and often not connected across systems. Scientific concepts reverse this idea and focus on real numbers in the larger sense first, with specific cases found in natural, whole, rational, and irrational numbers at the same time. Davydov (1966) conjectured that a general to specific approach in the case of the scientific concept was much more conducive to student understanding than using the spontaneous concept approach.

Children find themselves in places where there is a need to learn a formal representation or create a way of representing statements. The teacher in these instances first acts as the problem poser, giving students problems that present situations or create opportunities for cognitive dissonance. As students try to rectify the dissonance, they 'create' mathematics. In this sense *guided re-invention* is an aspect of MU. Students are literally developing the mathematics through their explorations. The links between the physical, symbolic, and other representations that use diagrams or models are carefully designed so that students can raise important mathematical questions.

The explorations could not be productive if there was not an extensive amount of interaction and discussion. This does, however, raise questions for the teacher about how to negotiate such encounters so that students can articulate the mathematics that was embedded in the task or lesson.

5.5 Concluding Remarks

Using the toolkit has enabled us to begin to tease out the similarities and differences between three approaches to algebra that relate to the work of four members of the Working Group. None of these approaches could be categorised as being wholly a generalisation approach, a problem-solving approach, a modelling approach, or a functional approach. Use of the toolkit has highlighted the ways in which the theoretical perspectives are different in each of these approaches, as are the teaching approaches, the ways of interacting with students, and the problem domains. For example, the Davydovian approach presents pupils with a generalisation (within a concrete context) and does not expect pupils to generalise from specific cases, which is what is practised within RME. Within RME and the Davydovian approach students work on pre-planned structured sets of problems, which is not the case in the community of inquiry approach of Section 5.3. We argue that within an international mathematics education community we do not always pay attention to such differences because a focus on mathematics and problem situations often leaves implicit what may be the more significant aspects of how such problems are approached in the classroom.

This chapter represents a collaborative way of working and writing. Many people, with different perspectives and from different cultures were members of the Working Group on Approaches to Algebra. We wanted the multiple and often different perspectives to have a voice within the group and to become represented within this chapter. Within this chapter multiple voices are also evident in the interpretations and re-interpretations of a piece of mathematics education history related to the question 'what does a generalisation approach mean?'

One of the aims of this chapter is to show that when international groups of researchers and teachers come together they can experience agreement when their conversations remain at a meta-level. It is only when specific examples, grounded in classroom practice are discussed that people begin to realise that they are thinking in different ways. It is as if ways of working at a classroom level embody particular approaches to algebra. However an international mathematics education community has to move beyond the level of particular narratives and examples and this is why we offer the toolkit for analysing different approaches to algebra. The toolkit, as presented in this chapter, is only a starting point and already our analysis of the three cases begins to show ways in which the toolkit could be developed.

By exposing some of the tensions and differences of opinion between members of the APPA group we hope that researchers and teachers will become more aware of how their theories, beliefs, and practices relate to the issues raised in this chapter.

5.6 References

Aczel, J. (2001). Towards a theoretical synthesis of research in the early learning of symbolic algebra. In H. Chick, K. Stacey, J. Vincent, & J. Vincent (Eds.), *The future of the teaching and learning of algebra* (Proceedings of the 12th ICMI Study Conference, pp. 13-20). Melbourne, Australia: The University of Melbourne.

Bazzini, L., Boero, P., & Garuti, R. (2001). Revealing and promoting the students' potential in algebra: A case study concerning inequalities. In H. Chick, K. Stacey, J. Vincent, & J. Vincent (Eds.), *The future of the teaching and learning of algebra* (Proceedings of the 12th ICMI Study Conference, pp. 53-60). Melbourne, Australia: The University of Melbourne.

Bazzini, L., & Tsamir, P. (2001). Research based instructions: Widening students' perspective when dealing with inequalities. In H. Chick, K. Stacey, J. Vincent, & J. Vincent (Eds.), *The future of the teaching and learning of algebra* (Proceedings of the 12th ICMI Study Conference, pp. 61-68). Melbourne, Australia: The University of Melbourne.

Bednarz, N., Kieran, C., & Lee, L. (1996). *Approaches to algebra: Perspectives for research and teaching*. Dordrecht, The Netherlands: Kluwer Academic.

Boero, P. (2001). Transformation and anticipation as key processes in algebraic problem solving. In R. Sutherland, T. Rojano, A. Bell, & R. Lins (Eds.), *Perspectives on school algebra* (pp. 99-120). Dordrecht, The Netherlands: Kluwer Academic.

Britt, M. (2001). Linear equations and introductory algebra. In H. Chick, K. Stacey, J. Vincent, & J. Vincent (Eds.), *The future of the teaching and learning of algebra* (Proceedings of the 12th ICMI Study Conference, pp. 103-109). Melbourne, Australia: The University of Melbourne.

Brown, L. (with Coles, A.). (1997). Being true to ourselves. Teacher as researcher: Researcher as teacher. In V. Zack, J. Mousley, & C. Breen (Eds.), *Developing practice: Teachers' inquiry and educational change* (pp. 103-111). Geelong, Australia: Deakin University Press.

Brown, L., & Coles, A. (1997). The story of Sarah: Seeing the general in the particular. In E. Pehkonen (Ed.), *Proceedings of the 21st annual conference of the International Group for the Psychology of Mathematics Education* (Vol. 2, pp. 113-120). Lahti, Finland: PME Committee.

Brown, L., & Coles, A. (1999). Needing to use algebra: A case study. In O. Zaslavsky (Ed.), *Proceedings of the 23rd annual conference of the International Group for the Psychology of Mathematics Education* (Vol. 2, pp. 153-160). Haifa, Israel: PME Committee.

Brown, L., & Coles, A. (2000). Complex decision-making in the classroom: The teacher as an intuitive practitioner. In T. Atkinson & G. Claxton (Eds.), *The intuitive practitioner: On the value of not always knowing what one is doing* (pp. 165-181). Buckingham, UK: Open University Press.

Brown, L., & Coles, A. (2001). Natural algebraic activity. In H. Chick, K. Stacey, J. Vincent, & J. Vincent (Eds.), *The future of the teaching and learning of algebra* (Proceedings of the 12th ICMI Study Conference, pp. 120-127). Melbourne, Australia: The University of Melbourne.

Cobb, P., & Bauersfeld, H. (Eds.). (1995). *The emergence of mathematical meaning: Interaction in classroom culture*. Hillsdale, NJ: Lawrence Erlbaum Associates.

Confrey, J. (1992). Using computers to promote students' inventions on the function concept. In S. Malcolm, L. Roberts, & K. Sheingold (Eds.), *This year in school science* (pp. 141-174). Washington DC: American Association for the Advancement of Science.

Davis, B., Sumara, D., & Luce-Kapler, R. (2000). *Engaging minds: Learning and teaching in a complex world*. London: Lawrence Erlbaum Publishers.

Davydov, V. V. (1966). Logical and psychological problems of elementary mathematics as an academic subject. In D. B. Elkonin & V. V. Davydov (Eds.), *Learning capacity and age level: Primary grades* (pp. 54–103). Moscow: Prosveshchenie.

Dougherty, B. (2001). Access to algebra: A process approach. In H. Chick, K. Stacey, J. Vincent, & J. Vincent (Eds.), *The future of the teaching and learning of algebra* (Proceedings of the 12th ICMI Study Conference, pp. 207-212). Melbourne, Australia: The University of Melbourne.

Dougherty, B., & Zilliox, J. (2003) Voyaging from theory to practice: The case of the dance of agency. In N. Pateman, B. Dougherty, & J. Zilliox (Eds.), *Proceedings of the 2003 joint meeting of PME and PMENA* (Volume 1, pp. 17-30). Honolulu, Hawai'i: Program Committee.

Filloy, E., Rojano, T., & Rubio, G. (2001). Propositions concerning the resolution of arithmetical-algebraic problems. In R. Sutherland, T. Rojano, A. Bell, & R. Lins (Eds.) *Perspectives on algebra* (pp. 155-176). Dordrecht, The Netherlands: Kluwer Academic.

Giles, G. (1972/1979). *Simple mappings*. (Booklet No. 1 from Number Patterns, Series No. 6 of Mathematics Workcard Booklets devised for the Fife Mathematics Project, University of Stirling, ISBN 0 05 003279 8). UK: Oliver and Boyd.

Gravemeijer, K. P. E. (1994). *Developing Realistic Mathematics Education*. Utrecht, The Netherlands: CD-Betá Press/Freudenthal Institute.

Hatch, G. (2001). Making algebra meaningful to pupils. In H. Chick, K. Stacey, J. Vincent, & J. Vincent (Eds.), *The future of the teaching and learning of algebra* (Proceedings of the 12th ICMI Study Conference, pp. 288-295). Melbourne, Australia: The University of Melbourne.

Henry, V. (2001). An examination of educational practices and assumptions regarding algebra instruction in the United States. In H. Chick, K. Stacey, J. Vincent, & J. Vincent (Eds.), *The future of the teaching and learning of algebra* (Proceedings of the 12th ICMI Study Conference, pp. 296-304). Melbourne, Australia: The University of Melbourne.

Horne, M., & Lindberg, L. (2001). A scaffolding for linear equations. In H. Chick, K. Stacey, J. Vincent, & J. Vincent (Eds.), *The future of the teaching and learning of algebra* (Proceedings of the 12th ICMI Study Conference, pp. 313-319). Melbourne, Australia: The University of Melbourne.

Kaput, J. (1995). Long-term algebra reform: Democratizing access to big ideas. In C. Lacampagne, W. Blair, & J. Kaput (Eds.), *The algebra initiative colloquium* (Vol. 1, pp. 33-49). Washington, DC: USA Department of Education.

Kaput, J. (1999). Teaching and learning a new algebra. In E. Fennema & T. Romberg (Eds.) *Mathematics classrooms that promote understanding* (pp. 133-155). Mahwah, NJ: Lawrence Erlbaum.

Mason, J. (1996). Expressing generality and routes of algebra. In N. Bednarz, C. Kieran, & L. Lee (Eds.), *Approaches to algebra: Perspectives for research and teaching* (pp. 65-86). Dordrecht, The Netherlands: Kluwer Academic.

Mason, J. (2001). On the use and abuse of word problems in moving from arithmetic to algebra. In H. Chick, K. Stacey, J. Vincent, & J. Vincent (Eds.), *The future of the teaching and learning of algebra* (Proceedings of the 12th ICMI Study Conference, pp. 439-430). Melbourne, Australia: The University of Melbourne.

Mason, J., Graham, A., Pimm, D., & Gowar, N. (1985). *Routes to/roots of algebra*. Milton Keynes, UK: The Open University Press

Mason, J., & Sutherland, R. (2002). *Key aspects of teaching algebra in schools*. Sudbury, UK: Qualifications and Curricula Authority.

McClain, K., Cobb, P., Gravemeijer, K., & Estes, B. (1999). Developing mathematical reasoning within the context of measurement. In V. L. Stiff & R. F. Curcio (Eds.), *Developing mathematical reasoning in Grades K–12, 1999 Yearbook* (pp. 93–106). Reston, VA: National Council of Teachers of Mathematics.

Nemirovsky, R. (1996). Mathematical narratives, modeling, and algebra. In N. Bednarz, C. Kieran, & L. Lee (Eds.), *Approaches to algebra: Perspectives for research and teaching* (pp. 197-220). Dordrecht, The Netherlands: Kluwer Academic.

Ng, S. F. (2001). Secondary school students' perceptions of the relationship between the model method and algebra. In H. Chick, K. Stacey, J. Vincent, & J. Vincent (Eds.), *The future of the teaching and learning of algebra* (Proceedings of the 12th ICMI Study Conference, pp. 468-474). Melbourne, Australia: The University of Melbourne.

Norton, T., Cooper, T., & Baturo, A. (2001). Same teacher, different teaching behaviours when students are using an ILS to learn algebra. In H. Chick, K. Stacey, J. Vincent, & J. Vincent (Eds.), *The future of the teaching and learning of algebra* (Proceedings of the 12th ICMI Study Conference, pp. 487-493). Melbourne, Australia: The University of Melbourne.

Pegg, J., & Redden, E. (1990a). Procedures for and experiences in introducing algebra in New South Wales. *Mathematics Teacher, 46*(2), 19-22.

Pegg, J., & Redden, E. (1990b). From number patterns to algebra: The important link. *Australian Mathematics Teacher, 83*(5), 17-33.

Pegg, J., & Tall, D. (2002). Fundamental cycles in learning algebra: An analysis. Paper presented to the 12th ICMI Study Conference on *The future of the teaching and learning of algebra*, Melbourne, Australia.

Rojano, T., & Sutherland, R. (2001). Arithmetic world - Algebra world. In H. Chick, K. Stacey, J. Vincent, & J. Vincent (Eds.), *The future of the teaching and learning of algebra* (Proceedings of the 12th ICMI Study Conference, pp. 515-522). Melbourne, Australia: The University of Melbourne.

Schoenfeld, A. H. (1996). In fostering communities of inquiry, must it matter that the teacher knows 'the answer'? *For the Learning of Mathematics, 16*(3), 11-16.

Stacey, K., & MacGregor, M. (2001). Curriculum reform and approaches to algebra. In R. Sutherland, T. Rojano, A. Bell, & R. Lins. (Eds.), *Perspectives on school algebra* (pp. 141-154). Dordrecht, The Netherlands: Kluwer Academic.

Sutherland, R. (2000). *A comparative study of algebra curricula.* London: Qualifications and Curricula Authority.

Sutherland, R. (2001). Algebra as an emergent language of expression. In H. Chick, K. Stacey, J. Vincent, & J. Vincent (Eds.), *The future of the teaching and learning of algebra* (Proceedings of the 12th ICMI Study Conference, pp. 570-576). Melbourne, Australia: The University of Melbourne.

Thompson, P. W. (1993). Quantitative reasoning, complexity, and additive structures. *Educational Studies in Mathematics, 25*(3), 165–208.

Treffers, A. (1987). *Three dimensions. A model of goal and theory description in mathematics education*. Dordrecht, The Netherlands: Kluwer Academic.

van Reeuwijk, M. (2001). From informal to formal, progressive formalization: An example on "Solving Systems of Equations". In H. Chick, K. Stacey, J. Vincent, & J. Vincent (Eds.), *The future of the teaching and learning of algebra* (Proceedings of the 12th ICMI Study Conference, pp. 613-620). Melbourne, Australia: The University of Melbourne.

Vygotsky, L. (1978). *Mind in society: The development of higher psychological processes.* Cambridge: Harvard University Press.

Vygotsky, L. (1986). *Thought and language.* Cambridge: MIT Press.

Wijers, M. (2001). How to deal with algebra skills in realistic mathematics education. In H. Chick, K. Stacey, J. Vincent, & J. Vincent (Eds.), *The future of the teaching and learning of algebra*

(Proceedings of the 12th ICMI Study Conference, pp. 649-654). Melbourne, Australia: The University of Melbourne.

Williams, A., & Cooper, T. (2001). Moving from arithmetic to algebra under the time pressures of real classrooms. In H. Chick, K. Stacey, J. Vincent, & J. Vincent (Eds.), *The future of the teaching and learning of algebra* (Proceedings of the 12th ICMI Study Conference, pp. 655-662). Melbourne, Australia: The University of Melbourne.

Wisconsin Center for Educational Research & Freudenthal Institute (Eds.) (1998). *Mathematics in Context: A connected curriculum for Grades 5-8.* Chicago: Encyclopedia Britannica Educational Corporation.

Yerushalmy, M., & Schwartz, J. (1993). Seizing the opportunity to make algebra mathematically and pedagogically interesting. In T. Romberg, E. Fennema, & T. Carpenter (Eds.), *Integrating research on the graphical representation of functions.* Hillsdale, NJ: Lawrence Erlbaum Associates.

Zazkis, R. (2001). From arithmetic to algebra via big numbers. In H. Chick, K. Stacey, J. Vincent, & J. Vincent (Eds.), *The future of the teaching and learning of algebra* (Proceedings of the 12th ICMI Study Conference, pp. 676-681). Melbourne, Australia: The University of Melbourne.

The Working Group on Technological Environments

Leaders: Carolyn Kieran and Michal Yerushalmy

Working Group Members:
Ferdinando Arzarello, Gary Asp, Michele Cerulli, Elisabeth Delozanne, David Haimes, Masami Isoda, Jean-François Nicaud, Henk van der Kooij, Jill Vincent and Rose Mary Zbiek.

The Working Group on Technological Environments. Seated (L to R): Masami Isoda, Henk van der Kooij, Carolyn Kieran, Elisabeth Delozanne, David Haimes. Standing (L to R): Jean-François Nicaud, Rose Mary Zbiek, Michal Yerushalmy, Ferdinando Arzarello, Michele Cerulli, Gary Asp. Absent: Jill Vincent.

The Working Group on Technological Environments had as its charge to deal with issues related to the fact that recent research, curriculum development, and classroom practice have incorporated a number of technologies to help students develop meaning for various algebraic objects, ideas, and processes. The technologies included function graphers, spreadsheets, programming languages, one-line programming on calculators, and other specific computer software environments. In order to prepare for the discussion of these issues at the conference, the members of the Working Group on Technological Environments contributed the following papers to the 12th ICMI Study Conference Proceedings. The authors (some with co-authors) and the titles of their papers are listed:

- Ferdinando Arzarello & O. Robutti: *From body motion to algebra through graphing* (pp. 33-40).
- Michele Cerulli & M. A. Mariotti: *L'Algebrista: A microworld for symbolic manipulation* (pp. 179-186).
- Alex Friedlander & M. Tabach: *Developing a curriculum of beginning algebra in a spreadsheet environment* (pp. 252-257).

- David Haimes & J. A. Malone: *Teaching algebra in a technology-enriched environment* (pp. 281-287).
- Masami Isoda: *Synchronization of algebraic notations and real world situations from the viewpoint of levels of language for functional representation* (pp. 328-335).
- Carolyn Kieran: *Looking at the role of technology in facilitating the transition from arithmetic to algebraic thinking through the lens of a model of algebraic activity* (pp. 713-720).
- Jean-François Nicaud, D. Bouhineau, & J.-M. Gélis: *Syntax and semantics in algebra* (pp. 475-486).
- Henk van der Kooij: *Functional algebra with the use of the graphing calculator* (pp. 606-612).
- Jill Vincent: *MicroWorlds™ and early algebra* (pp. 621-627).
- Michal Yerushalmy & D. Chazan: *Toward conceptual understanding and cognitively flexible fluency: Analyzing strategic decisions of curriculum developers* (pp. 668-675).
- Rose Mary Zbiek & M. K. Heid: *Dynamic aspects of function representations* (pp. 682-689).

Prior to the conference, specially prepared research briefs were circulated among the members of the Working Group (who may be contacted from their e-mail addresses listed at the back of the book). The authors and the titles of their briefs are:

- Ferdinando Arzarello: *Algebraic aspects of Cabri and other related dynamic geometry environments.*
- Michele Cerulli: *Review on symbolic manipulators to teach symbolic manipulation.*
- David Haimes: *The use of MBLs, CBLs, and probes in the learning of school algebra.*
- Masami Isoda: *Web-based environments and modeling environments.*
- Jean-François Nicaud: *Algebra in the AI-ED community.*
- Michal Tabach & Alex Friedlander: *Learning beginning algebra in a spreadsheet environment.*
- Henk van der Kooij: *Functional algebra with the use of the graphing calculator—Part 2.*
- Michal Yerushalmy: *Existing research on specially crafted software designed for the learning of algebra.*
- Rose Mary Zbiek: *Computer Algebra Systems in the learning and teaching of algebra: A brief summary of selected studies.*

This chapter, in its review of the current literature in the field, draws on the Working Group's discussions, as well as on some of the important ideas suggested in the prepared research briefs. The Group is grateful to Rose Mary Zbiek who graciously offered to take notes during discussions; these provided a valuable additional resource for the writing of this chapter. Particular thanks are extended to Carolyn Kieran and Michal Yerushalmy for their leadership and authorship roles.

Chapter 6

Research on the Role of Technological Environments in Algebra Learning and Teaching

Carolyn Kieran and Michal Yerushalmy
Université du Québec à Montréal, Canada, and University of Haifa, Israel

Abstract: As is suggested by the title, this chapter presents research on algebra learning and teaching that has been carried out in various technological environments, more specifically those where the focus has been either multiple representations in computer and graphics calculator environments, or dynamic control, or structured symbolic calculation. Discussion of these three areas of current research interest in technology-supported algebra education serves as a backdrop for reflecting on the duality of algebra with its multi-representational functional approaches, on the one hand, and symbol-based manipulation perspectives, on the other. A fundamental assumption of this chapter is that technological environments, if they are to support the learning of school algebra, ought to provide a bridge to algebraic symbolism.

Key words: Algebra, technological environments, computers, graphics calculators, multiple representations, functions, equivalent expressions, dynamic control, structured symbolic calculation, symbol manipulation, spreadsheets, programming

6.1 Introduction

Three main themes dominated the Working Group on Technological Environments discussions and have been developed into the main sections of this chapter: (a) multiple representations in computer and graphics calculator environments, (b) environments offering dynamic control, and (c) structured symbolic calculation environments. Computer Algebra Systems (CAS) environments, which were the specific focus of the Working Group on CAS (see Chapter 7 of this book), were thus excluded from the purview of this chapter.

Even though computers have been around since the late 1940s, it was only in the late 1960s that mathematicians and mathematics educators began to feel that computing could have significant effects on the content and emphases of school-level and university-level mathematics (Fey, 1984). Early visionaries soon saw the ways in which computing technology could be harnessed in order to more fully integrate the multiple representations of mathematical objects in mathematics teaching. The arrival of graphics calculators on the educational scene in the 1980s increased the accessibility of students to some of these new technologically-based emphases in mathematics programs. The interest in technology-supported multiple representations has not abated since its earliest days; in fact, it maintains its vigorous presence across several mathematical domains. Thus, much of the existing research literature on the role played by computer and graphics calculator technology in the teaching and learning of algebra has focused on the ways in which these technologies, by means of linked multiple representations, can enrich conceptual understanding of algebraic objects and processes.

As will become clearer in the next section of the chapter and as is also discussed in Chapter 5, this work of the past couple of decades integrates definitions of algebra that are broader than those formerly adopted in paper-and-pencil environments. A functional perspective on variables, expressions, and equations has been steadily finding its place within current views of algebra—views that emphasise more than the symbolic-calculation procedures for finding equation-solutions. However, the meaning of a functional perspective can be quite varied, as will be seen in the myriad ways in which multiple representations are used in technological environments. Thus, the next section of the chapter (Section 6.2), in its overview of learning and teaching research on multiple representations in computer and graphics calculator environments, includes also a discussion of the views of algebra inherent in these various environments.

The third section of the chapter deals with a more recently emerging theme in technology-related research in algebra. It describes examples of various types of environments offering dynamic control of mathematical representations, and includes, as well, the study of phenomena beyond algebra (with the control in both types of environments being provided by devices such as slidergraphs, dragging, and other tools), and the potential of these for developing algebraic thinking.

The fourth section of the chapter focuses on the transformational activity of algebra. While this work intersects to a certain extent with the multiple representation approaches described in the second section of the chapter, in particular those approaches designed to provide a foundation for conceptual understanding of algebraic symbols and symbolic calculation, the emphasis in the fourth section is on structured symbolic calculation. The environments focusing on structured symbolic calculation offer commands that transform expressions or equations and operate on the structure of these objects to preserve equivalences. In the context of recent theoretical work on the positive relationship between technical and conceptual development in technological environments, this stream of research with its focus on notions of equivalence offers a vehicle for thinking differently about the epistemic benefits that can be derived from symbolic manipulation.

6.1.1 Some earlier work involving technological environments

Because this conference had as its theme, "The future of the teaching and learning of algebra", the focus of this chapter is less on the early research literature related to technology-supported or technology-based work in algebra and more on recent developments and their potential for future work in algebra. However, in order to put the more recent research into context, we take a brief glimpse at some of these earlier studies. Among the clusters of past research on technology related to algebraic thinking and learning, two involved sustained effort, although from quite different sectors of the research community: the studies on programming, and those related to the use of intelligent tutors for equation-solving. As the latter are touched upon in Section 6.4, we restrict ourselves here to a look at some of the past work in programming, and discuss the extent to which the aims of this research were related to developing support for thinking about algebraic symbols and their transformations.

Some of the first research studies on the role of programming in learning mathematics were based on the notion that such activity would ultimately aid students in their mathematical problem solving. It was believed that "the activity of writing, processing, and studying the output of computer algorithms should promote the development of mathematical concepts and principles, computational skills, and problem-solving abilities of the students" (Hatfield & Kieren, 1972, p. 99), and support was indeed found for the idea that "such higher-order skills as problem solving, independent inquiry, and generalising can be enhanced with the design of computer algorithms" (p. 111).

However, it was with the Logo movement that a more direct link was made between programming and algebra. Feurzeig and Papert (1968) conceptualised Logo programming not only as a constructive problem-solving process but also as a procedural language whose definitions could be used to generate new commands and functions. That each procedure could be encapsulated as an object and then used within other procedures suggested to mathematics educators that Logo, and other such procedural languages (e.g., ISETL; see Dubinsky & Harel, 1992), could serve as a vehicle for developing algebraic thinking. As Love (1986) argued, algebra "concerns itself with those modes of thought that are essentially algebraic, for example, handling the as yet unknown, inverting and reversing operations, seeing the general in the particular; being aware of these processes, and in control of them, is what it means to think algebraically" (p. 49). In short, programming was viewed as an algebraic activity because it involved expressing mathematical ideas and processes in a general way with a particular language and accompanying syntax.

Furthermore, some studies (e.g., Clement, 1982) were showing that programming in fact facilitated the expressing of relationships because it permitted students to think about such relationships procedurally rather than structurally. A spate of Logo studies followed, many of them investigating various aspects of algebraic thinking: for example, students' conceptions of variables (e.g., Hoyles, Sutherland, & Evans, 1985), the process of formalisation (e.g., Noss, 1985), recursive thinking (e.g., Kieren, 1992), perceptual and analytical schemas (e.g., Kieran, Hillel, & Erlwanger, 1986), elements of group theory (e.g., Leron & Zazkis, 1992), and functional notions (e.g., Klotz, 1986).

However, the majority of the Logo studies did not go as far as making the connection with conventional algebraic concepts and notation. The algebra study of Noss (1985), for instance, which involved 10- and 11-year-olds, focused on investigating children's conceptual understanding of elementary algebraic ideas. Children constructed a relevant symbolism, à la Logo, to handle the algebra items they were presented with. The tasks were modified versions of questions from the Strategies and Errors in Secondary Mathematics project (Booth, 1984), where a rationale for the use of letters to write rules to solve whole classes of problems was provided by the idea of programming a virtual computer. Noss reported that the Logo experience "may have facilitated the understanding of the concepts of algebraic variable and elementary algebraic formalisation" (p. 434), even though the study did not include making a link with conventional algebraic symbols.

Another type of symbolic language for expressing mathematical ideas was featured in the spreadsheet research that followed shortly after the Logo studies of the 1980s. Like the Logo environments, spreadsheets were found to support pupils in moving their thinking from the specific to the general (Rojano, 1996), and to provide effective conceptual tools for students who had previously been unsuccessful with algebra (Sutherland, 1993). As spreadsheet environments are an integral part of the multi-representational work in the section that follows, their inclusion in the present discussion is intended merely to point to the role that they, like programming, are considered to play in the development of algebraic ideas—even if these ideas occur within the context of other types of symbolism. However, in the sections to come, it is not just the development of algebraic ideas that is considered important; particular attention will be paid to the ways in which the various technological environments support a bridging to algebraic symbolism.

6.2 Multiple Representations in Computer and Graphics Calculator Environments

6.2.1 The view of algebra that is the focus of this body of work

The ample number of studies on learning algebra with multiple representation technology (as reviewed by Zbiek in the brief prepared for the Working Group) suggests a range of types of *algebra*. Within these types, quite a few approaches to algebra that make intensive use of multiple representation technology define themselves as *functional approaches*. However, this common title does not fully capture the differences in intentions and approaches. For example, some have focused on the sorts of translations between representation systems of function that curricula ask students to master (Janvier, 1987). Others have focused on the ways in which functions can be conceptualised both as process and as object, and on the possible impact of various representations on such conceptualisations (Dubinsky & Harel, 1992; Sfard, 1992). Thus, the use of a multiple representation tool does not uniquely determine a functional approach to algebra.

Tabach and Friedlander, in the brief they prepared for the Working Group, reviewed studies of curricular sequences that use spreadsheets in algebra, and suggest that, in general, three factors allow for a functional approach in the teaching and learning of algebra: the potential to produce ample numerical tables, the need to use general expressions to create these tables, and the possibility of obtaining a wide variety of corresponding graphs. As an example, they refer to the *Compu-Math* curriculum, which makes heavy use of spreadsheets (*Excel*). Friedlander and his colleagues (Friedlander, 1999; Friedlander & Tabach, 2001a; Friedlander & Tabach, 2001b) describe *Compu-Math* as an intuitive functional approach based on complex, "real-life" or mathematical situations that lead to investigations of processes of quantitative variation and which involve generalising, justifying, modelling, and reflecting on thinking processes and solutions. The designers (Hershkowitz et al., 2002) state that, "our intention was to keep the formal aspect of the concept of function (definition, notation, mappings) at a minimal level and to require students to investigate variation, as expressed by numerical series, algebraic expressions, and graphs" (p. 679).

The *VisualMath* curriculum (Yerushalmy & Shternberg, 2001) is also a function-based curriculum that makes intensive use of multiple representation software. *VisualMath* uses specially designed software environments to support a variety of experiences with non-symbolic representations of functions as a way to establish understanding of multiple representations of functions that is not necessarily driven by analytic expressions. The dominant conception of letters is that of quantities that vary. Solving equations is conceptualised as a particular kind of comparison of two functions. Thus, this curriculum attempts to help students learn to do manipulations with an understanding of the qualities of the graphical and tabular meanings of these manipulations, as well as to obtain a sense of the purposes for which such manipulations are useful.

Another function-based curriculum is the *Core-Plus Mathematics Project* (CPMP), which approaches the mathematical ideas of the secondary school level through investigations of applied problems, using a variety of linked representations (Huntley, Rasmussen, Villarubi, Sangtong, & Fey, 2000). The connections among representations of functions are a major theme of the project materials (see Section 6.2.4.2 of this chapter for an extended description of *Core-Plus*). A similar orientation is embedded in the Dutch curriculum, which emphasises applied-problem situations as a major goal of algebra (see also Chapter 5 and Chapter 13 in this book that focuses on algebra around the world). The nature of the Dutch algebra (according to van der Kooij, in the brief prepared for the Working Group) involves bridging between the different subjects (e.g., the use of algebra when dealing with geometrical problems or the use of geometrical representations when doing algebra). The function orientation centres on *function as a process* at the junior high school and *function as an object* at the senior high school. Algebra in senior high school (15-18 years old) is mainly analysis of functions, including the use of derivatives and integrals. The focus in junior high school (12-15 years old) recently shifted from learning to manipulate algebraic forms (algorithmic routines) and the structure of number systems, to the interpretation and construction of formulas (and tables and graphs) that describe *relationships between quantities*.

For Dettori, Garuti, and Lemut (2000), the function orientation of spreadsheets allows for their effective use in the investigation of variation, but can cause difficulties in their implementation (e.g., solving equations or inequalities) and in performing algebraic manipulations. Thus, the approach to algebra via functions is, according to these researchers, limited to the introduction of algebra. When they investigated students' (13-14 years old) work on algebraic problems with spreadsheets, Dettori and colleagues were led to conclude that, "spreadsheets can start the journey of learning algebra, but do not have the tools to complete it" (p.206).

In other studies involving spreadsheets, Sutherland and colleagues (e.g., Filloy & Sutherland, 1996; Sutherland & Rojano, 1993) have described students' resulting problem-solving approaches as being of a *spreadsheet* type and have emphasised that the unknown (rather than the variable) is a central object of this approach—although the feel one gets from the description of the curricular sequences is that of a *functional approach* with a focus on variable.

Below we will look at how the various distinctions offered here about multiple representations and algebra are reflected in different studies.

6.2.2 Research related to creating meaning for the objects and processes of algebra

Functional dependencies that describe and allow us to make predictions about relationships between quantities can be found everywhere around us. In order to make these relationships the core of algebra courses, educators seek ways for students to look at relationships and functional dependencies, not according to their mathematical definitions, but rather by creating experiences in the real world around them by using technology that can lead to the acknowledgment of the centrality of these dependencies. They seek tools that will allow students to mathematically represent personal experiences as functions. Two styles of learning environments have recently been developed: one uses measurements and numerical data to represent dependencies by means of spreadsheets; the other uses the construction of graphs either directly from the situation or by graphical syntax. In the following two sections, we elaborate on research involving these two modes. Special attention is given below to the generation of algebraic symbols and the meaning they get in these two *bridging to functions* attempts.

6.2.2.1 Spreadsheet objects and operations as a bridging language between phenomena and algebraic models and methods

Problems in context that were usually approached as an application activity within formal algebra courses turn out to be an interesting subject for pre-algebra activities with the support of spreadsheets. Studies frequently follow whether and how the use of spreadsheet methods bridges between arithmetic and algebra and even supports the evolution of algebraic methods. Filloy, Rojano, and Rubio (2000) describe the *spreadsheet method* of solving an algebra word problem and how different it is from the other methods they observed students using without technology. The spreadsheet method involves getting a solution using spreadsheet symbolism, which is the solution to a family of problems rather than a solution to a specific problem. Filloy and colleagues suggest that the spreadsheet served as a bridging tool to algebra in that it helped to create conceptual meaning for algebraic objects and operations. Students were observed to move from focusing on a specific example to describing general relationships. Students learned to accept the algebraic idea of working with an unknown because the spreadsheet cell that represented the unknown number encouraged students to deal with unknown quantities.

The graphical representation is another central capacity of the spreadsheet, one that can support analysis of the generated data and perhaps even the generation of explicit correspondence rules to describe a relationship between columns. Ainley (1996) studied eight pairs of 11-year-old students (in their last year of primary school) within a larger project that aimed at finding out the role played by technology in giving meaning to formal notation by teaching it in meaningful contexts. After beginning the activity by collecting data, students used spreadsheets to describe numerically and graphically a situation. Ainley pointed to the active graphing facility—the shape of the graph and the evaluation of extreme points and other points of interest—as the major tool that facilitated moving from the practical and mental phases towards formalising using spreadsheet syntax.

However, Yerushalmy and Chazan (2002) argue that the types of conceptions that underlie symbolisation with a spreadsheet are a complicated affair:

> When students are working with symbols representing locations in the spreadsheet table, these symbols are neither unknowns, nor variables. They represent particular locations and in that sense seem too particular to be variables, though of course the values in cells to which they refer can change; the cells to which they refer either do or do not have values; when they do, it seems funny to call them unknowns. (p. 735)

The two ways to write formulas are also a complicated issue if one tries to determine the algebra these symbols represent:

> Copying rules down the spreadsheet's columns makes use of the tool's capacity to carry out recursive operations, yet the formula students must develop to compute one column from another is an explicit function rule on these variables. (p. 735)

Dettori, Garuti, and Lemut (2000), who studied students that were introduced to problems using spreadsheet formulations at the same time as traditional algebraic notation, pointed out similar complexities and argued that students do not grasp immediately what is a variable or an unknown:

> The sign of equality used in a spreadsheet is actually the assignment of a computed value to a cell. ... The inability to write relations in a spreadsheet means that it is not possible to use it to completely handle algebraic models. (p. 199)

While experimenting with different values in a cell leads to the feel of a variable, it is not always the case. When students are looking for a specific value that would set a specific result, the orientation is towards working with an unknown. However, at other times, the same spreadsheet operation would lead to thinking about variables and functions. Sutherland and Balacheff (1999), while arguing about the centrality of the teacher in directing students' actions toward different possibilities for viewing the subject matter, suggest a similar analysis of the complexity:

> A number in a cell can have several meanings, it can be a specific number or a cell representing a general number, or a cell representing an unknown number or a cell representing a relationship between numbers; the spreadsheet/algebraic approach is to view a cell as '*x*', either as an unknown or a general number and to express relationships with respect to this '*x*'. (p. 22)

However, most students approach it arithmetically, working from known numbers to unknown. Hershkowitz et al. (2002) have described how students tend, in the spreadsheet environment, to generalise recursively rather than explicitly. Designers of the *Compu-Math* project tried to deal with this difficulty in the following way:

> Students were asked to construct in a spreadsheet a sequence, by using the numbers in the column of the position index, or to express in words the weekly balance of a person's savings by using the number of weeks (and *not* the previous balance) as variable. (p. 684)

They also pointed to other distinctions that should be made in order to analyse the impact of the tool as a bridging environment to algebraic conceptualisation: (a) the type of modelled phenomenon (linear or not), (b) the method of presentation (consecutive or separate items of a sequence), and (c) the style of presentation.

Thus, spreadsheets can appropriately serve as a bridge to algebra and algebraic symbols if embedded in suitable sequences of activities. Tabach and Friedlander (in the brief prepared for the Working Group) have summarised the contribution and the delicacy of using spreadsheets along the following considerations:

- *Multiple representations*: Spreadsheet environments allow for an integrated use of numerical, graphical, and algebraic representations;
- *Requiring predictions*: Predictions in a spreadsheet activity can be made at the initial stage of getting acquainted with the problem situation and at the stage of transition from the numerical table to its graphical representation;
- *Generalisation by recursion and generalisation by position number*: Both methods have their advantages and disadvantages;
- *Spreadsheet formulas versus algebraic expressions*: The difference between the two is sometimes more than syntactic and can cause some conceptual difficulties;
- *Lack of transparency*: In a spreadsheet table, the formulas are "hidden" behind the resulting numbers. As a result, students can encounter both cognitive and technical difficulties in monitoring their work.

The above studies, which have illuminated the potential and the complexities of spreadsheets, point to the contributions that can be made by them and to the importance of appropriate activities for this type of environment.

6.2.2.2 Graphs as a bridging language from phenomena to algebraic models

Since the mid-1980s, Dugdale (1993) has been creating and studying multi-representational tools to promote students' understanding of graphs related to mathematical and non-mathematical situations. Her efforts have been directed to moving students beyond plotting and reading points to interpreting the global meaning of graphs and the functional relationships that they describe. In work at

TERC (Mokros & Tinker, 1987), which followed Dugdale's initial studies, students experienced real-time generation of graphs with probes and Microcomputer-Based Laboratories (MBLs). Use of MBLs allows students to gather data from an ongoing situation and to present the data graphically. These technologies bypass algebraic symbols (in this chapter, *algebraic symbols* will also be referred to as *symbolic language* and *symbolic representation of functions*) as the sole channel into mathematical representation and motivate students to experiment with the situation, and to analyse and reflect upon it, even when the situation is too complicated for them to approach symbolically. The visual analysis that emerges from work with MBL tools is different from that which arises from work with algebraic symbols or numerical tables. Attention is given to the shape of the graph and to the ways it relates to the qualities of the situation (Nemirovsky, 1994; Nemirovsky, Kaput, & Roschelle, 1998).

Another attempt at approaching the potential of graphs is not just as a representation of relationships in phenomena but also as a language in itself, which can then be used to bridge to symbolic language. Schwartz and Yerushalmy (1995) proposed an intermediate representation (demonstrated in Figure 6.1) that is between the complex natural language in which problems are often formulated and the dense and precise analytic and symbolic representation of the underlying mathematics. Similar uses of icons were suggested by Nemirovsky and Rubin (1992) and by Janvier, Girardon, and Morand (1993). Schwartz and Yerushalmy's intermediate linguistic representation, which is based on the function and its vocabulary, includes seven graphical icons that describe how the function and its rate of change both change.

Quantity Rate of change	Increasing	Decreasing	Constant
Increasing	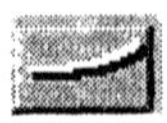		
Decreasing	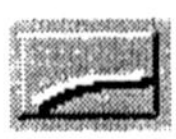		
Constant			

Figure 6.1. Seven graphical icons that show changes in the function and its rate of change.

In order to allow the mathematical construction to take place in the two channels in parallel—the iconic and the linguistic—the software environment, *The Function Sketcher* (Yerushalmy & Shternberg, 1994), was created. In a couple of studies, Yerushalmy and colleagues explored and described the nature of the transitions made to algebraic symbolism and the bridging functionality of the graphical objects with pre-algebra students. For example, Yerushalmy and Shternberg (2001) followed the learning of 7th graders during three phases of activities developed for use with *The Function Sketcher*—activities that preceded the learning of algebraic symbols in the *VisualMath* curriculum. The first phase involved building a graphical model of a situation (we will extend this example in Section 6.3.3.3). The second phase was intended to support the encapsulation of the process of graphing into objects: the seven graphical icons. To illustrate this phase, we use an example from Shternberg (2001a) who analysed the work of over 30 pairs of 7th graders using these bridging representations to solve the following "Motorcycle" problem.

> A motorcycle starts to move from rest. We use the graph to describe its distance from the starting point in the first 3 minutes of its motion. What is its speed in the second minute?

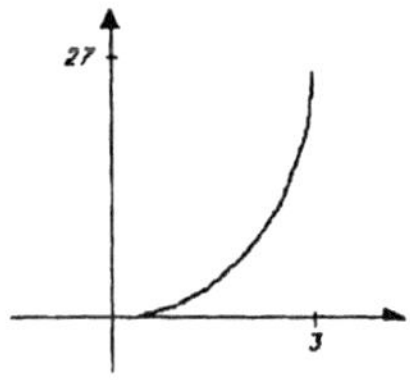

This problem challenges students' thinking about instantaneous speed and its possible representations. The students used *stairs* to describe the continuous graph of distance over time and to evaluate speed in small intervals. The extract below records their conversation, based on the *staircase* diagram they constructed (see Figure 6.2).

David: There is no speed - we have time and distance. Towards the end he started to drive faster. Let's do stairs - that would be the best. The distance increases in time and he starts to drive faster. But it is hard to say. Each stair starts slow and continues faster.

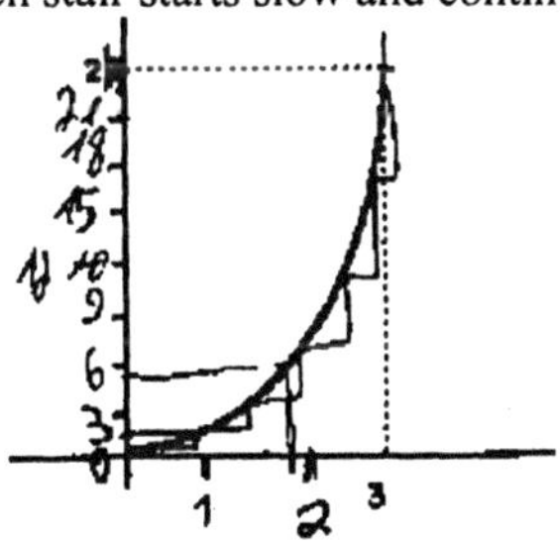

Figure 6.2. David's and Golan's working on the Motorcycle problem.

Golan: We would look at the distance change here, compare it to the previous one and then we will know the speed. ... so the speed in the second minute is 6.

Interviewer: So what do you mean saying that the speed is 6?

Golan: Oh, no! We should look at the difference between the first and the second minute and we will know how far he drove in the second minute.

The third phase led to another transition: from the iconic representations to the symbolic representation of the functions. In the environment *Escalators*, designed by Shternberg (2001b), expressions are constructed from (rather than given) and controlled by the graphs of rate of change (e.g., linear rate of change with starting value of 4 and difference of 2 is illustrated in Figure 6.3).

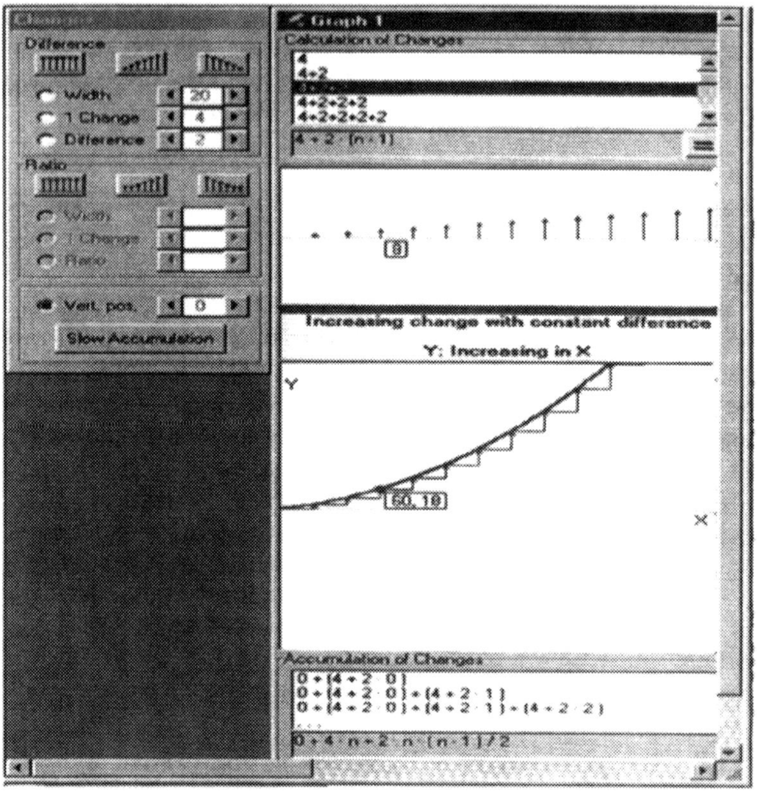

Figure 6.3. Using differences to construct a function's expression.

The upper windows describe the change (the height of the third *stair* is 4+2+2 and in general the function that describes the chosen rate of change here is $4 + 2(n - 1)$. The lower windows describe the accumulated quantities and the quadratic function constructed from the linear rate of change. The value of the function at the third *stair* is $18 = 4 + (4 + 2) + (4 + 2 + 2)$. (Note that the 60 of the ordered pair (60, 18) represents 20*3 where 20 is an arbitrary value for the width of the stair that is chosen by the user.) By generalising to n on the basis of the incremental calculations that are displayed, a quadratic expression, $4{*}n + 2{*}(n - 1)/2$ is constructed.

Although mathematical modelling cannot be fully accomplished by this qualitative sign system, the intermediate bridging language helped to form a mathematical construction with language that developed from acquaintance with physical scenarios. It supported the abstraction of everyday phenomena into a smaller set of mathematical signs that were manipulated with software as semi-concrete objects.

6.2.3 Research related to the transformational activities of algebra

Schoenfeld (2002) indicates that one of the major obstacles in the acceptance and implementation of *Standards* based curricula is the common assumption that, in mathematics, students have to master skills before using them for problem solving and applications. The NCTM *Standards* (1989, 1991, 1995, 2000) are a set of interrelated proposals for reform in content, teaching, and assessment of school mathematics K-12 (Pre-K-12 for the 2000 publication). While an underlying assumption of the reform is that students can develop mastery of skills through participation in problem solving, the mathematics community is debating the importance of students having to learn to carry out symbolic manipulations by hand. On the other hand, most algebra teachers cannot imagine an algebra course developing conceptual understanding rather than technique. Indeed, much of the power of algebra derives from the way it handles closed-form procedures (i.e., manipulation of symbols according to rules), which allows students to be less attentive to meaning. Much of the power of conceptual understanding lies in its attentiveness to meaning. Therefore symbolic manipulations are usually considered very remote from the type of bridging work with technology described above and, indeed, sometimes fluency in representations and links among them cause students to neglect algebraic symbols and manipulations (Hershkowitz & Kieran, 2001).

What could symbolic manipulations be in teaching algebra with multiple representation technology? Arguably, technology, such as graphics calculators, which links representations of functions, has the potential to provide students with visual feedback that emphasises various meanings of equivalence. Supporters have described, for example, its potential to help students understand visually both the equivalence of expressions and the equivalence of equations (as discussed in Schwartz & Yerushalmy, 1992). However, other studies suggest that there are reasons to be sceptical of the power of such linked representations, taken on their own, without any related changes to the curriculum or to modes of instruction. Goldenberg (1988) has argued (as well as others in Romberg, Fennema & Carpenter, 1993) that graphical representations are complex and require teaching and learning. For example, much needs to be made of the differences between graphing equations and graphing functions. One cannot simply expect that students will be able to read these different representations in the ways that they are intended. In this regard,

Yerushalmy and Chazan (2002) point to the tension between the standard curriculum and a technology-supported approach regarding the types of processes and actions represented by the equation (or by the equal sign). For instance, while equations in a single variable (e.g., $2x + 3 = 4x + 1$) are most commonly taken to represent a statement about unknown numbers, with graphing technology the linked representations are of expressions taken as the equations of functions ($y = 2x + 3$ and $y = 4x + 1$). Graphing $2x + 3 = 4x + 1$ could thus be interpreted as representing a comparison of two functions ($y = 2x + 3$ and $y = 4x + 1$) or as the solution set of a system of two equations in two unknowns; while its interpretation as an equation in a single variable would suggest the graph of a point on the x-axis where $x = 1$.

A broad view on the issues connected to students' accomplishments regarding algebraic calculations is given by the comparative study of CPMP (Huntley, Rasmussen, Villarubi, Santong, & Fey, 2000): "Students were not as proficient as control students at manipulation of symbolic expressions by hand; they had apparently learned a variety of alternative, calculator based strategies for accomplishing the same goals" (p. 354). This has been found to be a common trend in many current reform projects that use technology (Schoenfeld, 2002).

6.2.3.1 Operations with expressions

The alternate approaches taken by students in technology environments to the manipulation of algebraic expressions have been studied by Dugdale (1993). Working within the *Green Globs* environment (Dugdale & Kibbey, 1986), students were asked to tell for what value of the variable certain expressions were undefined. Rather than mechanically identifying a particular problematic x-value, the students were using the behaviour of the entire function, applying graphical transformations, and analysing immediate graphical feedback. According to Dugdale, the specific technological environment, along with appropriate learning sequences, played a major role in encouraging mathematical reasoning with the multiple representation tool rather than developing mere proficiency at solving specific types of problems.

Yerushalmy and colleagues have for over a decade been pursuing this issue of conceptual, transformational activities by crafting special-purpose software and designing a curriculum to incorporate proficiency in relational understanding of algebraic manipulations (e.g., Gafni, 1996; Yerushalmy, 1991; Yerushalmy & Gafni, 1992; Yerushalmy & Schwartz, 1993). They have described how different pieces of specially designed software provide an arena for conceptual, transformational activities and for developing different approaches related to explaining manipulations.

The role of technological environments in providing conceptual support for symbol manipulation was studied by Gafni (1996) who asked students to select expressions equivalent to $(3x - 3)/5$ from a list of four expressions including $(3x - 3)^2/25$ and $((3x - 3) + 8x)/(5 + 8x)$. Gafni analysed the performance of 44 *VisualMath* 9th graders (14-15 years old) who learned with graphing software, but which was not available during their exams. She compared their explanations to those of two control groups: a similar group (55 students) who learned with symbolic feedback and another one that did not use technology in algebra. She identified three strategies used by students in all three groups; however, in each group, different strategies led to different success rates and were used with different frequencies. The first strategy—the *syntactic strategy*—included arguments that quote a known rule regarding the eligible operations on expressions. Only 24% of the arguments in the multiple representation group used this strategy with 46% success. The second strategy—the *semantic strategy*—is one in which arguments that consider the type of operation are included (e.g., "squaring is not a legitimate operation because the numerator and the denominator were multiplied by different numbers or expressions"). All three groups used this strategy in about 30% of the arguments. The third strategy—the *graphic strategy*—is one where the explanations were mentally based on the functions' graphical properties (e.g., "the operation caused the line to turn to be a parabola, so they can't be equivalent"). It was found that 43% of the arguments of the multiple representation students were of this sort and 91% were correct. None of the other students used this strategy. Examples of this sort suggest that the use of multiple representation technology does not at all omit the structural ideas of expressions from the study of algebra but rather introduces a new perspective on activities that have a chance at introducing important ideas.

6.2.3.2 Solving and manipulating equations

The benefits and complexities that arise from viewing an equation as a comparison of two functions are often discussed in relation to problems in context. There are fewer opportunities to learn about equation solving as a pure analytic activity. Gafni (1996) studied the three groups described above on their approaches to solving equations when they either knew or did not know an algorithm for solving. A typical activity that Gafni used is given in Figure 6.4.

Students solved the problem with paper-and-pencil, with no technology present. She found that 41% of the experimental group (who had learned with graphing software) used the given graph to offer a range of constant functions that would cross the given graph at a single point and 100% of the attempts were correct. Another 21% tried to substitute numbers and only 56% of the attempts were successful. This example and many others in this study suggest that interpreting equations as a comparison of functions and learning to inspect the solution by reading the x-coordinate of the intersection is an intelligible process.

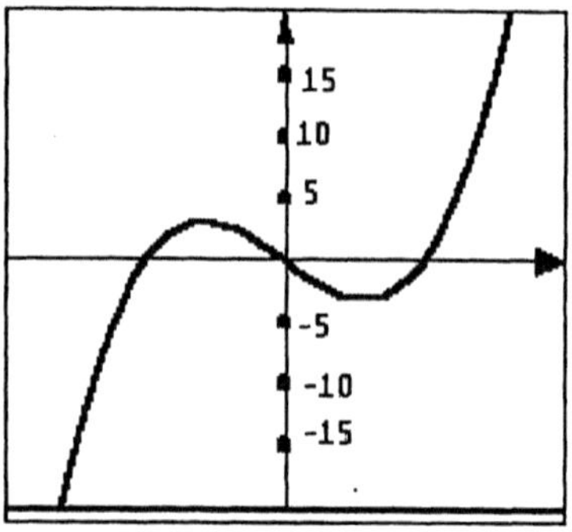

Here is a graph of the function: $f(x) = x^3 - 4x$.
Suggest a value for the parameter **a** such that the equation $x^3 - 4x = \mathbf{a}$ will have a single solution.

Figure 6.4. Solving a problem related to the graph of the function $f(x) = x^3 - 4x$.

Further evidence related to the role that graphical representations can play in meaningful equation solving comes from a study carried out by Kieran and Sfard (1999), who developed a 30-day module aimed at introducing algebraic objects and their transformations via graphs. One month after the students (about 13 years of age) had completed the instructional module, several were interviewed. Following is an excerpt with one student, Jer, who was as yet still unfamiliar with algebraic techniques of equation solving.

Interviewer:	*Can you solve $7x + 4 = 5x + 8$?*
Jer:	Well, you could, see, it would be like start at 4 and 8, this one would go up by 7, hold on, 8, 8 and 7, hold on, no, 4 and 7, 4 and 7 is 11. ...They'd be equal, like, 2 or 3 or something like that.
Interviewer:	*How are you getting that 2 or 3?*
Jer:	I'm just like graphing it in my head.

Yerushalmy and Chazan (2002) further discuss the meaning of having a solution rather than solving. Students who were used to viewing an equation as representing a comparison of two functions in a single variable were asked to describe the solution of an equation of two variables. The study analysed whether *VisualMath* students had internalised differences between functions in one variable and equations of two variables, and whether they were sensitive to the somewhat major conceptual changes that simple symbolic manipulations (such as *shifting* a term from one side to the other) lead to. Yerushalmy and Chazan describe a teacher of 9th graders who, as part of an introductory phase to teaching a system of equations in two variables, asked the students to focus on the following question: "How would you describe the equation $x + y = 2x - y$?" They indicate three strategies used by students to visualise an equation in two variables. The first strategy, which sought consistency in the representation of the equation as a comparison of two functions, viewed both x and y as independent variables and the equation as an object in three-dimensional space. Another strategy involved a delicate attempt to view one of the letters as a parameter and to describe families of functions. Students decided to think of x as an independent variable and of y as a parameter that could take on a range of values. A third strategy used the syntax and representations of function to give meaning to the *shift from side to side* manipulation: By using symbolic manipulations, they reduced the question of finding the solution of an equation in two variables (e.g., $x + y = 2x - y$) to a question about the zeros of a single function in one variable (e.g., $y = x/2$).

6.2.4 Research on integrating multiple representations with existing algebraic knowledge

Multiple representation software introduces many changes to the curriculum. In Section 6.2.2 we reported on ways in which the use of technology could support approaching algebraic symbols using intermediate lexicons such as graphs and symbols that do not result from symbolic expressions. In Section 6.2.3 we reviewed work that studied how to underpin the learning of symbolic manipulations with graphical representations of functions. Our review continues by looking at studies of students who are already familiar with multiple representations of functions and with algebraic syntax (symbols and manipulation rules.) These studies look mainly at how students implement this familiarity with multiple representations in conceptualising function and in solving contextual problems in algebra.

6.2.4.1 Conceptualising function within multiple representations

The construction of the function concept, which was normally part of calculus or pre-calculus courses, is now widely considered to be part of the knowledge of algebra. Schwarz and Hershkowitz (1999) characterise the concept images of function that arise in an interactive environment based on multi-representational software, and discuss technology-supported methods of conceptualisation of function in the light of previously known misconceptions and complexities. Students who learn about functions in an interactive environment: (a) often use prototypic functions (linear and quadratic) but do not consider them as exclusive, (b) use prototypes as levers to handle a variety of other examples, (c) articulate justifications often accounting for context, and (d) understand the attributes of functions. Several studies have tried to document what it is in the tool that actually leads to new ways of understanding.

Such a study was carried out with pre-university students using the graphics calculator in their mathematics course (Streun, Harskamp, & Suhre, 2000). Three classes used the graphics calculator throughout the entire year; five classes used the calculator only for one topic during the year, a topic that took six weeks to cover;

and four classes (the control group) did not use the calculator at all during the year. It was found that students who used the technology for the entire school year outperformed their counterparts in the control group on the post-test; however, no improvement was observed for the classes that used the calculator for only six weeks. A secondary analysis was able to confirm that the students who used the graphics calculator for the entire year learned to integrate graphing approaches with their previous algebraic methods and thus enriched their conceptual understanding of functions. However, those who used the graphics calculator for the shorter period of time merely replaced their algebraic or guess-and-test procedures with graphing methods and, as a result, did no better on the post-test than the students of the control group.

Ruthven's (1990, 1992a) *Graphic Calculators in Mathematics* project involved classes of students who had access to graphics calculators throughout their advanced-level mathematics course. One of the cases provided by Ruthven in his project reports deals with a session involving double-angle trigonometric identities. Students were asked to consider two possible equivalents for $\cos 2a$, namely, $2\cos a$ and $\cos^2 a$. The most productive work began when students started to compare graphs of the different possibilities. In the ensuing whole-class discussion, they found that not all had developed the same algebraic form of relationship. It was valuable for the students to be able to use their manipulative skills to establish which formulations were equivalent. According to Ruthven, "this example highlights the way in which it is possible to approach a new mathematical idea graphically, rather than symbolically, bringing a visual understanding of the idea which can be exploited to guide and support a symbolic treatment at a later stage" (Ruthven 1992a, p. 32).

Schwarz and Dreyfus (1993) attempted to track the ways in which students used particular representations. The investigation involved a detailed analysis of the work of 43 students, four of whom are reported in the article. The researchers used the multi-representational software *TRM* and a tutoring facility that helped obtain quantitative information about students' integration of representations. Schwarz and Dreyfus identify "parallelism of representations" (i.e., that representations should constitute a semantic field for the concept under consideration) to be a necessary condition (built into the software) to allow for successful work with multiple representations. However, they conclude that even the task and the software tools do not guarantee successful integration, and therefore it is important to construct tools that assist teachers in knowing the different methods of using multiple representations employed by their students. Hershkowitz and Kieran (2001), who also studied the ways in which 16-year-old students used the multiple representation and regression tools offered by graphics calculators, suggest that one of the crucial questions to be considered is *how much* and *in what way* we would like the tool "to do the work" for the students. Thus, the properties of the software and the situation in which it is being used can lead to the emergence of different types of knowledge.

6.2.4.2 Solving contextualised problems using multiple representations

A major agenda of algebra teaching is equipping learners with tools to mathematise their perception of situational contexts. As has been suggested up to now in this chapter (see also Heid, 1995; Nemirovsky, 1996), the concept of function can provide a set of tools for mathematical modelling that enlarges the time-honoured symbolic repertoire. More importantly, a multi-representational approach has the potential to shift the focus of solving even traditional word problems from assigning and solving for an unknown to analysing the various processes and relations among those processes. Thus, the integration of multiple representations of function opens up opportunities for developing a wider range of solution methods to traditional algebra problems. Huntley, Rasmussen, Villarubi, Santong, and Fey (2000) included in their analysis of CPMP students 49 specific questions in which students were asked either to formulate an algebraic model or to interpret results stemming from the use of a given algebraic model. CPMP students, as expected, outperformed control students on 44 of the questions. But beyond the mathematical formulation, the researchers looked at how students (a) evaluate the expressions that they formulate, (b) transform the expressions to equivalent ones, and (c) solve the equations involved in the contextual problems. Differences were expected because, in curricula that make use of numeric and graphic tools, "students have several more options available for answering such questions; furthermore, in *Standards*-based curricula that make heavy use of real-world contexts for teaching algebraic ideas, students are encouraged to use contextual metaphors as guides to thinking about algebraic tasks" (p. 346). Results show that the CPMP students outperformed the control students when they were able to use context clues and had graphics calculators available. However, Huntley and colleagues suggest some caution in interpreting these results, for it is not entirely clear whether students simply demonstrated a wider repertoire of problem-solving strategies when using the graphics calculator, or whether they developed a deeper understanding of the mathematics embedded in the contextual problems.

Zooming-in to the use of the graphics calculator, Doerr and Zangor (2000) found five patterns and modes of graphics calculator tool use that emerged within a study of pre-calculus students solving problems: computational tool, transformational tool, data collection and analysis tool, visualising tool, and checking tool. By *transformational tool*, Doerr and Zangor refer to the way in which the tool can transform tedious computational tasks into interpretation tasks. For example, in order to find a function to describe the vertical position of a point on a rotating ferris wheel, relative to the hub of the wheel, students linked the interpretation of the resulting equation with the experienced phenomena and shifted the focus of the task from pair-wise computations of values to a global interpretation of the function.

The impact of the range of representations on students' solving of complex contextual problems using spreadsheets is demonstrated by Molyneux-Hodgson, Rojano, Sutherland, and Ursini (1999). In this cross-national study, the multiple representations of the spreadsheet supported different approaches within the cultures of the schools: It brought Mexican students to appreciate and use graphical and numerical representations, and English students to make sense of algebraic representations.

Another aspect that has recently emerged from studies on the integration of graphing technology within contextual problem solving is the heightened performance and enhanced thinking processes of students who were previously unsuccessful with algebra in general and with solving contextual problems in particular. Huntley, Rasmussen, Villarubi, Santong, and Fey (2000) discuss the solving of contextual problems by lower and higher achievers and suggest that "new approaches to algebra might well be enabling traditionally unsuccessful students to gain access to the problem-solving power of the subject" (p. 357). They found that students who were not strong in symbol-manipulation skills could outperform symbolically-capable students when the tasks required formulation and interpretation of situations.

Yerushalmy (2000) describes a longitudinal observation of a lower-achieving pair of students who studied algebra for three years using the *VisualMath* function approach that included intensive use of graphing technology. Yerushalmy analysed their problem-solving processes using two schemes: (a) the phases related to various problem-solving strategies, and (b) the mathematical resources (categorised by the different representations) used at each phase. An overview based on interviews throughout the three years of learning algebra offers trends in the learning and development of the students' skills using multiple representations. The time spent on each problem was long: Students seemed to have developed a facility with all the representations and they needed time to manoeuvre. Traditional problem-solving phases such as verification and implementation became indistinguishable, as both were integrated into the solution process with the graphing software. Thus, lower achievers seemed to benefit from learning higher-order skills that involved the use of multiple representation tools when solving problems in context.

6.2.5 Tensions involved in teachers' conceptualisations of school algebra when using multiple representation technology

The increasing availability of multiple representation technology and of curricula involving new approaches to school algebra have begun to be reflected in a few of the studies of teacher thinking and knowledge. Some studies have considered teachers' attempts in implementing reform-oriented curricula, and their beliefs and attitudes towards different types of curriculum (e.g., Heid, Blume, Zbiek, & Edwards, 1999; Kendal & Stacey, 2001; Zehavi, 1996). Although multiple representations of functions has turned out to be a widely-used approach among algebra teachers using graphics calculators inside the standard curriculum, or even as part of a reformed curriculum, many teachers experience tension between familiar sources and methods and the use of technology, or between the curriculum and their educational conceptions (e.g., Chazan, Larriva, & Sandow, 1999; Haimes, 1996; Lloyd, 1999; Monaghan, 2001; Wilson & Krapfl, 1994).

For example, Wilson and Krapfl (1994) analysed teachers' use of graphics calculators in their classroom and stressed the lack of studies on the tension between traditional methods of solving equations and those offered with graphics technology. They found that teachers often do not think about mathematics and mathematics teaching in ways that would enable them to use the reform recommendations. In contrast, Doerr and Zangor (2000) who studied how graphics calculators were used to support students in exploratory processes of modelling suggested that the role, knowledge, and beliefs of the teacher influenced the emergence of rich usage of the graphics calculator. The teacher's role in encouraging interpretation and explanation led to meaningful constructions of mathematics. In fact, the importance given to meaning-making led the teacher to devalue regression equations as solutions and limited the "black-box" use of the graphics calculator. But, in another study, Haimes (1996) observed teachers who used the new functional approach with the best of intentions to implement it as designed; however, they failed to realise the spirit of the innovative curriculum.

In a four-year, technology-implementation study in Mexico involving fifteen secondary schools (7th, 8th, 9th grades, 12-15 years old), Cedillo and Kieran (2003) found that the introduction of TI-92s into the teaching of algebra promoted not only improved student learning among both the stronger and weaker students, but also a much more positive attitude toward mathematics. This, in turn, affected the teachers. Among the teachers taking part in the project were those whose mathematical background was quite strong and who seemed initially very difficult to convince about the potential benefits of using technology. Those with a long experience in teaching showed themselves to be particularly reluctant. But once they witnessed that their students were using strategies that they themselves had not taught, and how well their students were progressing, they changed their attitudes and rapidly became themselves avid learners of the new possibilities provided by technology. These teachers changed their teaching style too. They abandoned lecturing at the front of the classroom. They started to produce new materials on their own to supplement the activities that had been provided by the project designers. They began to respect students' self-pacing of learning and to be more alert to helping students on an individual basis.

Similar findings have been reported by Ruthven (1992b), who described the reactions of some initially-sceptical teachers involved in the *Graphic Calculators in Mathematics* project to their students "forever making new discoveries and devising ingenious strategies for solving problems" (p. 100), with the aid of their new technological tool. As a result, these teachers became much more enthusiastic and, in fact, disclosed that, "the introduction of the graphic calculator revolutionised [their] approach to the teaching of many mathematical topics" (p. 100). See also

Chapter 7 for discussion relating to graphics calculators with symbolic algebra capabilities.

The complexity involved in teachers' thinking about the transition towards an algebra that is heavily based on multiple representations of functions is vividly reflected in their writing of the experience in a study conducted by Slavit (1996). Having introduced a novel pedagogical aid (graphics calculator) into a somewhat traditionally-taught Algebra II classroom, a teacher was worrying about students' tendencies to focus on graphing-strategy solutions, but without actually reflecting on what they were doing—a familiar dilemma, in a novel context. He was also struggling with fitting a new tool into a traditional textbook presentation that made limited use of the main features of the tool. Related dilemmas have been described by Chazan (1999, 2000), who tells about the concerns he experienced in making the transition from algebra as the study of symbols and manipulations to algebra as the study of phenomena outside of mathematics—a perspective that is often embedded in *functional approach* curricula. Chazan, who was able to make the algebra relevant to his students by having them look at the world around them, did not however feel that his students had developed sufficient facility with symbols. But what does sufficient mean? And according to what standards and goals? Chazan (2000) answers as follows:

> Of course, the matter of just what constitutes sufficient facility with symbols is a matter of debate; ... there is a dilemma that pits my desire to help my students see mathematics as meaningful activity against my desire to make many possible future trajectories open and available for them. (p. 98)

6.3 Environments Offering Dynamic Control

The previous section has focused on the roles played by multiple representation environments as a bridge to algebraic symbolism. Several of these environments offered dynamic control of representations and of phenomena, but elaboration of that aspect was reserved for this section. Dynamic control involves the direct manipulation of an object or a representation of a mathematical object. The manipulation can be of a continuous nature, or it can be discrete. Dynamic control can be achieved by means of several devices, for example, slidergraphs, sliders, dragging facilities, and so on—a variety which in itself opens up interesting and, as yet, unexplored research questions.

Before looking at examples of various environments offering direct manipulation and the nature of the algebra learning that can occur in these environments, we note that such environments are of two types. If the manipulations are related solely to the three mathematical representations of a function, then the environment is of the first type: Within Functional Representations (WFR). If, on the other hand, the environment includes phenomena from outside algebra, such as physical or geometric objects, then it is of the second type: Phenomena Beyond Algebra (PBA). Our discussion of these two types of environments begins by distinguishing between discrete and continuous control within a framework of multiple representations of functions.

6.3.1 Discrete versus continuous control

The issue of control involves two questions: *what* is being controlled and *how*. When one remains within a single representation of a function (the *what* part of the question), there are three easily identifiable elements: an input, some transformation carried out on that input, and the output (which is the transformed input). For example, staying within the graphical representation of a function, the input could be the graph of a linear function passing through the origin with a slope of 2 (we deliberately refrain here from employing a letter-algebraic symbolism to describe this function), the transformation could be a vertical shift of 3 units, and the output would then be the graph of the linear function passing through (0, 3) and with a slope of 2. The transformation (the *how* part of the question) could have been carried out by a continuous manipulation of the graph or by discrete steps using a translation button.

But the situation becomes more complex when one is dealing with multiple representations of a function. For instance, suppose there are two representations of a function on the screen, the graphical and the symbolic. Let us assume for this example that the environment has been designed in such a way that the "driving" input is the symbolic one. The graphical representation is quite present, but it is being "driven" by the symbolic representation. As the driving input is the symbolic one, the transformations that are to be carried out are executed on the symbolic representation, but these transformations are also mirrored in the driven representation in one of two ways—either simultaneously if the transformation is a continuous one, or at the completion of the transformation if it is a discrete one. The output that results from the transformation is reflected in both representations: the symbolic, which was the driver, and the graphical, which was being driven.

Four examples of direct manipulation of functions are provided in the paragraphs below—all of them involving multiple representations: two operate on the graphical representation, and two on the expression representation. In each group of two, one of the transformations is discrete and the other continuous.

6.3.1.1 Manipulating the graph

The first example involves transforming graphs by acting on icons in discrete mode (e.g., horizontal translation or stretch, vertical translation or stretch, reflection). Such environments were part of a few studies that analysed students' mathematical constructions when working on tasks in algebra. (e.g., Borba, 1993; Schwartz & Yerushalmy, 1992). Figure 6.5 illustrates how the discrete (stepwise) transformation of the function can be explicitly read from the expression when linked to the traces of the graphical transformation process.

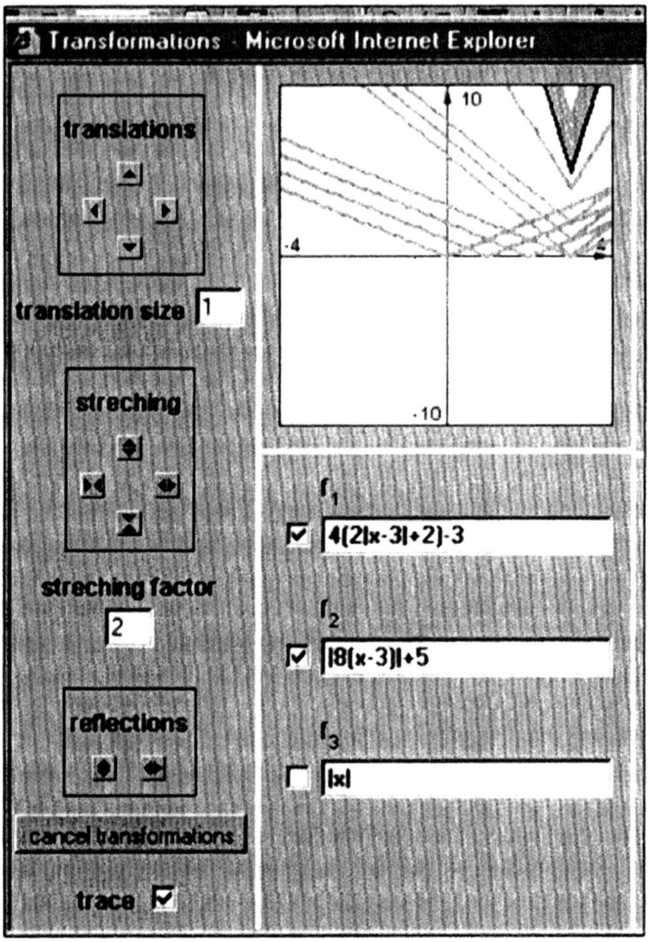

Figure 6.5. Discrete manipulation of the function graph.

The task is part of a newer environment of the *VisualMath* series (http://www.cet.ac.il/math/function/english) and it involves describing how the function $|x|$ was transformed to the function f_2 ($|8(x-3)| + 5$). The process that is recorded in the expression $4(2|x-3| + 2) - 3$ suggests that one way in which this could be done would be to perform 3 steps of 1 unit of horizontal translation, a stretch by a factor of 2, a vertical translation of 2, a stretch by 4, and a vertical translation downwards of 3. The point here is of course not the specific process but rather the transparency of the structure of the expression reflecting the discrete transformations.

The second example related to graph manipulation involves direct continuous manipulation of the graph itself. Another piece of software from the same *VisualMath* series (Schwartz & Yerushalmy, 1996) allows the user to go beyond a direct stepwise manipulation of the graph to its direct continuous manipulation. In Figure 6.6, the task is to fit the function $f(x) = x$ to the four non-linear points.

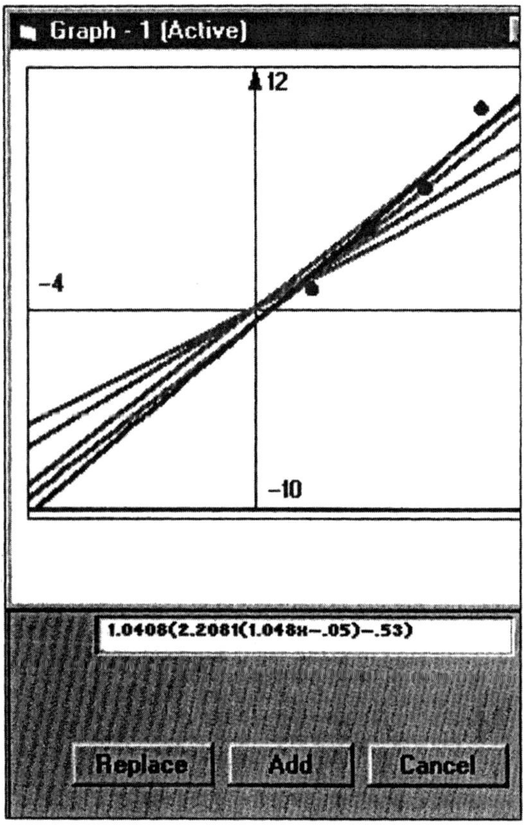

Figure 6.6. Direct continuous manipulation of the function graph.

The design of the environment allows the use of continuous translations to relocate the graph in a specific position, as well as the use of stretch to change the shape of the graph—both without any pre-determination of the step size or the direction. While this might be a new channel to visualise graphs, the multiple representation linkage is more delicate here. The changes in the expressions and in the table of values are sensitive to the manual inputs and are quite difficult to read and interpret. Although the structure of the expression reflects the chronological order of the transformations on the function $f(x) = x$, we assume that asking a question similar to that in Figure 6.5 would not be as appropriate with this environment and other more suitable tasks would have to be designed for such continuous processes.

6.3.1.2 Manipulating the expression

The first example below, which is also drawn from *VisualMath*, illustrates the manipulation of the parameters of an expression directly by fixing the step size for a desired change. By means of the arrow buttons, the user transforms the step size for a parameter of the given expression. Once the arrow button has been set to a particular value, the expression changes to reflect that transformation and the graph is redrawn (see Figure 6.7).

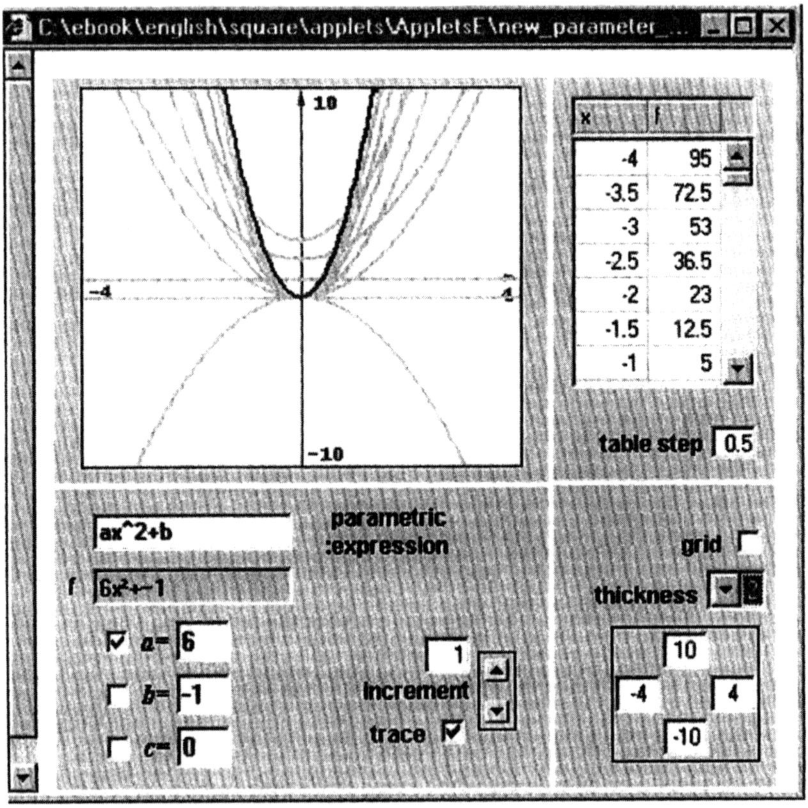

Figure 6.7. Discrete manipulation of the function expression parameters.

In contrast to the discrete transformations that are an integral part of the environment above, the second example of this pair illustrates a continuous transformation that is obtained by means of the dynamic control of a "slidergraph". Zbiek and Heid (2001) provide an example involving a family of functions whose general member is represented by $f(x) = a / (1 + be^{cx}) + d$, where a, b, c, and d are real numbers. As seen in Figure 6.8, the screen display includes both the expression for the family of functions and a graphical representation. The lower portion of the screen contains the slidergraph for changing the values of the four parameters, while the upper portion displays the resulting value of the expression. The slidergraph works by sliding the point corresponding to a given parameter along its horizontal line at the bottom of the screen. At the same time as the slidergraph moves along, the graphical representation responds to and reflects the various values that the functional expression takes on. In the example shown in Figure 6.8, dragging the point b to the left causes the graph to split suddenly as the value of parameter b changes from positive to negative.

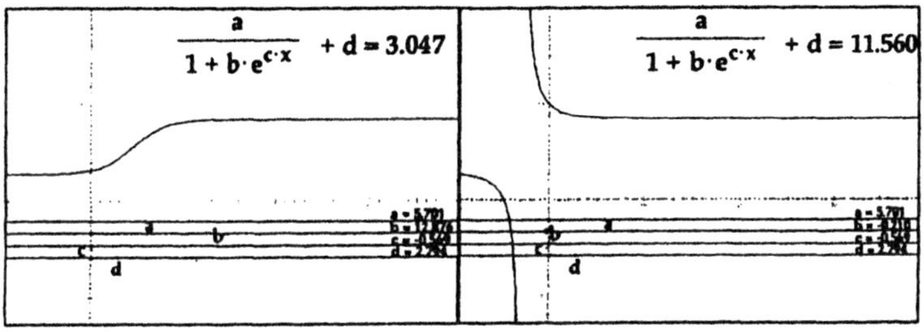

Figure 6.8. Direct continuous manipulation of the function expression parameters.

Research on the impact of directly controlling a change in one representation to effect changes in another (or the same one) remains relatively underdeveloped. As will be seen from the discussion below, based on a sample of the types of environments offering dynamic control, emerging findings of this growing body of research suggest that the kinesthetic relation between the user and the phenomenon being controlled plays a crucial role in developing a deeper understanding of the phenomenon—be it a property of a mathematical object, or of a mathematical representation of some change in a physical object.

6.3.2 Staying within the mathematical representations of a function

Within-Functional-Representations (WFR) environments are those that involve some combination of graphs, algebraic expressions, and numerical values (usually in tabular form). The environment featured in the Zbiek and Heid (2001) study, seen above, illustrates how the direct continuous manipulation of elements of a function's expression can be reflected in the continuous movement of the graphical representation. Zbiek and Heid have pointed out that the ease with which the slidergraph, unlike static graphing utilities, allows students to vary a single parameter quickly and consistently leads to surprising observations and is, in fact, the key to launching student reasoning.

The dynamic change that is the key feature of this slidergraph environment can be viewed functionally in the following way. The transformation that the students apply involves changing the values of the parameters of the input expression by moving the slidergraph for one of the given parameters. This transformation is simultaneously reflected in the graphical representation that changes all the while that the slidergraph is being moved along. The "final" output, as determined by the final resting place of the slidergraph, is the expression with the most recent value of the modified parameter, along with the most recent form of the graph. But in actual fact, there have been a stream of outputs because the transformation was a direct continuous one. As the slidergraph moved along the horizontal band, a series of new values of the given parameter—and thus a set of new values for the expression of the function—was generated, and the graphs corresponding to those functions were traced. In their attempt to make better sense of the given family of functions (expressed symbolically), students in the Zbiek and Heid study used the slidergraph to produce a seemingly continuous set of graphical correspondences. Each parameter could be controlled individually, in combination with particular values of the others. The object of study here was the family of functions expressed symbolically, but one where the graphical "output" served to help the students in creating sense for the symbolic representation. Based on the findings of their research, Zbiek and Heid (2001) argue that this tool "facilitates the study of function families involving larger numbers of parameters and more intricate relationships among the effects of parameter values" (p. 684).

The new methods of interacting with expressions and graphs, which are made possible by slidergraph environments such as this one, seem promising ones to explore in further research. These environments permit the study of questions such as, "How do student interpretations of transformations with computers affect the way in which they think about and construct their algebra?", "How do students perceive the differences between continuous and discrete change of parameters in expressions?", and "What is the nature of the conjectures that emerge as students control mouse-based transformations in a continuous way?"

Borba (1993) studied the roles of visualisation and direct actions on graphs in the environment, *Function Probe* (*FP*) (Confrey, 1991). With *FP*, a function can be transformed into another by an action on the graph using what Borba calls "*direct action*"—the mouse clicks on one of three icons (and not on the graph itself) that controls changes in the equation by inputting values for the step size of the desired transformation. Borba's findings, which cover a wide range of aspects of multiple representation software and in particular the study of manipulations of graphs, shed light on two issues related to direct control of graphs: one involves the impact of the design of the software and the other, the impact of the task.

Regarding the design of the environment, Borba argued that the software influenced the mathematics that was studied and produced. With respect to the impact of the task, he claimed that the "static" and the "dynamic-process" styles of thinking that are elicited in *FP* are very much dependent on the nature of the tasks that students work with in a given environment. To illustrate the point, Borba commented that, "most of the time, they [the students] saw transformations as a process rather than as a static (beginning and end) feature, and they could connect particular cases such as $y = f(2x)$ and $y = f(3x)$ to a general 'sub family' $y = f(\mathrm{b}x)$" (p.366). Conversely, when they worked on the problem, "How would you investigate the relationship between coefficients in the equation of a function and the graph of a function?" they "saw transformations as a 'static change' focusing on a comparison of a beginning and an ending state." Borba argued that when the goal of the task was to determine what happens to the graph when the coefficient changes, it invited a static view of transformations. It would be an interesting analysis to study students' work on the same task with the environment described by Zbiek and Heid and to see whether the continuous manipulation of the coefficients with the slidergraph changes the mode of thinking on this task.

6.3.3 Environments that include phenomena from beyond algebra

Dynamic environments that include phenomena from beyond algebra (previously referred to as PBA environments) tend to involve geometric constructions or interactions with various physical phenomena. The former range, for example, from the work of Hazzan and Goldenberg (1997), who explored the types of understanding of function that emerge from operations in dynamic geometry (e.g., "What are the difficulties in understanding constant function?"), to the work of Arcavi and Hadas (2000) on the role played by dynamical graphical representations (and their symbolic counterparts) in developing a deeper understanding of geometrical relationships (e.g., the relation between the area of a triangle and its altitude). In the dynamic geometry environment studied by Arcavi and Hadas, properties of the geometric object are graphed as a function (e.g., length of an altitude serving as the independent variable and area of the triangle as the dependent variable). With activities such as the one reported by Arcavi and Hadas, the dynamic of the movement of the geometric shapes creates dynamic changes in the graphs.

In another category of somewhat different geometric constructions are those created in the Logo *Microworlds Pro* programming environment. The dynamic control afforded by this environment is not that of acting directly on the geometric figure by means of the mouse, but rather using a slider to control numerical values of the variables. Vincent (2001) has been conducting studies with 5th to 8th graders, in an environment where Logo procedures are displayed alongside the work area containing the sliders (Logo Computer Systems, 2000). She has observed students constructing procedures for, say, regular polygons, which contain two variables—one for the number of sides and one for the length of the sides—and has noted that, by adjusting the position of the slider for one variable at a time (e.g., the number of sides), pupils are provided with an opportunity to explore the various roles that this variable plays in the relationships of the polygon situation—for example, the measures of the exterior and interior angles.

The next major group of dynamic environments that deal with phenomena from outside algebra are those that feature physical objects and their change. Such environments generally involve the modelling of the physical situation by means of graphical and symbolic representations (as reviewed by Isoda, in the brief prepared for the Working Group); however the opposite process—starting with the graphical representation and using it to control the phenomenon—is an important part of the environments that will be discussed below. In the brief he prepared for the Working Group, Arzarello—whose ideas helped in the development of some of the ideas contained in this section—has highlighted the difference between the dynamic-geometry and physical-phenomena environments in terms of virtual and physical movement. Accordingly, he classifies as environments that offer virtual movement those that feature sliders or dragging in some form, as in Cabri and other dynamic geometry environments, or in Excel, Logo MicroWorlds Pro, and so on. In contrast, environments that offer direct physical movement are those that involve, for example, motion sensors, such as calculator-based rangers (CBRs) hooked up to graphics calculators (see, e.g., Arzarello & Robutti, 2001), or other various devices where the computer presents a graphical reflection of some outside physical activity (e.g., the environments created by Nemirovsky and his collaborators [e.g., Nemirovsky & Noble, 1997; Nemirovsky, Tierney, & Wright, 1998; Schnepp & Nemirovsky, 2001] involving minicars, where the motion of the cars is controlled by the user). This dichotomy proposed by Arzarello does not, however, distinguish between those environments that stay within the three mathematical representations of a function and those that venture beyond in order to include phenomena from outside algebra.

6.3.3.1 Two examples from dynamic geometry environments

The quintessential environments offering dynamic control are the dynamic geometry environments (Cabri, Geometer's Sketchpad, Geometry Inventor, etc.). Because the connection of these environments with algebra may not be obvious, the first example discussed below was chosen not only to illustrate the nature of the direct interaction between user and environment but also to suggest how such environments may be viewed as algebraic.

Hazzan and Goldenberg (1997), in a study involving dynamic geometry environments, have argued that, while students are typically found to encounter difficulty in analysing or describing function behaviour, this difficulty tends to disappear when functions are represented dynamically (see also Goldenberg, Lewis, & O'Keefe, 1992):

> When movements of the student's hand (computer mouse) represent (control) movements of the function variable, and when the function value is observable as a separate coordinated movement, students spontaneously notice and invent ways of talking about and classifying different kinds of change. (Hazzan & Goldenberg, 1997, p. 274)

In student activities involving, for example, parallelograms inscribed in a quadrilateral, Hazzan and Goldenberg noted that dynamic change in a geometry construction can be viewed functionally in at least two different ways:

> From one perspective, the movement of the user's hand from a specific point in R^2 to another specific point in R^2 along a particular path is the function that transforms the object on the screen from one state (pre-image) to another state (image). The input of a function of this kind is a construction (complete with its placement in R^2) – one element drawn from the space of all possible constructions that could be on the screen.
>
> From another point of view, the rules that determine a dependency, captured in a construction, define a function. Such functions relate values in R^2 – a position of a point – to other values that may also be in R^2 (the position of another point), or in R (a measurement performed on some feature), or even in higher dimensions (e.g., the position, slope, and length of a segment). The output varies as the input is dragged, according to the dependencies determined as the construction was built. (p. 269)

Thus, if as in dynamic geometry environments, not only is the notion of a function expanded, but also the nature of the sets that can serve as domain and range, then it seems reasonable to conclude that in other dynamically controlled environments new ways of thinking about functions and their inputs and outputs are conceivable.

The second example from dynamic geometry, which is presented just below, has a completely different feel to it from the Hazzan and Goldenberg study above. It involves the relationship between two components of a geometric object that are first measured, then represented as coordinates of a point in the plane, and finally manipulated so as to create a graphical trace of the functional relation.

Arcavi and Hadas (2000) have described how, within the given sequence of activities, students from 9th and 10th grades, as well as teachers attending a professional development workshop, explored triangular relationships, in particular, the areas of an isosceles and of a nearly isosceles triangle as functions of the length of the base and of the altitude to the base. In the first part of the activity, after constructing an isosceles triangle with equal sides of 5 units, students were encouraged to drag the vertex C so as to change the length of the base AC. They were asked to consider questions such as, "What changes and what stays the same?" The study of the variation of the area as a function of AC was then proposed.

Soon the participants constructed a Cartesian graph of the variation of the area of the triangle ABC as AC changes.

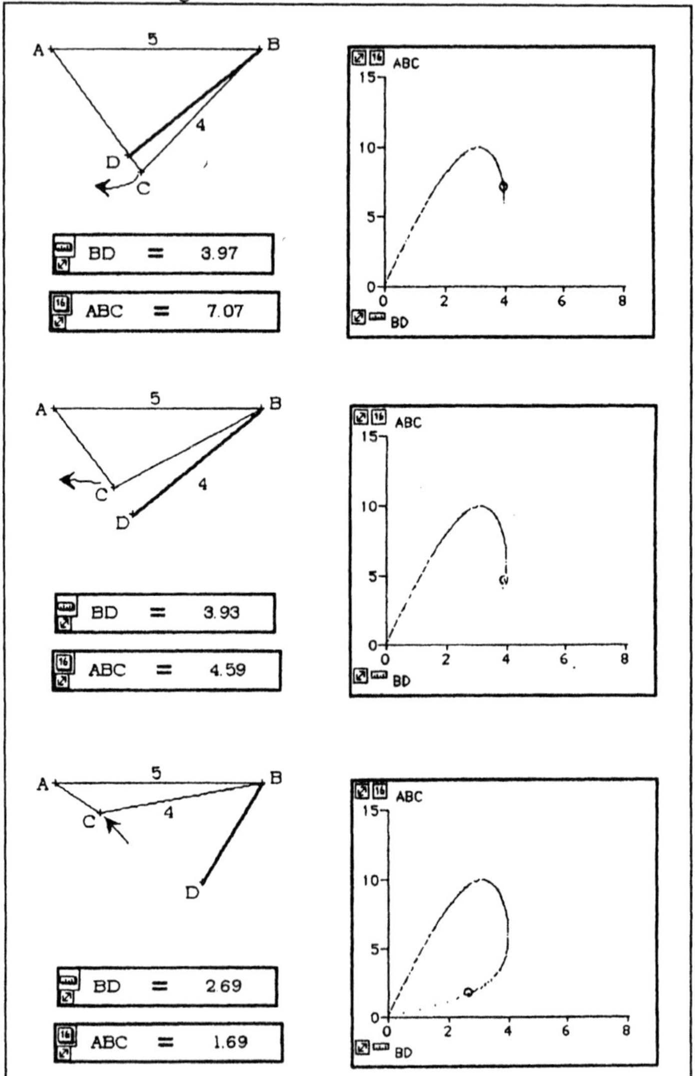

Figure 6.9. Area of a nearly isosceles triangle plotted as a function of the altitude BD
(From Arcavi & Hadas, 2000. Copyright by Kluwer Academic Publishers. Reprinted by permission)

This was followed by related activities with a second triangle where one of the two equal sides was changed from 5 units to 4 units. A surprise arrived when the area of the nearly isosceles triangle as a function of the altitude BD was plotted. When the altitude jumped from inside the triangle to outside, the graph "went backwards" (see Figure 6.9).

This totally unexpected graph created a strong need in participants to understand the situation. Representing the situation with algebraic symbols added, according to the researchers, another layer of meaning that helped students to see why, for the same altitude BD, one obtains two different values for the area:

Area = (BD.AC)/2 = [BD (AD ± DC)]/2, and when BD = x then

$$\text{Area} = [x(\sqrt{25-x^2} \pm \sqrt{16-x^2})]/2.$$

6.3.3.2 Examples involving direct control on functions as a way to manipulate physical phenomena

We shift now from environments involving geometric objects to environments featuring physical phenomena. These environments are first examined through the lens of direct control of the functional representation in order to manipulate the physical object—a point of view that is the opposite of a modelling perspective. For example, an environment has been developed at TERC for the study of the mathematics of change (Schnepp & Nemirovsky, 2001). It consists of graphing software and hardware that link a computer to miniature cars on parallel linear tracks and to a miniature stationary bike. The software allows for two vantage points from which to study the relationships between physical motion and motion graphs: "line becomes motion" (LBM), and "motion becomes line" (MBL). With MBL, as with many modelling tasks of this sort, the motion is the object of study and mathematical tools are used to construct a graphical representation of it. However, with LBM, the user constructs a graph on a computer, which in turn communicates with a motor that moves the mechanical device according to the graphical specifications. This feature thus permits a shift in focus to the study of the mathematical function, as was confirmed by Schnepp and Nemirovsky, when they analysed the conversations that took place among the 12th grade calculus students who participated in their study.

A second example of an environment involving direct control of the function as a means of manipulating physical phenomena is drawn from the *SimCalc* Project (http://www.simcalc.umassd.edu/simcalcframe.html). In one of the activities designed for use with this environment (see Figure 6.10), students try to adjust a velocity graph to enable the baby duck to catch up with its mother. The baby duck starts out several metres behind the mother, and must go quickly to catch up, but then must slow down at the right time to match the mother's motion. In another mode of the software, students adjust the velocity graph by directly dragging and the position graph changes at the same time.

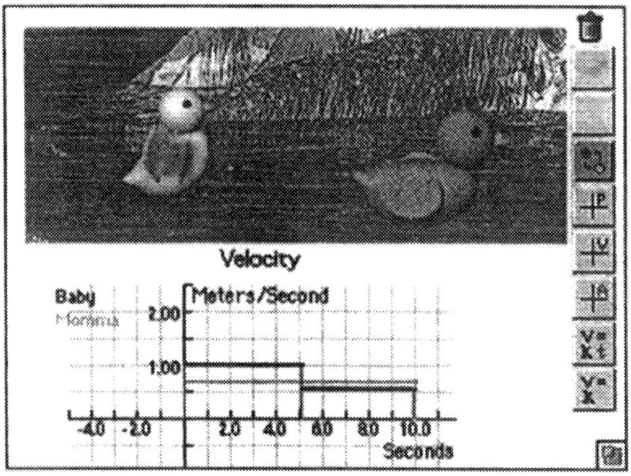

Figure 6.10. Example from the *SimCalc* environment involving direct control of a function to manipulate a physical phenomenon.

From their research with various activities of the *SimCalc* environment, Kaput and Roschelle (1997) have argued for "the important role of physical motion in understanding mathematical representations ... [whereby] students confront subtle relations among their kinesthetic sense of motion, interpretations of other objects' motions, and graphical, tabular, and even algebraic notations" (p. 106). They have also emphasised the need for students to use various representations, especially graphical ones, to control phenomena, not just to interpret them.

6.3.3.3 Examples involving direct dynamic control on the phenomenon as a way to change its algebraic representation

In the two examples of this particular subsection, another perspective will be taken, that of direct control of some physical phenomenon in order to effect a change in its algebraic representation. The control is of a continuous nature, and the change is reflected first in graphical or tabular representations, which then serve as a conceptual basis for producing algebraic representations. The first example is drawn from the *VisualMath* curriculum used with 7th graders (Yerushalmy & Shternberg, 2001). The researchers describe the use of an authentic motion situation involving a ball (which is at first disconnected from the computer), the trajectory of which is sketched directly on an electronic coordinate plane and which simultaneously provides the graphs of the process as it changes over time.

The task described by the researchers is as follows: "Hava loves to play with the ball. She throws it to the ground. The ball hits the ground and then it hits the wall" (Yerushalmy & Shternberg, 2001, p. 255). The students—using the *Function Sketcher* software (Yerushalmy & Shternberg, 1994)—drew freehand with the mouse their own interpretation of the path of the ball. Figure 6.11 illustrates examples of their dynamic drawing, accompanied by the simultaneous appearance of various graphs of the parameters of motion over time.

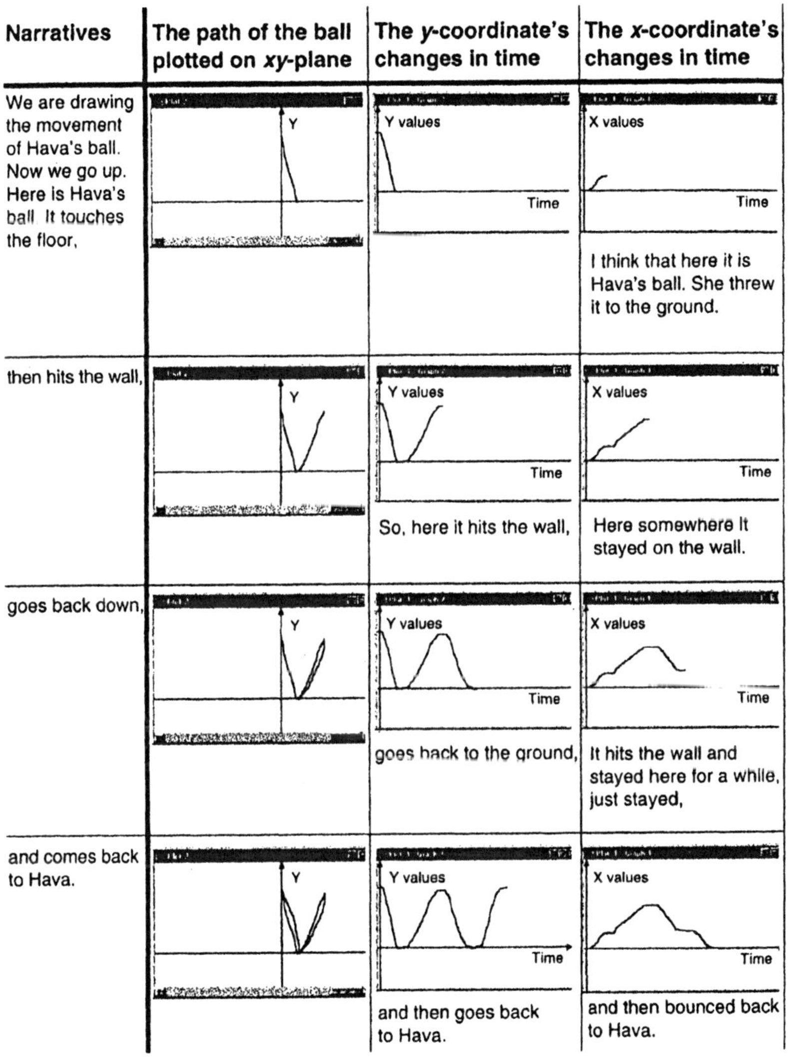

Figure 6.11. Dynamic drawing accompanied by the simultaneous appearance of graphs of motion over time (Yerushalmy & Shternberg, 2001). *(Reprinted with permission from* The Role of Representation in School Mathematics: 2001 Yearbook of the National Council of Teachers of Mathematics, *copyright 2001 by NCTM. All rights reserved.)*

The simultaneity of the graphs in response to the students' dynamic drawing of the ball's movement faced them, according to the authors, "with the challenge of explaining the differences between the drawing and the graphs, an activity that led them to elaborate on the meaning of the graphic representation. As the students drew the ball's path on the basis of their personal experience, they were discussing their view of the motion of the ball at the same time that they were receiving graphs of the path's *x*- and *y*-coordinates over time" (p. 255). Once students became proficient with visual representations, these representations served as a foundation for constructing symbolic representations (as described in Section 6.2.2.2).

Microcomputer-based laboratories and calculator-based rangers (CBRs) also offer direct control over phenomena, and thus a vehicle for relating the changes in them to their algebraic representations. A study by Arzarello and Robutti (2001) illustrates an environment focusing on changes in bodily motion. These researchers developed a research project that was carried out with a classroom of 9th grade students, working in groups of three-to-four pupils, and equipped with CBRs. The task was as follows:

> Walk or run in the corridor in order to make a uniform motion; when you arrive at the red line, come back with the same motion. The CBR will record your position with respect to time and will collect the data in a graph and in a table. The data are expressed in seconds (s) and in metres (m) respectively. Each 1/10 second, a couple of data (time and position) are collected.
>
> (a) Describe the kind of motion you made in the corridor.
>
> (b) Using the graph and the table, describe how space changes with respect to time (increase, decrease, ...).
>
> (c) Analyse the graph. Is it like a line? Is it like a curve? Does the curve increase? Does the curve decrease? ... Consider the ratio: $m = (s_2 - s_1) / (t_2 - t_1)$ and use it to describe mathematically the graph of your motion (t_1 and t_2 are two subsequent time data and s_1 and s_2 are two subsequent position data). (Arzarello & Robutti, 2001, p. 35)

Initially students were encouraged to try various running patterns in order to create different graphs. The continuous nature of the CBR graphing allowed students to test conjectures in a direct manner, controllable by their own physical movement. The authors observed that, "students' cognitive activity passes through a complex evolution, which starts in their bodily experience (namely, running in the corridor), goes on with the evocation of the just lived experience through gestures and words, continues connecting it with the data representation, and culminates with the use of algebraic language to write down the relationships between the quantities involved in the experiment" (p. 39).

6.3.4 Summary remarks

This section has presented examples of environments offering dynamic control—environments that permit students to manipulate directly objects on the screen or outside it, in order to try to grasp either connections between representations, or properties of the object being represented, or the relation between change in an object and the mathematical representation of this change. As suggested by the studies mentioned herein, the feature of dynamic control in these environments is creating fresh opportunities for the development of algebraic reasoning. Such possibilities open up new ground for research.

6.4 Structured Symbolic Calculation Environments[1]

6.4.1 What is structured symbolic calculation?

The meaning of *structured symbolic calculation* can perhaps best be understood by distinguishing it from multi-representational approaches. Multi-representational approaches to symbolic work use two or more representations (e.g., from among symbolic, graphical, and tabular representations) to create meaning for algebraic objects or processes. In other words, according to Cerulli (in the brief prepared for the Working Group), "the transformation rules that correspond to algebraic principles are viewed as transformations that keep some invariant in the multiple representations of expressions or equations." For example, two expressions might be considered to be equivalent if they have the same graphs or the same tables of values (Kieran, 1994). In Yerushalmy (1989), for instance, the user enters one expression, and then a transformed expression; the computer plots not only the graph of the two expressions but also the difference graph. If the difference graph corresponds to $y = 0$, then the transformation that has been applied to the initial expression is a valid one. Another example is based on an environment described by Lesh and Herre (1987): Given the equation A[X] = B[X], the software plots y = A[X] and y = B[X]; when the equation is transformed into A[X] + C[X] = B[X] + C[X], the software plots the new functions y = A[X] + C[X] and y = B[X] + C[X], thus making it possible to see how the set of solutions to the equation remains constant when the equation is transformed.

Not only are structured symbolic calculation environments different from environments based on multi-representational approaches, they do not use multiple representations. They offer commands that transform expressions or equations by operating on their structures and preserving equivalencies (see Cerulli's brief). One of the earlier examples of such an approach to constructing meaning for the transformational activity of algebra was the work of Thompson and Thompson (1987). Their study was based on the use of *EXPRESSIONS*, "a special computer program ... that enabled students to manipulate expressions, but which constrained them to acting on expressions only through their structure." The software presents expressions as trees and the user can operate on these expressions by clicking on an action (for instance, distribute, commute, etc.), and then on the head of the branch of the represented expression to be transformed. The software thus allows step-by-step transformations, which contrasts it (and other examples of structured symbolic calculation that may also include a tutoring facility) with Computer Algebra Systems (CAS). Most CAS handle algebraic transformation processes (such as factor, solve, simplify) in one step, with black-box algorithms that produce solutions that are, at times, harder to understand than the original problem (e.g., Stacey, 1997; see also Pimm, 1995, for discussion of issues related to symbolism, manipulable dynamic microworlds, and CAS "black boxes").

Other examples of earlier work in this domain include the intelligent computer tutors for solving algebraic equations that appeared in the 1980s. The tutoring system developed by McArthur and his colleagues (McArthur, 1985; McArthur, Stacz, & Hotta, 1987) incorporated higher-level reasoning processes (such as "eliminate fractions") as well as the lower-level simplifications required to effect these higher-level goals. In this environment, the student could ask the "resident expert" if a particular step were correct, or for some assistance with respect to the next step to take. Another such environment was *Algebraland* (Brown, 1985), in which the student had only to decide which operator (such as isolate, collect, group, split, or simplify) to apply and what to apply it to; the system then performed the operation. Other intelligent tutoring systems for equation solving included the *Algebra Tutor* (Lewis, Milson, & Anderson, 1987) and the *Algebra Workbench* (Roberts, Carter, Davis, & Feurzeig, 1989). However, most of the above systems are no longer being researched and further developed. The exceptions include the *Algebra Tutor* (which has since been modified and integrated into the *Cognitive Tutor Algebra I*) and *Aplusix*, which are discussed in the subsections below (see also Nicaud, Delozanne, & Grugeon, 2002, for further examples).

6.4.2 Examples of structured symbolic calculation environments

A structured symbolic calculation environment is one that, broadly speaking, is a system that in the past might have been referred to as an intelligent tutoring system. However, recent advances in the field have led to such a variety in these systems that there is no straightforward characterisation applicable to all. For example, many will not execute an operation that is incorrect; others will give some form of feedback or a detailed error message; others will permit pupils to make errors, but make available a checking mechanism, should the pupil so wish. Because systems for "algebraic reasoning" (as described by Nicaud in the brief prepared for the Working Group) generally take charge of part of the calculation process, they can be analysed in terms of who selects the rule and the sub-expression to which it is to be applied, as well as who applies the rule. From the tutoring point of view, Nicaud has also pointed out that systems can be distinguished according to whether they provide hints on rules, matching, and strategies, as well as on the nature of their feedback on errors. The considerable variety inherent in these systems is reflected in the examples below.

6.4.2.1 L'Algebrista

L'Algebrista (Cerulli & Mariotti, 2001, 2002) is a "microworld of algebraic expressions (totally under the user's control) where the user can transform expressions on the basis of the fundamental properties of operations." Once the user has entered an expression into the workspace of the environment, he/she chooses a subpart of the expression to be transformed, along with a button depicting the axiom to be applied (such as a commutativity button or a distributivity button, the icons of which show both the name and the structure of the axiom) (see Figure 6.12). However, *L'Algebrista* will not carry out a transformation that is invalid. For example, if the user entered 2*a + 3 into the workspace and selected the subexpression a + 3 as the one to be transformed, *L'Algebrista* would automatically extend the selection to 2*a + 3. According to the authors, "this feature corresponds to the fact that the expressions of this microworld incorporate a fundamental algebraic characteristic of mathematical expressions: their tree structure." As the user's algebraic knowledge increases, users may create their own buttons; however, *L'Algebrista* does not check the mathematical correctness of these new transformation rules. According to the didactical choices made by the authors, the pupil is responsible for the validation of a new theorem or transformation rule.

The basic aim of the design of this environment is to introduce students to symbol manipulation as an activity of theory development, using the concept of equivalence relation as the basic principle underlying symbol manipulation. Cerulli and Mariotti's (2001) research thus far with 9th grade classes has included work involving the following notion: Two expressions are equivalent if it is possible to transform one into the other using the given buttons. Once this equivalence relation is accepted, pupils are asked to compare expressions. The equivalence of two expressions is considered proved if one expression is transformed into the other using the axioms. Cerulli and Mariotti have found, in pupils' paper-and-pencil algebra, that *L'Algebrista* plays the role of semiotic mediator, as evidenced by the iconography of the buttons appearing off to the side in their written work on a comparison of expressions task. The authors argue that, "a single mathematical concept can be represented using very different signs that make explicit different aspects of the same concept; but the crucial point remains the effectiveness of such signs as instruments of semiotic mediation" (p. 186). We return to this issue in Section 6.4.3.

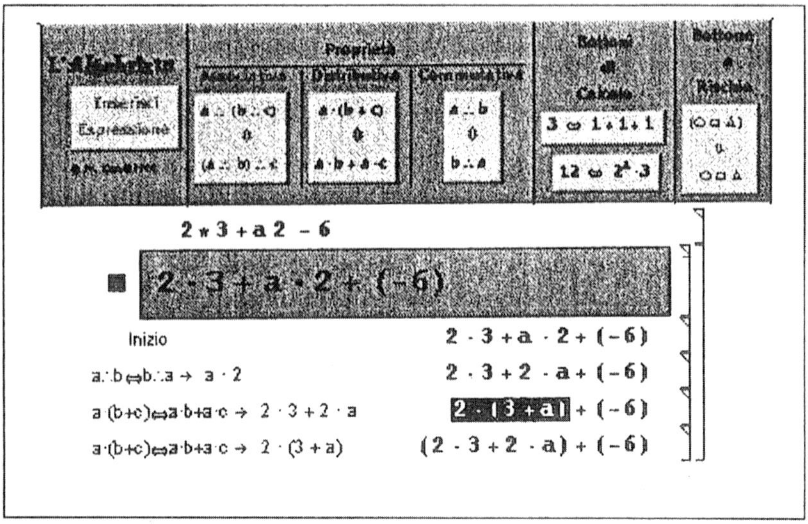

Figure 6.12. Transforming expressions in the *L'Algebrista* environment (Cerulli & Mariotti, 2001).

6.4.2.2 Aplusix

Aplusix (see http://aplusix.imag.fr) is another structured symbolic calculation environment that, according to Nicaud (in the brief prepared for the Working Group; see also Bouhineau, Nicaud, Pavard, & Sander, 2001; Nguyen-Xuan, Nicaud, Bastide, & Sander, 2002), is "an educational environment for algebraic reasoning, [where] the student solves exercises step by step; a step is performed by choosing a transformation rule in a menu, then selecting with the mouse the subexpression on which to apply it." An example is provided in Figure 6.13. Several levels of use are permitted: from novice through intermediate, all of which are under the complete control of the user. The environment provides hints when requested, as well as feedback on errors. Studies carried out from 1990 to 2000 with this environment have produced findings such as the following: (a) Students benefit more from scanty hints but very detailed feedback (Nguyen-Xuan, Nicaud, & Gélis, 1997), and (b) Learning by observation is better for very beginners while learning by doing is better for novices who are working at becoming more expert (Nguyen-Xuan, Bastide, & Nicaud, 1999). Nicaud (in the brief prepared for the Working Group) emphasises that this structured symbolic calculation environment is one that promotes algebraic reasoning in that it facilitates students' solving of algebraic tasks by developing in them a step-by-step reasoning.

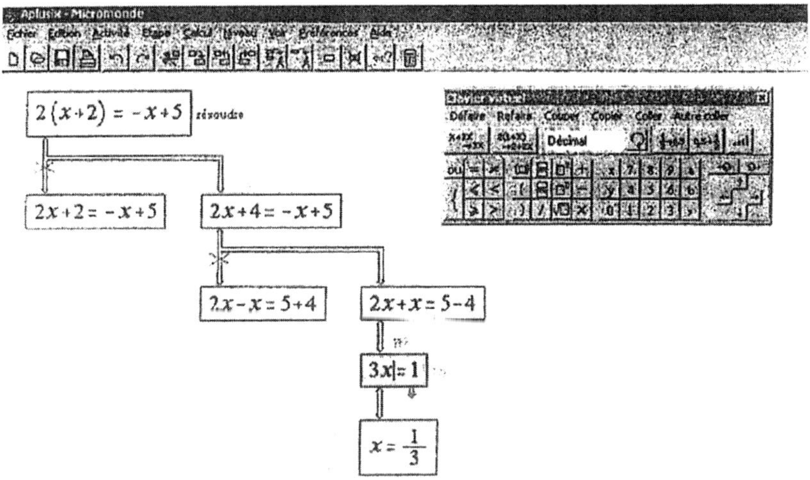

Figure 6.13. A student's attempt at solving an equation in the *Aplusix* environment.

6.4.2.3 Cognitive Tutor Algebra 1

The *Cognitive Tutor Algebra 1* (Carnegie Learning, 1998) includes its equation-solving system within a broader multiple-representation, curricular environment, and thus is not exclusively a *structured symbolic calculation environment* (see Figure 6.14 for a sample equation-solving trace). After entering an equation into the Solver window, the student chooses from a menu offering the following options: (i) Add to both sides, Subtract from both sides, Multiply both sides, Divide both sides; (ii) Combine Like Terms, Multiply, Reduce Fractions; (iii) Distribute; and (iv) Erase last step. In (i), an additional window opens to ask what number is to be added (or whatever operation has been selected) to both sides, and then the system carries out the operation. In (ii) and (iii), the system asks whether the operation is to be carried out on the left side, on the right side, or on both sides. Hints are available if the student asks for them. Research with the *Cognitive Tutor Algebra 1* environment (Koedinger, Anderson, Hadley, & Mark, 1997) has provided evidence of positive results of its structured symbolic calculation environment within the broader curriculum.

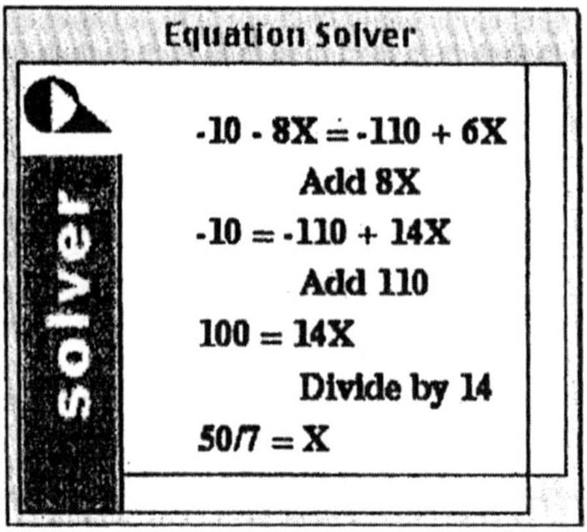

Figure 6.14. An equation being solved using the *Cognitive Tutor Algebra I.*

6.4.3 Additional remarks

In the example seen above with *L'Algebrista,* Cerulli and Mariotti emphasised the mediating role played by the software in subsequent paper-and-pencil work that involved the step-by-step transformation of an expression into an equivalent form. Students were clearly thinking about the computer environment and its representations and methods while they were doing similar transformations later on paper. But they were also thinking about the structural underpinnings of the transformations they were carrying out. What the emerging findings of studies involving structured symbolic calculation are suggesting is that the interaction between student and technological tool within these environments can lead to deep conceptual understanding of the structure of expressions and notions of equivalence. Recent theoretical work on the instrumental genesis of mathematical concepts has yielded a new vision of the relations among technological tool, technique, and algebraic thinking. For example, Lagrange (1999), who has drawn on Verillon and Rabardel's (1995) ideas of instrumental genesis to think about how students develop their own mental schemes for working with advanced calculators in mathematics, emphasises "technique" as a rational elaboration of a set of methods, that is, as the means for thinking beyond the algorithm. Zbiek (in the brief prepared for the Working Group) has described a task prepared by Lagrange and his team involving the production of by-hand and by-CAS derivatives, along with the question of explaining why these two different-appearing expressions are equivalent. According to Zbiek, Lagrange's analysis addresses a "complex, student-specific connection among conceptual understanding and algorithmic skills in concert with technology use, formal reasoning, and reflection on appropriate aspects of mathematical and technological events." Further research on tasks of the sort used by Lagrange could lead to shedding more light on the co-emergence of proficiency and conceptual understanding in algebra-based technology environments.

6.5 Reflection and Discussion

In her Plenary Lecture (see Chapter 2 of this book), Carolyn Kieran presented a threefold model of algebraic activity consisting of (a) generational activity, (b) transformational activity, and (c) global / meta-level activity. Generational activity involves creating meaning for the objects of algebra, such as expressions and equations. Transformational activity, which deals especially with expression simplification and equation solving, focuses on equivalence and the notion of solution of an equation. The global / meta-level activity of algebra comprises those activities for which algebra is used as a tool, but which are not exclusive to algebra and which could be engaged in without using any algebra at all, such as problem solving, modelling, noticing structure, justifying, proving, and predicting. Much of the work discussed in the present chapter adopted a "functional approach" for developing student proficiency within the threefold activities of algebra. This was most apparent in Section 6.2 where multiple representation technology served as a basis for bridging to algebraic symbolism. As well, Section 6.3, with its emphasis on dynamic control, involved continuous or discrete movement of objects and multi-representational tools. However, such a focus was clearly absent in Section 6.4 with its theme of structured symbolic calculation, involving a single representation. Thus, we are reminded that the presence of technology does not determine a single approach to algebra learning and teaching. We showed clearly the diversity of approaches that exists in algebra, such diversity that it calls for further work on distinctions that could help to better describe how one might plan a learning sequence, what the tasks or activities are about, what we might expect students to do with the tools, which ways students might interpret them, and how student work might be assessed.

This last observation on diversity of approaches introduces the concluding section of this chapter, a brief section devoted to reflection and discussion. In looking at some of the principal ways in which technology is being used to support algebra learning, the preceding sections have touched upon many issues. The aim of this last section is to draw out some of these, pointing to areas in need of rethinking and of further research effort.

6.5.1 Algebraic technique is a substantial mathematical activity within technological environments

Sections 6.2 and 6.4 presented different approaches for using technology with symbolic manipulation. The examples shown are substantive mathematically because they deal with central objectives of mathematics; they allow students to relate the mathematical activity to broader ideas, content (in and out of the mathematics), and representations; and they promote thinking in different ways about carrying out the same task (tasks that were usually designed to generate proficiency in a single procedure). Not only is the activity substantive, it also allows students to inject meaning into their work with algebraic symbols. Furthermore, the presence of technology does not eliminate symbolic manipulation from algebra, but it does change it.

Mathematics education has tended, in the past, to separate conceptual from skill learning, in the belief that the practice of procedures led to skill development but had no impact on conceptual growth. However, manipulation activity within technological environments is leading to a shift in this view. Students are questioning the meanings of the unexpected expression produced by the choice of a particular button, or the new position of a graph when they slide the control for a parameter of a given expression. They are also learning to use consciously the items of an algebra menu and to decide whether expressions are equivalent, as well as anticipate the output of a given transformation. We are thus seeing *conceptual learning evolve within technique-oriented / transformational activity.* This leads to a myriad of questions that have never been asked before on the relation between technical and conceptual learning, as well as to *the search for appropriate research tasks* that can help us to identify the nature and the circumstances of such learning in technological environments.

6.5.2 A variety of "algebraic" notations may make conventional notations meaningful

Traditionally we formulate instruction so that it proceeds from the symbolic to the numerical (tables of values) and from there (sometimes) to the graphical. This path requires students to first master algebraic rules and transformations, and as a result is not usually successful in having them create meaning for symbols. Using different sorts of multiple representation technology, it is possible to begin with any of the representations and to proceed to any other representation. For example, one can start with situations described in natural language and mathematise them using qualitative graphs whose properties are put forward in natural language. The necessity for more precise language then leads to introducing symbols. Another path would start with numerical problems and build the need for a symbolic set that would support the generation of ample examples and the formulation of numerical patterns—as is carried out with spreadsheets. Such patterns can be represented graphically and the analysis of these graphs can then lead students to the need for another formal symbolic representation. Representations of functions, and mainly *"visual notations", have the potential to change the assumed notations and to make them more accessible*, but they also present their own new complexities that call for further research.

6.5.3 The use of tools challenges us to rethink what we assumed to be natural regarding the growth of knowledge

In this chapter we mainly discussed tools for representing functions and exemplified how novel research explores what becomes *natural*. This research offers results very different from what cognitive research previously assumed. For example, we reviewed studies that describe how the graphing of two-variable equations as two 3D surfaces or as a parametric family turned out to be a natural extension of the graphic methods used to solve an equation in a single variable. *Such phenomena call for research of a didactic nature* with different kinds of learning sequences.

We also reviewed studies involving dynamic control of the manipulation of an object or a representation of a mathematical object. In some studies, direct control made the phenomena—the process and its graphical representations—become the manipulable leading object rather than a representation of the symbolic model. The existing dynamic-control technology and its use in algebra calls for research on what can be accessible, meaningful, and for whom. *In which ways do continuous changes of a representation lead to different cognitive processes than those based on discrete changes?* How do direct manipulations change the sequence of learning?

And more generally: *Do our assumptions of what tools can do now lead research and what we envision can be done?* What kind of methodology could we set up that would have an impact on existing and future technology? By trying to define more generally what should be done with, for example, symbolic manipulators, graphic manipulators, and dynamic control of phenomena inside and outside of mathematics, could research on conceptual understanding of algebra support the development of ideas that could become "standards" of software for algebra? Could this research then devise interesting questions and appropriate methodologies for each of those types of technology environments?

6.5.4 Conventions regarding the algebraic performance of low-performing students are to be reconsidered

Research on the long-term learning of algebra with technology shows that it is not only high-performing students who can excel with the aid of technology; students who tend to be low-performers in algebra do better when they can move part of the responsibility to the tool (e.g., if they solve equations using graphical intersections) or when they can work on manipulations in context. The organisation of the algebra curriculum around a few central objectives, and the availability of visual and numerical representations that support the work of students as independent learners, make it possible to construct part of the algebra curriculum based on open-ended, larger scale, activities. Such learning environments require the active participation of students in making conjectures and refutations, in discussing, and in engaging in expressive and exploratory modelling in ways that were not accessible in traditional algebra. A large part of the learning is built upon students' ideas, and in cooperation with their peers and the teacher. How does this shift in goals and habits work out for lower performers in mathematics? We should *seek appropriate tools to study and identify ways to describe the work of low performers when they have access to technology in algebra.* (See Chapter 12 in this book for further discussion by Mollie MacGregor related to this issue).

6.5.5 We need to rethink the necessity of learning algebraic concepts and skills by-hand before using technology

Research studies that have shown that technologically-based approaches to algebra are enabling traditionally unsuccessful students to gain access to the problem-solving aspects of algebra have also suggested the following: Students who have learned, without the aid of technology, to be good at symbolic manipulations of algebra in the traditional sense are not necessarily as good at conceptual tasks as those who are long-time users of various technological approaches and who may have learned algebra primarily with technology. In other words, students who learn algebraic skills in various technological environments are *also* developing a deeper conceptual knowledge of the objects they are manipulating. But research needs to be able to tell us more about how this happens. Do students need to use the technology all of the time, or just some of the time, and when? Related to this question is another that is more value-laden: How much, in what ways, and at what times would we like the tool "to do the work" for the students? There is a need for research on the *roles and responsibilities that we assume the tool should take.* Although technology allows shifting most of the responsibility for generating symbolic descriptions of data to the machine, a major bulk of studies suggests that we need to continue to seek ways to make meaningful the processes of symbolic construction and manipulation.

Related to the question of algebraic skills is their assessment. How do we design activities for testing (and for different assessment modes other than testing), which would fit with what we teach with technology? We reviewed here some studies that especially looked at explanation. We believe that an appropriate strategy for identifying conceptual understanding of algebra is to look for tasks that would promote explanations both with and without the use of technology. The difficulty with this strategy is that explanations in algebra are quite rare. Moreover, in research projects, one would want to pose problems that would encourage different types of argumentation and to devise ways in which it could be assessed. What kinds of explanations should be considered conceptual as a result of using software? How would that be assessed with software? Whom would it make sense to assess without the tools with which they learned?

6.5.6 Multiple representations in algebra change the views of algebra and, for many teachers, lead to a reconsideration of the goals of the algebra course

The use of technology, and mainly the use of graphing technology, introduces other representations besides symbols and equations into school algebra. Studies of teachers who teach with technology reveal the challenges and the complexity of rethinking traditional sequences. We doubt whether research can help to decide which particular sequences would be the best ones. It seems that such decisions are rooted mainly in values, beliefs, and political considerations (see also Chapter 12 in this book). The current round of mathematics education reform seeks to develop classrooms in which students' mathematical ideas are integral to instruction. It requires that teachers overcome their own mathematical biases in order to listen to the different perspectives that students may bring. It also requires that teachers transcend the linearity of textbooks and develop the capacity to use curricular materials flexibly in order to be responsive to their students' ideas.

There is also a need to establish a different relationship between theory and practice regarding teaching within new environments—one where theory and practice refine each other. Because almost everything is new—the sequences, the tasks, the assessments, the focus—and because many studies have shown that the technology and the designed sequence give rise to both a myriad of different interpretations by teachers and unexpected reactions on the part of students, there is a need to create learning environments and settings where teachers can experiment according to their own beliefs and practical experience. Collaborative work involving algebra teachers with their intimate knowledge of the classroom setting, in conjunction with a deep awareness of the mathematical and pedagogical affordances of the tool, is vital for progress. In this way, practice and research can work together to ensure the improved algebra learning of students.

6.6 References

Ainley, J. (1996). Purposeful contexts for formal notation in a spreadsheet environment. *Journal of Mathematical Behavior, 15*, 405-422.

Arcavi, A., & Hadas, N. (2000). Computer mediated learning: An example of an approach. *International Journal of Computers for Mathematical Learning, 5*, 25-45.

Arzarello, F., & Robutti, O. (2001). From body motion to algebra through graphing. In H. Chick, K. Stacey, J. Vincent, & J. Vincent (Eds.), *The future of the teaching and learning of algebra* (Proceedings of the 12th ICMI Study Conference, pp. 33-40). Melbourne, Australia: The University of Melbourne.

Booth, L. R. (1984). *Algebra: Children's strategies and errors.* Windsor, UK: NFER-Nelson.

Borba, M. C. (1993). *Students' understanding of transformations of functions using multi-representational software.* Unpublished doctoral dissertation, Cornell University, Ithaca, NY.

Bouhineau,, D., Nicaud, J.-F., Pavard, X., & Sander, E. (2001). Un micromonde pour aider les élèves à apprendre l'algèbre [A microworld to help students learn algebra]. *Actes de EIAO2001*. Paris: Hermès.
Brown, J. S. (1985). Process versus product: A perspective on tools for communal and informal electronic learning. *Journal of Educational Computing Research, 1*, 179-201.
Carnegie Learning, Inc. (1998). *Cognitive Tutor® Algebra I©* [Computer software]. Pittsburgh, PA: Carnegie Learning.
Cedillo, T., & Kieran, C. (2003). Initiating students into algebra with symbol-manipulating calculators. In J. T. Fey (Ed.), *Computer algebra systems in secondary school mathematics education* (pp. 219-239). Reston, VA: National Council of Teachers of Mathematics.
Cerulli, M., & Mariotti, M. A. (2001). L'Algebrista: A microworld for symbolic manipulation. In H. Chick, K. Stacey, J. Vincent, & J. Vincent (Eds.), *The future of the teaching and learning of algebra* (Proceedings of the 12th ICMI Study Conference, pp. 179-186). Melbourne, Australia: The University of Melbourne.
Cerulli, M., & Mariotti, M. A. (2002). L'Algebrista: un micromonde pour l'enseignement et l'apprentissage de l'algèbre [L'Algebrista : a microworld for the teaching and learning of algebra]. *Sciences et techniques éducatives, 9*, 149-170.
Chazan, D. (1999). On teachers' mathematical knowledge and student exploration: A personal story about teaching a technologically supported approach to school algebra. *International Journal of Computers for Mathematical Learning, 4*, 121-149.
Chazan, D. (2000). *Beyond formulas in mathematics and teaching: Dynamics of the high school algebra classroom*. New York: Teachers College Press.
Chazan, D., Larriva, C., & Sandow, D. (1999). What kind of mathematical knowledge supports teaching for "conceptual understanding"? Preservice teachers and the solving of equations. In O. Zaslavsky (Ed.), *Proceedings of the 23rd International Conference for the Psychology of Mathematics Education* (Vol. 2, pp. 193-200). Haifa, Israel: Program Committee.
Clement, J. (1982). Algebra word problem solutions: Thought processes underlying a common misconception. *Journal for Research in Mathematics Education, 13*, 16-30.
Confrey, J. (1991). *Function Probe©* [Computer software]. Ithaca, NY: Cornell University.
Dettori, G., Garuti, R., & Lemut, E. (2000). From arithmetic to algebraic thinking by using a spreadsheet. In R. Sutherland, T. Rojano, A. Bell, & R. C. Lins (Eds.), *Perspectives on school algebra* (pp. 191-207). Dordrecht, The Netherlands: Kluwer Academic.
Discussion Document for the 12th ICMI Study. (2001). In H. Chick, K. Stacey, J. Vincent, & J. Vincent (Eds.), *The future of the teaching and learning of algebra* (Proceedings of the 12th ICMI Study Conference, pp. 1-6). Melbourne, Australia: The University of Melbourne.
Doerr, H. M., & Zangor, R. (2000). Creating meaning for and with the graphics calculator. *Educational Studies in Mathematics, 41*, 143-163.
Dubinsky, E., & Harel, G. (1992). The nature of the process conception of function. In G. Harel & E. Dubinsky (Eds.), *The concept of function: Aspects of epistemology and pedagogy* (MAA Notes, Volume 25, pp. 85-106). Washington, DC: Mathematical Association of America.
Dugdale, S. (1993). Functions and graphs - perspectives on students' thinking. In T. A. Romberg, E. Fennema, & T. P. Carpenter (Eds.), *Integrating research on the graphical representation of functions* (pp. 101-130). Hillsdale, NJ: Lawrence Erlbaum.
Dugdale, S., & Kibbey, D. (1986). *Green globs and graphing equations* [Computer software]. Pleasantville, NY: Sunburst Communications.
Feurzeig, W., & Papert, S. (1968). Programming languages as a conceptual framework for teaching mathematics. In *Proceedings of the NATO Conference on Computers and Learning* (pp. 19-42). Nice, France.

Fey, J. T. (Ed.). (1984). *Computing and mathematics: The impact on secondary school curricula.* College Park, MD: The University of Maryland.

Filloy, E., Rojano, T., & Rubio, G. (2000). Propositions concerning the resolution of arithmetical-algebra problems. In R. Sutherland, T. Rojano, A. Bell, & R. C. Lins (Eds.), *Perspectives on school algebra* (pp. 155-176). Dordrecht, The Netherlands: Kluwer Academic.

Filloy, E., & Sutherland, R. (1996). Designing curricula for teaching and learning algebra. In A. J. Bishop, D. H. Clements, C. Keitel, J. Kilpatrick, & C. Laborde (Eds.), *International handbook of mathematics education* (pp. 139-160). Dordrecht, The Netherlands: Kluwer Academic.

Friedlander, A. (1999). Cognitive processes in a spreadsheet environment. In O. Zaslavsky (Ed.), *Proceedings of the 23rd International Conference for the Psychology of Mathematics Education* (Vol. 2, pp. 337-344). Haifa, Israel: Program Committee.

Friedlander, A., & Tabach, M. (2001a). Promoting multiple representations in algebra. In A. A. Cuoco & F. R. Curcio (Eds.), *The roles of representation in school mathematics* (Yearbook of the National Council of Teachers of Mathematics, pp. 173-185). Reston, VA: NCTM.

Friedlander, A., & Tabach, M. (2001b). Developing a curriculum of beginning algebra in a spreadsheet environment. In H. Chick, K. Stacey, J. Vincent, & J. Vincent (Eds.), *The future of the teaching and learning of algebra* (Proceedings of the 12th ICMI Study Conference, pp. 252-257). Melbourne, Australia: The University of Melbourne.

Gafni, R. (1996). *Using multiple representation of function to improve semantic understanding of algebraic expressions and equations.* Unpublished doctoral dissertation (in Hebrew), Haifa University, Israel.

Goldenberg, P. E. (1988). Mathematics, metaphors, and human factors: Mathematical, technical and pedagogical challenges in the educational use of graphical representations of functions. *Journal of Mathematical Behavior, 7,* 135-173.

Goldenberg, P. E., Lewis, P., & O'Keefe, J. (1992). Dynamic representation and the development of a process understanding of function. In G. Harel & E. Dubinsky (Eds.), *The concept of function: Aspects of epistemology and pedagogy* (MAA Notes, Vol. 25, pp. 235-260). Washington, DC: Mathematical Association of America.

Haimes, D. H. (1996). The implementation of a "function" approach to introductory algebra: A case study of teacher cognitions, teacher actions, and the intended curriculum. *Journal for Research in Mathematics Education, 27,* 582-602.

Hatfield, L. L., & Kieren, T. E. (1972). Computer-assisted problem solving in school mathematics. *Journal for Research in Mathematics Education, 3,* 99-112.

Hazzan, O., & Goldenberg, E. P. (1997). Students' understanding of the notion of function in dynamic geometry environments. *International Journal of Computers for Mathematical Learning, 1,* 263-291.

Heid, M. K. (1995). The impact of technology, mathematical modeling, and meaning on the content, learning, and teaching of secondary-school algebra. *Journal of Mathematical Behavior, 14,* 121-137.

Heid, M. K., Blume, G. W., Zbiek, R. M., & Edwards, B. S. (1999). Factors that influence teachers learning to do interviews to understand students' mathematical understanding. *Educational Studies in Mathematics, 37,* 223-249.

Hershkowitz, R., Dreyfus, T., Ben-Zvi, D., Friedlander, A., Hadas, N., Resnick, T., Tabach, M., & Schwarz, B. (2002). Mathematics curriculum development for computerized environments: A designer-researcher-teacher-learner activity. In L. D. English (Ed.), *Handbook of international research in mathematics education* (pp. 657-694). Mahwah, NJ: Lawrence Erlbaum.

Hershkowitz, R., & Kieran, C. (2001). Algorithmic and meaningful ways of joining together representatives within the same mathematical activity: An experience with graphing calculators. In M. van den Heuvel-Panhuizen (Ed.), *Proceedings of the 25th International Conference for the*

Psychology of Mathematics Education (Vol. 1, pp. 96-107). Utrecht, The Netherlands: Program Committee.

Hoyles, C., Sutherland, R., & Evans, J. (1985). *A preliminary investigation of the pupil-centred approach to the learning of Logo in the secondary school mathematics classroom, 1983-1984.* London, UK: University of London, Institute of Education.

Huntley, M. A., Rasmussen, C. L., Villarubi, R. S., Sangtong, J., & Fey, J. T. (2000). Effects of Standards-based mathematics education: A study of the Core-Plus Mathematics Project Algebra and Functions Strand. *Journal for Research in Mathematic Education, 31*, 328-361.

Janvier, C. (Ed.). (1987). *Problems of representation in the teaching and learning of mathematics.* Hillsdale: NJ: Lawrence Erlbaum.

Janvier, C., Girardon, C., & Morand, J.-C. (1993). Mathematical symbols and representations. In P. S. Wilson (Ed.), *Research ideas for the classroom: High school mathematics* (pp. 79-102). New York: Macmillan.

Kaput, J. J., & Roschelle, J. (1997). Deepening the impact of technology beyond assistance with traditional formalisms in order to democratize access to ideas underlying calculus. In E. Pehkonen (Ed.), *Proceedings of the 21st Conference of the International Group for the Psychology of Mathematics Education* (Vol. 1, pp. 105-112). Lahti, Finland: Program Committee.

Kendal, M., & Stacey, K. (2001). Influences on and factors changing technology privileging. In H. Chick, K. Stacey, J. Vincent, & J. Vincent (Eds.), *The future of the teaching and learning of algebra* (Proceedings of the 12th ICMI Study Conference, pp. 360-367). Melbourne, Australia: The University of Melbourne.

Kieran, C. (1994). A functional approach to the introduction of algebra: Some pros and cons. In J. P. da Ponte & J. F. Matos (Eds.), *Proceedings of the 18th International Conference for the Psychology of Mathematics Education* (Vol. I, pp. 157-175). Lisbon, Portugal: Program Committee.

Kieran, C., Hillel, J., & Erlwanger, S. (1986). Perceptual and analytical schemas in solving structured turtle-geometry tasks. In C. Hoyles, R. Noss, & R. Sutherland (Eds.), *Proceedings of the Second International Conference for Logo and Mathematics Education* (pp. 154-161). London, UK: University of London, Institute of Education.

Kieran, C., & Sfard, A. (1999). Seeing through symbols: The case of equivalent expressions. *Focus on learning problems in mathematics, 21*(1), 1-17.

Kieren, T. E. (1992). Mathematics in a Logo environment: A recursive look at a complex phenomenon. In C. Hoyles & R. Noss (Eds.), *Learning mathematics and Logo* (pp. 223-246). Cambridge, MA: The MIT Press.

Klotz, F. S. (1986). When is a program like a function? In C. Hoyles, R. Noss, & R. Sutherland (Eds.), *Proceedings of the Second International Conference for Logo and Mathematics Education* (pp. 14-24). London, UK: University of London, Institute of Education.

Koedinger, K. R., Anderson, J. R., Hadley, W. H., & Mark, M. A. (1997). Intelligent tutoring goes to school in the big city. *International Journal of Artificial Intelligence in Education, 8*, 30-43.

Lagrange, J.-b. (1999). Complex calculators in the classroom: Theoretical and practical reflections on teaching pre-calculus. *International Journal of Computers for Mathematical Learning, 4*, 51-81.

Leron, U., & Zazkis, R. (1992). Of geometry, turtles, and groups. In C. Hoyles & R. Noss (Eds.), *Learning mathematics and Logo* (pp. 319-352). Cambridge, MA: The MIT Press.

Lesh, R., & Herre, J. (1987). Dienes revisited: Multiple embodiments in computer environments. In J. C. Bergeron, N. Herscovics, & C. Kieran (Eds.), *Proceedings of the 11th International Conference for the Psychology of Mathematics Education* (Vol. I, pp. 211-220). Montreal, QC: Program Committee.

Lewis, M. W., Milson, R., & Anderson, J. R. (1987). Designing an intelligent authoring system for high school mathematics ICAI: The TEACHER'S APPRENTICE Project. In G. P. Kearsley (Ed.), *Artificial intelligence and instruction: Applications and methods*. Reading, MA: Addison-Wesley.

Lloyd, G. M. (1999). Two teachers' conceptions of a reform-oriented curriculum: Implications for mathematics teacher development. *Journal of Mathematics Teacher Education, 2*, 227-252.

Logo Computer Systems, Inc. (2000). *MicroWorlds Pro©* [Computer software]. Montreal, QC: LCSI.

Love, E. (1986). What *is* algebra? *Mathematics Teaching, 117*, 48-50.

McArthur, D. (1985). Developing computer tools to support performing and learning complex skills. In D. Berger, K. Pedzek, & W. Ganks (Eds.), *Applications of cognitive psychology*. Hillsdale, NJ: Lawrence Erlbaum.

McArthur, D., Stasz, C., & Hotta, J. Y. (1987). Learning problem-solving skills in algebra. *Journal of Educational Technology Systems, 15*, 303-324.

Mokros, J. R., & Tinker, R. F. (1987). The impact of microcomputer-based labs on children's ability to interpret graphs. *Journal of Research in Science Teaching, 24*, 369-383.

Molyneux-Hodgson, S., Rojano, T., Sutherland, R., & Ursini, S. (1999). Mathematical modeling: The interaction of culture and practice. *Educational Studies in Mathematics, 39*, 167-183.

Monaghan, J. (2001). Teachers and computer algebra systems. In H. Chick, K. Stacey, J. Vincent, & J. Vincent (Eds.), *The future of the teaching and learning of algebra* (Proceedings of the 12th ICMI Study Conference, pp. 462-467). Melbourne, Australia: The University of Melbourne.

National Council of Teachers of Mathematics. (1989). *Curriculum and evaluation standards for school mathematics*. Reston, VA: NCTM.

National Council of Teachers of Mathematics. (1991). *Professional teaching standards for school mathematics*. Reston, VA: NCTM.

National Council of Teachers of Mathematics. (1995). *Assessment standards for school mathematics*. Reston, VA: NCTM.

National Council of Teachers of Mathematics. (2000). *Principles and standards for school mathematics*. Reston, VA: NCTM.

Nemirovsky, R. (1994). On ways of symbolizing: The case of Laura and velocity sign. *Journal of Mathematical Behavior, 13*, 389-422.

Nemirovsky, R. (1996). Mathematical narratives, modeling, and algebra. In N. Bednarz, C. Kieran, & L. Lee (Eds.), *Approaches to algebra: Perspectives for research and teaching* (pp. 197-220). Dordrecht, The Netherlands: Kluwer.

Nemirovsky, R., Kaput, J. J., & Roschelle, J. (1998). Enlarging mathematical activity from modeling phenomena to generating phenomena. In A. Olivier & K. Newstead (Eds.), *Proceedings of the 22nd International Conference for the Psychology of Mathematics Education* (Vol. 3, pp. 287-294). Stellenbosch, South Africa: Program Committee.

Nemirovsky, R., & Noble, T. (1997). Mathematical visualization and the place where we live. *Educational Studies in Mathematics, 33*, 99-131.

Nemirovsky, R., & Rubin, A. (1992). *Students' tendency to assume resemblances between a function and its derivative*. Unpublished manuscript. Available from first author at TERC, 2067 Massachusetts Ave., Cambridge, MA 02140.

Nemirovsky, R., Tierney, C., & Wright, T. (1998). Body motion and graphing. *Cognition and Instruction, 16*, 119-172.

Nguyen-Xuan, A., Bastide, A., & Nicaud, J.-F. (1999). Learning to solve polynomial factorization problems: by solving problems and by studying examples of problem solving, with an intelligent learning environment. *Proceedings of the Conference on Artificial Intelligence in Education* (pp. 215-224). IOS Press.

Nguyen-Xuan, A., Nicaud, J.-F., Bastide, A., & Sander, E. (2002). Les expérimentations du projet [APLUSIX Experiments in the APLUSIX project]. *Sciences et techniques éducatives, 9,* 63-90.

Nguyen-Xuan, A., Nicaud, J.-F., & Gélis, J.-M. (1997). Effects of feedback on learning to match algebraic rules to expressions within an intelligent learning environment. *Journal of Computers in Mathematics and Science Teaching, 16,* 291-322.

Nicaud, J.-F., Delozanne, E., & Grugeon, B. (Eds.). (2002). Logiciels pour l'apprentissage de l'algèbre [Software for the learning of algebra]. Special issue of *Sciences et Techniques Éducatives, 9*(1-2).

Noss, R. (1985). *Creating a mathematical environment through programming: A study of young children learning Logo.* Published Ph.D. thesis, Centre for Science and Mathematics Education, Chelsea College, University of London.

Pimm, D. (1995). *Symbols and meanings in school mathematics.* London: Routledge.

Roberts, N., Carter, R., Davis, F., & Feurzeig, W. (1989). Power tools for algebra problem solving. *Journal of Mathematical Behavior, 8,* 251-265.

Rojano, T. (1996). Developing algebraic aspects of problem solving within a spreadsheet environment. In N. Bednarz, C. Kieran, & L. Lee (Eds.), *Approaches to algebra: Perspectives for research and teaching* (pp. 137-146). Dordrecht, The Netherlands: Kluwer Academic.

Romberg, T. A., Fennema, E., & Carpenter, T. C. (Eds.). (1993). *Integrating research on the graphical representation of functions.* Hillsdale, NJ: Lawrence Erlbaum.

Ruthven, K. (1990). The influence of graphic calculator use on translation from graphic to symbolic forms. *Educational Studies in Mathematics, 21,* 431-450.

Ruthven, K. (1992a). Calculator strategies. *Micromath, 8*(1), 31-32.

Ruthven, K. (1992b). Personal technology and classroom change: A British perspective. In J. T. Fey & C. R. Hirsch (Eds.), *Calculators in mathematics education* (Yearbook of the National Council of Teachers of Mathematics, pp. 91-100). Reston, VA: The Council.

Schnepp, M., & Nemirovsky, R. (2001). Constructing a foundation for the fundamental theorem of calculus. In A. A. Cuoco & F. R. Curcio (Eds.), *The roles of representation in school mathematics* (Yearbook of the National Council of Teachers of Mathematics, pp. 90-102). Reston, VA: NCTM.

Schoenfeld, A. A. (2002). Making mathematics work for all children: Issues of standards, testing, and equity. *Educational Researcher, 31*(1), 13-25.

Schwartz, J. L., & Yerushalmy, M. (1992). Getting students to function in and with technology. In E. Dubinsky & G. Harel (Eds.), *The concept of function: Aspects of epistemology and pedagogy* (MAA Notes, Vol. 25, pp. 261-289). Washington, DC: Mathematical Association of America.

Schwartz, J. L., & Yerushalmy, M. (1995). On the need for a bridging language for mathematical modeling. *For the Learning of Mathematics, 15*(2), 29-35.

Schwartz, J. L., & Yerushalmy, M. (1996). *Calculus unlimited* [Computer software]. Tel-Aviv, Israel: Center for Educational Technology.

Schwarz, B. B., & Dreyfus, T. (1993). Measuring integration of information in multirepresentational software. *Interactive Learning Environments, 3,* 177-198.

Schwarz, B. B., & Hershkowitz, R. (1999). Prototypes: Brakes or levers in learning the function concept? The role of computer tools. *Journal for Research in Mathematics Education, 30,* 362-389.

Sfard, A. (1992). Operational origins of mathematical objects and the quandary of reification: The case of function. In E. Dubinsky & G. Harel (Eds.), *The concept of function: Aspects of epistemology and pedagogy* (MAA Notes, Vol. 25, pp. 59-84). Washington, DC: Mathematical Association of America.

Shternberg, B. (2001a). *Effect of analyzing the change of a function on the understanding of the mathematical meaning of phenomena and of the concept of a function.* Unpublished doctoral dissertation (in Hebrew), University of Haifa, Israel.

Shternberg, B. (2001b). *Escalators©* [Computer software]. Tel-Aviv, Israel: Center for Educational Technology.

Slavit, D. (1996). Graphing calculators in a "hybrid" Algebra II classroom. *For the Learning of Mathematics, 16*(1), 9-14.

Stacey, K. (1997). Mathematics – what should we tell the children? *The International Journal of Computer Algebra in Mathematics Education, 4*(4), 387-390.

Streun, A., Harskamp, E., & Suhre, C. (2000). The effect of the graphic calculator on students' solution approaches: A secondary analysis. *Hiroshima Journal of Mathematics Education, 8,* 27-39.

Sutherland, R. (1993). Symbolising through spreadsheets. *Micromath, 10*(1), 20-22.

Sutherland, R., & Balacheff, N. (1999). Didactical complexity of computational environments for the learning of mathematics. *International Journal of Computers for Mathematical Learning, 4,* 1-26.

Sutherland, R., & Rojano, T. (1993). A spreadsheet approach to solving algebra problems. *Journal of Mathematical Behavior, 12,* 353-383.

Thompson, P. W., & Thompson, A. G. (1987). Computer presentations of structure in algebra. In J. C. Bergeron, N. Herscovics, & C. Kieran (Eds.), *Proceedings of the 11th International Conference for the Psychology of Mathematics Education* (Vol. I, pp. 248-254). Montreal, QC: Program Committee.

Verillon, P., & Rabardel, P. (1995). Cognition and artifacts: A contribution to the study of thought in relation to instrumented activity. *European Journal of Psychology of Education, X,* 77-101.

Vincent, J. (2001). MicroWorlds® and early algebra. In H. Chick, K. Stacey, J. Vincent, & J. Vincent (Eds.), *The future of the teaching and learning of algebra* (Proceedings of the 12th ICMI Study Conference, pp. 621-627). Melbourne, Australia: The University of Melbourne.

Wilson, M. R., & Krapfl, C. M. (1994). The impact of graphics calculators on students' understanding of function. *Journal of Computers in Mathematics and Science Teaching, 13,* 252-264.

Yerushalmy, M. (1989). The use of graphs as visual interactive feedback while carrying out algebraic transformations. In G. Vergnaud, J. Rogalski, & M. Artigue (Eds.), *Proceedings of the 13th International Conference for the Psychology of Mathematics Education* (Vol. 3, pp. 252-260). Paris: Program Committee.

Yerushalmy, M. (1991). Effects of computerized feedback on performing and debugging algebraic transformations. *Journal of Educational Computing Research, 7,* 309-330.

Yerushalmy, M. (2000). Problem solving strategies and mathematical resources: A longitudinal view on problem solving in a functional based approach to algebra. *Educational Studies in Mathematics, 43,* 125-147.

Yerushalmy, M., & Chazan, D. (2002). Flux in school algebra: Curricular change, graphing technology, and research on student learning and teacher knowledge. In L. D. English (Ed.), *Handbook of international research in mathematics education* (pp. 725-755). Mahwah, NJ: Lawrence Erlbaum.

Yerushalmy, M., & Gafni, R. (1992). Syntactic manipulations and semantic interpretations in algebra: The effect of graphic representation. *Learning and Instruction, 2,* 303-319.

Yerushalmy, M., & Schwartz, J. L. (1993). Seizing the opportunity to make algebra mathematically and pedagogically interesting. In T. A. Romberg, E. Fennema, & T. C. Carpenter (Eds.), *Integrating research on the graphical representation of functions* (pp. 41-68). Hillsdale, NJ: Lawrence Erlbaum.

Yerushalmy, M., & Shternberg, B. (1994). *The Function Sketcher*© [Computer software]. Tel-Aviv, Israel: Center for Educational Technology.

Yerushalmy, M., & Shternberg, B. (2001). Charting a visual course to the concept of function. In A. A. Cuoco & F. R. Curcio (Eds.), *The roles of representation in school mathematics* (Yearbook of the National Council of Teachers of Mathematics, pp. 251-268). Reston, VA: NCTM.

Zbiek, R. M., & Heid, M. K. (2001). Dynamic aspects of function representations. In H. Chick, K. Stacey, J. Vincent, & J. Vincent (Eds.), *The future of the teaching and learning of algebra* (Proceedings of the 12th ICMI Study Conference, pp. 682-689). Melbourne, Australia: The University of Melbourne.

Zehavi, N. (1996). Challenging teachers to create mathematical projects with DERIVE. *International DERIVE Journal, 3*(2), 1-16.

---

[1] We acknowledge the contributions contained in the research briefs of Michele Cerulli, Jean-François Nicaud, and Rose Mary Zbiek, many of which were instrumental in the framing of ideas in this section (see Working Group report).

The Working Group on CAS and Algebra

Leaders: Barry Kissane and Mike Thomas

Working Group Members:
Lynda Ball, Roger Brown, Paul Drijvers, David Driver, Peter Flynn, Kathleen Heid, Margaret Kendal, Ivy Kidron, Jean-baptiste Lagrange, David Leigh-Lancaster, Giora Mann, John Monaghan, and Robyn Pierce.

The Working Group on CAS and Algebra. Front: Peter Flynn; Middle (L to R): Mike Thomas, Giora Mann, Robyn Pierce, Margaret Kendal, Lynda Ball; Back (L to R): John Monaghan, Paul Drijvers, Barry Kissane, Jean-baptiste Lagrange, Kathleen Heid, David Driver, Roger Brown. Absent: Ivy Kidron, David Leigh-Lancaster.

Prior to the Conference, the members of the Working Group on CAS communicated with each other via email discussions that were coordinated by Barry Kissane. Each member also reviewed a particular aspect of the current literature and these reviews were distributed to other members during the conference. In addition, each member prepared a paper for the 12th ICMI Study Conference Proceedings. These papers reflected members' expertise and prior experience in teaching, curriculum development, and researching algebra using Computer Algebra Systems (CAS). The authors (sometimes with co-authors) and the titles of their papers are listed:

- Lynda Ball: *Solving equations: Will a more general approach be possible with CAS?* (pp. 48-52).
- Roger Brown & B. Nielson: *What algebra is required in "high stakes" system wide assessment? A comparison of three systems* (pp. 128-135).

- Paul Drijvers: *The concept of parameter in a computer algebra environment* (pp. 221-227).
- David Driver: *The trial of algebraic calculators in senior mathematics by distance education* (pp. 228-237).
- Margaret Kendal & K. Stacey: *Influences on and factors changing technology privileging* (pp. 360-367).
- Ivy Kidron: *Teaching Euler's algebraic methods in a calculus laboratory* (pp. 368-375).
- Jean-baptiste Lagrange: *Tasks for students using computer algebra: A study of the design of educational computer applications* (pp. 376-383).
- David Leigh-Lancaster & M. Stephens: *Responding to the impact of increasing accessibility of Computer Algebra Systems (CAS) on national and other school systems at the senior secondary level* (pp. 400-404).
- John Monaghan: *Teachers and computer algebra systems* (pp. 462-467).
- Robyn Pierce: *Algebraic insight for an intelligent partnership with CAS* (pp. 732-739).
- Michael Thomas: *Building a conceptual algebra curriculum: The role of technological tools* (pp. 582-589).
- Michael Thomas & D. Tall: *The long-term cognitive development of symbolic algebra* (pp. 590-597).
- R. M. Zbiek & Kathleen M. Heid: *Dynamic aspects of function representations* (pp. 682-689).
- N. Zehavi, L. Wasserteil, & Giora Mann: *From a word problem to a family of word problems* (pp. 690-696).

After the initial meeting, the members of the Working Group elected to work in either the subgroup that focused on functional issues of curriculum (tasks, assessment, and teaching), led by Barry Kissane or the subgroup that focused on the processes and outcomes of student learning, led by Mike Thomas. Pairs of members explored particular issues within each of these subgroups (e.g., assessment, teaching with CAS, etc.). They examined the relevant literature, thoroughly explored important current issues, proposed areas for future research, and finally presented their work to all members of the Working Group on CAS. Barry Kissane and Mike Thomas coordinated all of these reports into a power point presentation that was delivered to all of the conference participants towards the end of the conference.

This chapter was developed from the ideas of all of the participants of the Working Group and the work of each member is gratefully acknowledged. Individual members can be contacted using their e-mail addresses listed at the back of this book.

Thanks are extended to Barry Kissane and Mike Thomas for their capable leadership of the Working Group on CAS. Finally, Mike Thomas is particularly thanked for undertaking the authorship of this chapter with John Monaghan and Robyn Pierce who graciously undertook to help Mike at very short notice.

Chapter 7

Computer Algebra Systems and Algebra: Curriculum, Assessment, Teaching, and Learning

Michael O.J. Thomas, John Monaghan, and Robyn Pierce
University of Auckland, NZ, University of Leeds, UK, and University of Ballarat, Australia

Abstract: Computer algebra systems (CAS) were originally designed for mathematicians, scientists, and engineers, and their implementation into education, and especially into schools, is still very much in its infancy. This situation is reflected in the research on CAS and hence our ability to describe the influence of CAS on student learning in algebra. This chapter synthesises key results from research related to using CAS in algebra from two broad perspectives: firstly, the issues of curriculum, assessment, and teaching; and secondly, that of student learning. Our analysis supports the view that CAS has much to offer the teaching and learning of algebra, but that real benefits may accrue only from thoughtful and structured approaches which take into account the perspectives of the student and teacher, and the intricacies of the relationships between student, teacher, and CAS. In writing the chapter we are mindful that many more questions than answers have emerged and we have included a significant number of these in the hope that they may give impetus and direction to research in the area.

Key words: Algebra, CAS, Computer Algebra Systems, calculators, curriculum, assessment, teaching, learning, symbolic manipulation

7.1 Introduction

This chapter arose from the discussions of the Working Group on Computer Algebra Systems (CAS), reported on in the two introductory pages to this chapter, with the aim of presenting what we know about CAS and algebra at present and where we think research might go in the future. There is no doubt in the minds of those involved in that forum that in the hands of teachers and learners CAS have the potential to change the teaching and learning of algebra, and possibly to do so in radical ways. However, it is far less clear exactly what these ways are and how such change might best be accomplished.

Computer Algebra Systems (CAS) were originally designed for mathematicians, scientists, and engineers. Although they vary in capabilities and can be used on calculators or computers, all CAS are powerful mathematical systems which include at least facilities for symbolic manipulation of standard real and complex variables, function graphing, and working with numerical tables and lists. Of particular interest in this chapter is the impact on teaching mathematics of the symbolic manipulation facility. Research on CAS use is tightly connected with the research on using graphics calculators, the term we use to describe hand-held calculators with function graphing and numerical tables facilities, but not symbolic manipulation.

Our discussion started with many questions, and finished with even more. One reason for this is that *CAS in education* is in its infancy. To employ a road metaphor, we are a little way down a road but we don't know where it is leading and we don't know what other roads we've overlooked. There is, then, no suggestion that we can provide definitive or even highly reliable answers to questions. A second reason is that individuals with different backgrounds and work interests, such as practical teaching, education research, and curriculum/assessment management, will have differing perspectives. The interests, for example, of a classroom teacher are likely to be different from those of someone who is employed by an assessment authority. It is good for various partners in mathematics education to come together to talk, but it is also important to note that motives for entering such a dialogue may differ. Particular differences are evident between radical constructivists and socio-culturalists with regard to what learning mathematics in a CAS environment entails.

However, in an attempt to describe the major issues and implications of CAS use in algebra learning as we see them, we will consider two broad perspectives. The first looks at issues of curriculum, assessment, and teaching (espousing a narrower interpretation of the word curriculum than usual), while the second looks at the processes and outcomes of student learning. In this discussion we will look at some of the recent research in each area and consider the implications for future research and possible ways forward in the implementation of CAS in the learning of algebra.

7.2 Curriculum, Assessment and Teaching Matters

The *on the curriculum* subgroup was charged with addressing the impact of CAS from three viewpoints: curriculum, assessment, and teaching. Subdividing mathematics education into discrete parts is a dubious activity. We do this purely for reporting purposes with an awareness that CAS work in the classroom, lecture theatre, or computer room integrates these discrete elements in an environment where learning (the focus of the second subgroup) is, we hope, paramount.

7.2.1 The algebra curriculum

7.2.1.1 Extensions, additions, and changed emphasis

The availability of CAS can provide a catalyst for a fundamental review of our traditional algebra curricula (see e.g., Stacey, Asp, & McCrae, 2000). The sections below provide an overview of some of the issues, possible revisions, and positive change of emphasis that the facility of CAS may afford our programs of study.

A strength of CAS is the ability to perform algorithmic algebraic routines quickly and correctly. This facility, along with ready access to symbolic, graphic, and numeric representations opens up the possibility of curriculum change. This possibility can be viewed as an opportunity or a threat, but either way we must be prepared to justify the choices that we make to use or not use CAS in the classroom. It has been claimed that the support of CAS can facilitate the implementation of a curriculum that places less emphasis on manipulation skills and more emphasis on conceptual understanding and symbol sense (see for example Heid, 1988; Repo, 1994). However, this claim is challenged by others such as Lagrange (1999a, 1999b) and Artigue (2002). A counter-claim is succinctly made by Lagrange (2000, p. 14):

> The opposition between concepts and skills masks an essential point. There is a technical dimension to the mathematical activity of students which is not reducible to skills. When technology is used, this dimension is different, but it retains its importance in giving students understanding.

This debate is, at the time of writing, ongoing in CAS research. Whatever the resolution of this debate, in the non-CAS classroom, students' success with algebra has often been measured by their ability to perform appropriate by-hand routines that have commonly been memorised as a set of narrowly applicable rules. While the availability of CAS apparently devalues the learning of rules, it offers the possibility to focus on the selection of appropriate techniques, the anticipation of general patterns, and the interpretation of results and application to real life problems. In addition, explicit provision must be made in the curriculum for students to learn to use CAS effectively, and this will be discussed below. Some new curriculum possibilities are summarised in the section below. These include topics specifically related to using CAS and a new emphasis required in foundational algebra teaching.

7.2.1.2 CAS supports emphasis on concepts

The availability of CAS presents an opportunity to build a conceptual algebraic curriculum, emphasising the study of conceptual objects of mathematics through a number of different representations. Thomas (2001) has commented that, from the students' perspective, the focus of the algebra curriculum has been oriented to routine processes. CAS offers the potential to shift away from an almost total reliance on symbolic representation and its consequent emphasis on procedural thinking. However, Thomas also warns that the use of technology does not make this change simple or automatic. The ability to use a calculator or computer software does not necessarily develop or indicate mathematical understanding. Guin and Trouche (1999) emphasise that potential benefits from CAS use require acknowledgment, in the curriculum, of the importance of the task of integrating CAS with other mathematical practices. This requires the full involvement of teachers, discussed later in this chapter. A CAS-available algebra curriculum would allocate less time to memorising routines and more to identifying structure, patterns, and the linking of representations in order to enhance conceptual understandings.

7.2.1.3 CAS supports emphasis on generalisation

An important aspect of algebra is building mental, abstract models and manipulating these models to find new relationships and understandings that are not always apparent in particular cases. CAS affords the opportunity to explore symbolic representations and their connections. Through guided explorations of multiple examples, multiple representations, or general expressions, students may find patterns and observe links. Cnop (2001) notes that doing mathematics is essentially a creative process that most of the time involves lots of experimentation. Many educators, he says, want to replace the traditional teaching paradigm of

definition → theorem → proof → corollary (and then perhaps application)

with a discovery approach of

problem → experiment → conjecture (and then perhaps proof).

CAS allows students to experiment in a way that can lead to reasonable, well-supported conjectures. For example, the solution of large numbers of sets of equations, moving from numerical examples to those with letters, and exploring the impact of different parameters both allows and develops students' understanding of generality (Kidron, 2001).

On the other hand CAS also facilitates the possibility of introducing general principles and then moving to particular examples. Ball (2001) for example discusses a more generalised approach to teaching students techniques for the solution of equations. She suggests that students should consider general forms, such as $af(x) + b = c$ in parallel with specific examples, such as $2(x + 3)^2 + 5 = 12$ or $2\sin^2 x + 5 = 15$. With CAS, the generality of the method of solution can be stressed.

Current CAS cannot solve all mathematically soluble equations instantly. A revised curriculum needs to include not only the use of built-in CAS features but strategies for rewriting expressions into appropriate forms. In a new world of shared cognition, students need to be given sufficient directed experience to enable them to make sensible choices about what to do themselves, on paper, or in their head, and what to delegate to the machine. A student's decision to use CAS or not is most commonly based on their perception of the time the problem may take and their likelihood of making a manipulation error (Pierce & Stacey, 2001a). A new range of problems, which may now be included in the curriculum, would without CAS be considered too time consuming by hand, and perhaps too complicated in terms of the routines required for simplification and solution.

7.2.1.4 CAS supports mathematical modelling

Using CAS can support multiple methods for the solution of equations. Without CAS students need to spend much time trying to learn to identify problem types and implement appropriate solution strategies. For many problems the resources of school algebra offer only one correct solution method. The multiple representations offered by CAS provide both a facility for both exploring a problem in order to choose a solution method and a set of alternative solution paths. Kissane (1999, 2001a) emphasises that students may take a solution path which employs CAS at none, some, or all stages and may choose to use one or more representations in order to make progress towards a satisfactory outcome. Kissane (2001a) also reminds us that exact solutions exist for only a limited range of problems. CAS technology will continue to increase the emphasis on numerical approximations and to support the inclusion in our curricula of situations where exact solutions do not exist.

CAS can provide a stimulus for, and access to, more sophisticated algebra. For example CAS can support the use of both different types of functions and combinations of functions that, as Thomas (2001) and Zehavi, Wassertel, and Mann (2001) argue, provide the facility to build problems around modelling real life situations. In current curricula, students often become fluent at performing mathematical processes that are, in fact, meaningless to them. The opportunities that CAS can provide to choose, and work with, one or more representations may help students to build a conceptual understanding of mathematically complex systems. CAS affords the opportunity to tackle harder and more realistic problems, to extend topics, and to connect topics. However, achieving good results from these new curriculum opportunities is in no way an automatic consequence of the availability of CAS, an issue that is taken up in the second part of this chapter

7.2.1.5 Key curriculum issues still in question

In our well-established algebra curricula we have classic teaching examples. A new CAS curriculum, with new emphases, also requires good examples. It is vital that mathematics educators share their experiences so that we can build a portfolio of simple, elegant teaching examples that will form the skeleton of a new curriculum in practice.

Along with the excitement of new challenges still being explored, for example, the opportunity to focus on conceptual understanding rather than the memorisation of routines to which students attach no meaning, a nagging doubt remains. A persistent question which has fundamental implications for our curriculum and has been raised by many researchers (see for example Flynn, Berenson, & Stacey, 2002), is the role of by-hand and mental working in the learning process. Relevant questions include:

- What is learnt by physically writing out all the terms of an expansion? Is this a necessary experience for learning and understanding? To what extent are these traditional activities part of the essence of doing algebra?
- Is it important that students plot points, write out the steps involved in solving an equation, write fractions over a common denominator before simplifying, repeat a routine until they have memorised the process? Does the action of undertaking these tasks build different neural connections than those developed by working with the technology and does this matter to students' long-term ability to understand and use algebra?
- What are the essentials of the techniques of algebra? Can these be achieved with CAS and if so is this a sensible thing to do or is by-hand work more efficient or effective?

A CAS-active school algebra curriculum may privilege different aspects of algebra from the traditional curriculum. In this way CAS may provide more access to algebraic thinking and techniques for a greater number and variety of students, but may present yet another barrier for other students. There is still much to learn about the possible impact of CAS on the algebra curriculum.

7.2.2 Assessment

A survey of assessment appropriate to a CAS learning environment needs to consider both timed examinations and portfolio work (or coursework as it is known in some countries). Wurnig and Townend (1997) provide portfolio examples which, they believe, show how "assessment can move away from unrealistic, mechanistic problems to more interesting exercises ..." (p. 83). Lumb, Monaghan, and Mulligan (2000), however, cite an example of coursework for which CAS was suitable but for which students also chose instead to use spreadsheets, graph plotters, and graphics calculators. Lumb, Monaghan, and Mulligan's example raises two immediate issues. First, viewing CAS in isolation from other mathematical tools represents a somewhat limited viewpoint. Second, the suitability of CAS (or not) for assessment tasks is not something that can be determined solely by teachers or curriculum developers. The students-using-tools will decide in practice on the suitability or not of the tool for the task. For studying these phenomena, the unit of analysis should not be the agent or the tool but the agent-using-the-tool.

Wurnig and Townend's (1997) paper is notable for the care and attention they pay to scaffolding (*scaffolding* in the sense of Wood, Bruner, and Ross (1976), not in the sense of Kutzler (1994)) associated with the work of 'weaker' students. Wurnig and Townend's comments (pp. 82-83) on the intention of this assessment item to help develop students' understanding of odd and even functions could be interpreted as the assessment item being intended to meet the student at the zone of proximal development (Vygotsky, 1978). This is, perhaps, one of the most important affordances for new tools such as CAS in mathematics education but it is also an area which receives scant consideration with regard to school mathematics due, we believe, to the dominance of high stakes, timed examinations. Such examinations are usually written by a Chief Examiner (unknown to school personnel) and the examination questions are not seen prior to the examination by students or teachers. This form of assessment can do little more than statically examine what students can reproduce under such examination conditions.

High stakes examinations extend CAS assessment issues beyond teacher-student interaction and into policy considerations. At the time of writing CAS calculators are permitted in examinations in France, Advanced Placement Calculus in the USA, in Danish upper secondary education, and in the Victorian Certificate of Education in Australia. The International Baccalaureate is also commencing a pilot project with CAS-allowed examinations—although it is intended that examination questions will be set so that the students who use CAS will not be advantaged.

Leigh-Lancaster and Stephens (2001) appraise the situation from an examination authority point of view. They note the need to manage change responsibly with due regard to the stakeholders. Key policy issues they raise concern equity, teacher development, and the integrity of assessment procedures. They consider various models that examination authorities can adopt:

A *no change now model* affords minimal disturbance in the short term but may underestimate the pace of future change.

A *dual approach* permits CAS and by-hand work through parallel CAS and non-CAS questions. An advantage of this model is that teachers may can adopt CAS when they are confident. A disadvantage is that preparing examinations that purport to offer no advantage either way is problematic.

A *pilot curriculum and assessment approach* permits a cohort of schools/classes to follow a CAS route whilst the majority follow the traditional route. This allows time for specialist curriculum and teacher development but has the disadvantage of attempting to prepare two cohorts of students for further study.

CAS-permitted or CAS-required models provide a clear endorsement for CAS use and encourage both teachers and students to consider the possibilities and constraints of CAS and non-CAS solutions. A disadvantage is possible inequities for students arising from their teachers' expertise, or not, with CAS. (*CAS-permitted* questions are likely to be written to confer no advantage to CAS users whereas *CAS-required* questions are likely to be written to make positive use of CAS.)

Brown and Nielsen (2001) report on a comparative study of algebraic skills assessed by three examination boards that allowed graphics calculators for examinations, but their considerations are relevant for CAS. They examined the extent to which the call by some authors, for example Heid, Choate, Sheets, and Zbiek (1995), for more emphasis on conceptual understanding and less on the execution of algorithms, was evident in the examinations. Brown and Neilson conclude that most questions focused on standard manipulative algebra and did not encourage innovative, non-procedural uses of technology. We consider this *inertia of tradition* an important point to note with regard to calls for radical changes. What technology reformers want (such as investigative questions requiring substantial time) may not be achievable in high stakes examinations.

Brown and Nielsen (2001) highlight differences between *a priori* and *a posteriori* considerations. At the time of this Algebra ICMI study (2001) most papers on CAS and assessment were *a priori* considerations of general issues, that is they were based in theoretical analysis. Since this study, *a posterior* analyses based on empirical results related to high stakes examination issues have raised new issues. The several analyses below draw attention to some of the different issues that have emerged.

Monaghan (2000) reports on *a priori* considerations of several UK assessment Working Groups from the 1990s which contemplated future CAS-permitted examinations. Findings from these Working Groups include the notions that CAS can automate (*trivialise*) many traditional questions and that adapting questions for CAS use may make questions more difficult for low attaining students. An example of trivialisation is in factorising $x^3 - 4x^2 + 4x - 1$. There are many ways to do this. One way is to spot that $x = 1$ makes the value of the expression 0, so $(x - 1)$ is a factor and then divide $(x - 1)$ into the expression. This requires knowledge of the factor theorem, spotting the substitution, and the ability to do polynomial division. None of this knowledge is needed with a CAS—all one must do is enter $Factor(x^3 - 4x^2 + 4x - 1)$ in the syntax of the specific CAS. This is, however, an extreme case, because there is a direct one step mapping of the mathematical idea onto the CAS command. Most problems do not map so directly, and others are very difficult to map. At the other extreme of difficulty, getting a current basic CAS to calculate r given that $r = s + t$ and $r + s + t = 30$ is beyond most people's ability although it is relatively easy to do by hand (see Recio, 1997).

Flynn (2002, 2003), however, in an *a posteriori* analysis of students factorising with CAS notes the need for student algebraic insight in apparently trivialised questions. Some students-with-CAS factorised $(x^3 - 9xy^2)^2 + 3(3x^2y - 3y^3)^2$, for example, but omitted the leading bracket and, instead of the desired $(x^2 + 3y^2)^3$, obtained $x^2(x - 3y)^2(x + 3y)^2 + 3(3x^2y - 3y^3)^2$. Flynn notes that only two of the five students who omitted the bracket exhibited adequate algebraic expectation (i.e., recognising this as the sum of two terms and hence not a fully factorised version of the original) and entered the expression again to obtain the perfect cube form.

A priori reasons for believing that questions in a CAS-allowed examination may become more difficult for low attaining students arise from two assumptions: that questions may move from those which examine skills to questions of a more conceptual nature and that skill questions are the lifeline of low attaining students. But are these assumptions correct? What is known from *a posteriori* studies is non-conclusive. The comparative study of Brown and Nielsen (2001) suggests a move to conceptual questions in technology-allowed examinations is not a foregone conclusion. Leigh-Lancaster's (2003) comparison of CAS and non-CAS students' examination results, "in broad terms" indicate that CAS students were not disadvantaged on common questions (but the fact that these were common questions in *CAS-permitted* and CAS-*not permitted* examinations is important to note). Finally a study by Berry, Johnson, Maull, and Monaghan (1999) indicates that all students score substantially more marks on routine questions, but lower attaining students do obtain proportionally more of their marks on routine questions. Because of this they may take to procedural use of CAS readily, but as Hong, Thomas, and Kiernan (2000) have shown there is then a danger that CAS may undermine their learning, since they came to rely totally on it, rather than learning either the concepts or the procedures.

The upshot of this discussion is recognition that *a priori* beliefs must be treated with scepticism and that further studies from real implementations are needed. Now, as we write (2003), is an exciting time as real student data is coming in from the Melbourne CAS-CAT project (see http://www.edfac.unimelb.edu.au/DSME /CAS-CAT/). Such data not only allows us to interrogate prior beliefs, it also raises emergent themes we hitherto did not consider. From this project, Ball (2002, 2003) considers the RIPA (***Reasons–Inputs–Plan–(some)Answers***) rubric for students' written solutions and examined how students' written examination solutions changed (see discussion in 7.3.1.5). Flynn (2003) has analysed how students' abilities to demonstrate mathematical results (i.e., problems to prove) may be tested in CAS-permitted examinations. While the results are not conclusive, they raise important issues for question designers and researchers. In addition, he has studied the challenges of constructing examinations that are fair to users of different CAS (for three different CAS calculators), and has created guidelines for setting examination questions that are fair and accessible to users of different CAS.

7.2.3 Teaching and CAS

Very little is known about the issues which arise when teachers use CAS in their classrooms. An early review of CAS research, Mayes (1997), noted that the majority of published research focused on student learning and did not mention teachers *per se*. Zehavi (1996) and Lachambre and Abboud-Blanchard (1996) focus on in-service training which is sensitive to tasks and the classroom culture in which these tasks may be used. However, out of class training on its own has difficulties enabling teachers to effect innovative change in their classes (Bolam, 1982). We do not have solutions to the problem of how to assist teachers to make better use of CAS but feel that: teacher professional development should involve them doing a great deal of algebra with CAS; teachers' experiences in their own learning of algebra with CAS should be positive; and teachers should be allowed to create problems and solutions.

Heid (1995) and Zbiek (1995) present case studies where CAS has threatened teachers' perceived command of their subject knowledge and, in Zbiek's paper, document a teacher changing her original plans to ensure she was in command of her subject knowledge in lessons. This is understandable as CAS can, without warning, confront the depth of teachers' understanding of algebra with unexpected responses. Our experience is that everyone, whether they have strong or weak understanding, becomes a learner of algebra when using CAS. Teachers' subject knowledge and, indeed, teachers' knowledge in all its forms (technical, pedagogical, and pedagogical content knowledge) are clearly issues to be addressed in the preparation of teachers for using CAS, and they need to be addressed sensitively. However, although we expect a positive correlation between teachers' subject knowledge of algebra and good teaching with CAS this should not be assumed *a priori*.

The most important issues, however, are arguably those centred on teachers' practice. Practice, what teachers and students actually do in their (CAS) classrooms, is influenced by cognitive functioning, but analyses of practice cannot be reduced to analyses of cognitive functioning. Other influences include social relations, institutional norms, tools used, and tasks set (see Kendal & Stacey, 2002). Practice is a whole experience and is more than the sum of parts which influence it. One way in which this practice has been described is through the concept of a *didactic contract* (Brousseau, 1986). Delos Santos and Thomas (2002) investigated the didactic contract of a teacher during function and limit concept lessons using graphics calculators, and concluded that forming a didactic contract, especially with technology present, is not an easy task. They maintain that the teacher has to be open to new approaches, be willing to work around constraints, be open to learning, and be able to reflect "on the tension created between valuing a formal, primarily algebraic approach to mathematics and an investigative style of teaching" with CAS (p. 357).

Little is known about teachers' practices in algebra lessons. Kieran (1992) states that available research suggests that algebra teachers are primarily concerned with classroom management and covering material and view themselves as providers of information. In a paper on teachers' beliefs about the development of algebraic reasoning, Nathan and Koedinger (2000, p. 181) state that "textbooks have been identified as a primary resource—and often the *only* source—of the content planning performed by expert and novice high school mathematics teachers". Their study, moreover, showed a highly significant rank correlation between teachers' ranking of algebra item difficulty and its curricular sequencing in the textbook they used.

The conflicts that can arise when teachers who rely on textbooks as their primary resource consider implementing CAS were highlighted in a study of Lumb, Monaghan, and Mulligan (2000). They report on the work two teacher-researchers did with senior high school classes. One teacher relied on his own worksheets to "ensure control over students' work" and claimed that the textbook and CAS did not "fit" (he reflected that written mathematics and CAS mathematics were two different mathematics). The other teacher tried two quite distinct CAS-focused mathematics books (one *traditional* and one *investigational*). He found both of them of little use:

the traditional book had no lesson ideas whilst the investigational book presented activities that did not suit his curriculum goals.

Kendal's work has generated a number of papers that focus on both teachers and students and examines how teachers' *privileging* differentially affected their students' learning in CAS lessons, especially related to multiple representations. Kendal and Stacey (1999) report on the work of three teachers who helped design and teach an experimental introductory calculus course. The overall test results for classes were similar but the teachers differed in their privileging of symbols/graphs, concepts/skills and by-hand/by-technology which impacted on their classes' performance in these areas. For example Teacher A privileged technological and symbolic approaches. Class A students were able to compensate for weaker symbolic skills by using (and overusing) CAS. Class B and C students managed the symbolic items almost as well as Class A students but without the over-dependence on computer algebra. Teacher B emphasised by-hand symbolic approaches, and Class B students demonstrated the highest proficiency with by-hand symbolic manipulation. On occasions they failed to use CAS when its use would have been advantageous. Teachers B and C privileged conceptual understanding built from illustrating algebraic ideas graphically. In consequence, Class B and C's conceptual error rates were lower than for Class A.

The question of how teachers, who have learned algebra *without* CAS, make decisions about what students should do *with* CAS raises important issues. This question has many facets. Teachers design "routes to knowledge" (Lumb, Monaghan, & Mulligan, 2000) which are products of what they perceive counts as knowledge (Kendal, 2002). Issues may emerge in CAS classrooms (time spent mastering CAS syntax/commands and algebraically equivalent forms—see Lagrange, 1999a, 1999b) which may make teachers using CAS ask "Is this (useful) mathematics?" This is an area where *a priori* analysis must give way to *a posteriori* analysis of emergent issues for teachers using CAS in classrooms. The question also relates to teachers' use of new tools (mediational means). This links with (mainly) French work on instrumentation detailed in the next section of this chapter. Work on CAS instrumentation has, to date, focused on students but research on instrumentation and teachers is clearly important. There is scope for at least two directions in this work: parallel work on teachers and instrumental genesis (which is likely to be somewhat different to the instrumental genesis of students as teachers already know standard algebra) and how teachers *orchestrate* students' instrumental genesis (work on this area has already begun, see Guin and Trouche (2000)). As the subgroup reported to the Study Conference in its Plenary Presentation, "teachers knowing how to factorise, solve, and expand with CAS does not necessarily lead to knowledge of how to provide students with learning activities that provide students with algebraic insight".

CAS work in the classroom is currently mainly carried out by enthusiasts but, if indeed there is a place for CAS, *The Future of the Teaching and Learning of*

Algebra requires that all algebra teachers are CAS aware and CAS competent. Future research should examine teachers' motives for using CAS on particular tasks. It is possible to generate wonderful new tasks suitable for CAS work. Aldon's (1996) "factorise $x^n - 1$" task is one of a multitude of such tasks available in the literature and on web sites around the world. But such tasks must *fit* with the teachers' motive for the classroom activity (which will surely be different in a CAS-supported lesson) or else teachers are likely to ask that very awkward question "Should I be using class time to do this?"

Changing practice for a CAS future will take time training and continuing dialogue with teachers, schools, governments, students, parents, and researchers, because government and school policies, attitudes and beliefs of parents, students, teachers, and researchers impinge on the ways in which teachers actually use CAS for the teaching and learning of algebra.

7.3 Student Learning

The second key area of discussion of CAS use in algebra is that of student learning. We believe that there are three crucial questions to ask when considering implementation of CAS in the learning of algebra. These are:

- How does CAS use influence student conceptualisation?
- What is the nature of the students' development and ability to work with CAS as an instrument?
- How does the way students work on tasks by hand inform their work in a CAS environment and vice versa, and how does each of these interact with students' mathematical thinking and understanding?

We will look at some of the issues surrounding each of these key questions below.

7.3.1 Student conceptualisation

Students' ability to conceptualise the mathematics they are studying is influenced by a wide range of factors in a CAS-environment. Some of these factors are discussed below and include the impact of CAS on students' computational skills, their level of dependency on the tool, and learning opportunities available to help students improve their understanding.

7.3.1.1 Impact on by-hand computational skills

One of the first issues that must be addressed is whether using CAS is likely to be beneficial to student learning or not. Some, for example, are concerned that students' by-hand computational (i.e., symbolic manipulation) skills may suffer if CAS is used. In this regard, Keller (1994) found that CAS groups improved their performance on pattern spotting and other problem-solving approaches but performed worse on standard questions. Examining this same issue Heid (1988) and

Palmiter (1991) had previously both found that students who used CAS had a deeper conceptual understanding without losing traditional computational skills. A later study by Keller and Russell (1997) considered three types of problems: those that could be solved with CAS by direct entry (i.e., by a single CAS command), or by a standard set of CAS commands, or by non-standard symbolic procedures. In all three cases the students taught and tested with CAS were consistently more successful in using correct solution methods and at producing correct answers than the students without CAS. Hence Keller and Russell conclude there is no need to be concerned about possible negative effects on the development of by-hand computational skills in a CAS environment since "if an environment fosters the development of symbol sense, then the symbolic manipulation development proceeds naturally" (p. 93). Heid and Edwards (2001) support the view that when using CAS, symbol sense (Arcavi, 1994), symbolic reasoning, and symbolic disposition assume greater importance than manipulation of expressions. This is also supported by Heid, Choate, Sheets, and Zbiek (1995), who argue for more emphasis on conceptual understanding and less on algorithmic skills of symbolic manipulation. Heid (2002a) recently reviewed the arguments against CAS use in the secondary algebra classroom, including the idea that they lead to a loss of by-hand skills, and came to a contrasting view. In fact, rather than supplanting such skills, she asserts, "Today… the focus is on enhancing students' understanding of symbolic aspects of algebra—a focus that can be ably assisted by CASs." (p. 663). To support this stand she gives examples including how a consideration of the expansion of products of binomials could assist students to investigate their structure, to abstract the general case, or even to extend it, and how they can begin to understand the important role of parameters in functions such as $f(x)=a/(b + ce^x) + d$. In keeping with these expectations, research by Driver (2001) showed that students who used CAS attained a higher level of achievement than would otherwise be expected of them.

7.3.1.2 Dependency on CAS

Some aspects related to CAS use recorded in the literature appear negative in nature. Hunter, Marshall, Monaghan, Roper, and Wain (1993) used Derive with 14- and 15-year-old students who were working with factorising and expanding quadratic expressions. They found that although some students who used CAS were not motivated by using it they nevertheless became dependent on it. When not permitted to use CAS they did worse than the control group. A similar outcome is recorded by Hong, Thomas, and Kiernan (2001) who showed that lower attaining students may become reliant on the CAS as a problem-solving support in a manner that has a negative effect on their learning, since they fail to learn necessary procedures. Monaghan, Sun, and Tall (1994, p. 282) also record how, for some students, CAS use can become a mere button pushing process that can obscure deeper understanding. One reason for negative reactions of some students using CAS was recorded by Bergsten (1996) following CAS investigations on the convergence of sequences and Taylor expansions using Maple. He records that "It was very clearly stated by many of the students ...that to *learn* analysis you must do the hard work by hand" (p. 38). The students complained that since they did not see what happened inside the machine, they could not check it. Edwards (2001, p. 299) considers the awkwardness of the CAS output on calculators as well as the calculator's tendency to perform "too many steps" automatically may have contributed to the students' preference for by-hand methods. Goos, Galbraith, Renshaw, and Geiger (2001) found a similar attitude among some students, who expressed their feelings along the lines: "I just don't understand what I'm learning here. I mean all I have to do is ask the machine to solve the problem and it's done. What have I learned?" (p. 226). Bergsten (1996) stresses the importance of clarifying the role of computation in developing mathematical understanding, which he claims and we have noted above, is far from understood within mathematics education. While it is not always possible to separate out the influences in a learning environment there has been a suggestion that CAS may have limited extra value in solving complex problems in calculus, compared with the student-centred instructional techniques which technology use can promote. Cooperative learning and discussion formats were shown to have a positive overall effect on students' performance in calculus, when used alongside the TI-92 (Keller, Russell, & Thompson, 1999).

7.3.1.3 Opportunities for improved student learning with CAS

Clearly, with its valuable pedagogical elements (see for example, De Alwis, 2000), using CAS is likely to have a positive influence on student conceptualisation of mathematics, offering relevant representations for thinking that a student can internalise. Hence it is no surprise that there are many studies which find positive advantages for student learning with CAS use. The following section reports on specific ways of learning with CAS that may assist students' learning.

Use of multiple representations. CAS can assist students through offering multiple dynamically linked representations, which appear to provide good opportunities for learning (Hart, 1992). These representations may be simultaneously presented on screen so students may more readily link different representations of the same object, something which Lesh (2000) describes as essential to mathematics learning. Meel (1998), for example, has shown that students often use the CAS to scaffold their problem solving, building their solution by moving among various representations. Likewise Slavit (1996, p. 14) noticed how in high school algebra II classes (approximately 16 years old), the "multi-representational capabilities of the [graphics calculator] allowed additional aspects of a problem to be quickly analysed in a 'representationally-connected' fashion." Hong, Thomas, and Kwon (2000) investigated Korean students' conceptual thinking during the solving of linear algebraic equations using a TI-92 CAS calculator. They presented solution methods which encouraged students to think inter-representationally about the concept of a solution to an equation, asking them to solve a symbolically-presented equation in a graphical and a tabular domain. They found that while some students were assisted to make the conceptual links between the representations, this was patchy. Hong and Thomas (2001a, 2001b) have also used a CAS calculator to assist students in both New Zealand and Korea in their understanding of how the algebraic formula for the Newton-Raphson method works. The tangent function for $f(x)$ automatically generated in the graphical representation was copied and pasted to the symbolic home screen, and then used to solve the equation $f(x) = 0$ using the Newton-Raphson formula. They report that this gave students a richer conceptual perspective of the method, rather than a purely procedural one, and enabled them to understand the graphical foundation of the symbolic formulation.

Even though CAS provides a range of representational tools, use of multiple representations is not automatic with students not necessarily using a second representation to check their work until they encounter obstacles in their problem solving with the first representation. In fact Crowley (2000) found that students did not link graphical and symbolic representations unless specifically instructed to do so, and Weigand and Weller (2001, p. 109) found that "learners ... seldom have the patience to read representations on the screen and then to interpret and reflect on them. Computer generated representations are often viewed only as pictures." Hence, Lagrange (1999) suggests that students need adequate algebraic knowledge and flexible connections between various representations as a prerequisite for CAS use. Reviewing the potential benefits and pitfalls from graphics calculator use, Wilson and Krapfl (1994) encouraged more studies to address how and why improvements occur. In particular they suggest looking at "how students use graphic calculators to develop conceptual links between graphical, tabular, and algebraic representations of functions" (p. 261). This is still a crucial area for research since only when we know how this process proceeds can students be helped to understand graphs and other representations in a fundamentally different way, as tools to help interpret and solve mathematical problems.

It appears that many students may prefer graphical and numerical representations to reason from, and tend to work towards a symbolic representation rather than from one. Kendal and Stacey (2003) found that in their study, only the most capable students achieved the goal of developing facility with numerical, graphical and symbolic representations of functions and derivatives, which they thus felt might be an unrealistic goal for the majority of students. They propose that it may be necessary to introduce such multiple representations over time, rather than together, and to teach students explicitly to take full advantage of the numeric, graphic, and symbolic capabilities of their CAS. Heid (2002b, p. 104) considers theories of representation to be a key component in CAS use and makes some suggestions on possible research in this area, which we would endorse:

Future research is needed to further our understanding of the impact of technology on students' abilities to convert from one representation to another, students' abilities to interpret in a second representation what they learn in a first, students' reasoning from symbolic representations that were externally generated using technology, and students' abilities to capitalize on the opportunities the CAS affords.

Experimentation and generalisation. A further advantage of CAS in student learning is that they allow efficient variation of symbolic constraints and parameters so that students can generate and examine many examples and regard them with confidence. This capacity should enable and stimulate experimentation, generalisation, and pattern recognition (Trouche, 1998). The ability to deal with literal expressions in CAS may afford students the opportunity for more ready access to generality. An approach employing experimentation and generalisation was employed by Graham and Thomas (2000), using graphics calculators to assist construction of the concept of variable. Their method of supplying a mental model for a variable in terms of stores, which would also work well with CAS, provided a good basis for conceptual understanding of algebraic variables. Edwards (2001) has provided some evidence that high school algebra students have learned to build conjectures through experiences with TI-92s by generating numerous examples. Other aspects of CAS use that may promote learning include the greater reliability of results which may lead students to feel more confident in reasoning from them, and the immediate feedback which can bring confirmation (or otherwise) of their thinking.

Focus on functions and parameters. The capacity of CAS to allow students to operate on variables and functions either as "objects" or processes may promote deeper or more advanced understanding of concepts of variable, function, expression, and particular types of functions could result (Boers-van Oosterum, 1990; Heid, 1988; O'Callaghan, 1998; Repo, 1994). One crucial concept is the role of letter as parameter. Drijvers (2000, 2003) studied this using a TI-89 with 14- and 15-year-old students and concluded that CAS helped some students but not others. The opportunity CAS provides to treat letters in different roles, and especially the need to distinguish parameters from variables, may help develop a better understanding of the distinction between variable and parameter. Further, the option of using multiple-letter variable names may offer a semantic advantage, although there is a possibility that this notation may foster confusion in the development of understanding of variable and function. Slavit (1996) reports that using a graphics calculator for algebra problems allowed for several parameters to be used, increasing the breadth of investigations.

It has been argued that students can learn by-hand symbolic manipulation routines more quickly if their introductory algebra experiences use CAS to prioritise concepts of function, and variable in contextual situations (Heid, 1988, 1992).

O'Callaghan (1998) reports a better overall understanding of function, including modelling, interpreting, and translating components by students who were exposed to computer-intensive algebra, although there was no evidence of improved reification of function. This suggests that when students use the CAS as the primary generator of symbolic routines, their work can focus more intently on conceptual development than if they were not using the CAS. Thus their later by-hand symbolic work is supported by deeper than usual conceptual understanding. Research by Leinbach (2001) supports this contention. In his study students focused on understanding an algorithm for solving cubic equations, avoiding the need for difficult, and possibly error prone, symbolic manipulation. This approach requires an appreciation of the nature of polynomials and the role that their coefficients play in determining the shape of the graph. Such conceptual understanding is required of the user, and cannot be provided directly by the CAS, although this may be used to scaffold it by providing easy access to many examples. Weigand and Weller's (2001) study on student understanding of quadratic and trigonometric functions in a CAS environment reported no evidence of a better understanding of function when using CAS, but they did find a different understanding from by-hand working, including seeing functions as objects.

Other possible benefits of CAS use on student conceptualisation suggest that students might develop a deeper insight into the structure of formulas, a better symbol sense, and improved symbolic reasoning (Boers-van Oosterum, 1990; Pierce & Stacey, 2001a). In addition, learning with CAS gives students an opportunity to develop algebraic insight and to communicate their mathematical understanding in a new way. These new benefits are discussed in the following sections.

7.3.1.4 Opportunity to develop algebraic insight

Many routine algebraic simplifications, manipulations, and solutions may be assigned to CAS but the user must still be in control. The student must decide which routines are appropriate, enter expressions using appropriate syntax, and interpret results that are not always presented in a conventional format. It is not uncommon to meet teachers, who, having observed CAS capabilities but not worked with CAS in their own classroom, think that "we will no longer need to teach algebra". In contrast, there is agreement amongst experienced CAS teachers, researchers, and students that in fact working with CAS requires a sound understanding of the foundations of algebra (Björk & Brolin, 1998; Herget, Heugl, Kutzler, & Lehmann, 2001). Cnop (2001) noted that skills such as factorising and simplification of expressions may not be useful in themselves with CAS but are needed so that students retain an ability to control the output of the computer. This requires the student to gain algebraic insight. Students often express a preference for doing a few simple examples by hand before making use of CAS to explore variations or tackle more difficult problems. From this starter level of algebraic knowledge, strategic use of CAS can be used to help build algebraic insight.

Pierce and Stacey (2001b) have also proposed a framework for algebraic insight that might give a focus to the new, developing algebra curriculum. They characterise algebraic insight in terms of two key aspects: algebraic expectation and the ability to link representations. Guiding students' development of algebraic expectation requires a curriculum that, for example, emphasises the structure and key features of algebraic expressions. This helps recognition of equivalent expressions—an essential skill when working with CAS. Ability to link representations by identification of symbolic form and key features will also help students progress to the solution of algebraic problems. Such use of CAS requires a curriculum which both develops algebraic insight and technology skills.

7.3.1.5 Explaining results, communicating mathematics

Learning to write mathematics in a way that communicates to others is a core skill in any mathematics curriculum. The traditional conventions of written algebra may be challenged as we consider the questions of what is recorded and why when using CAS. Ball (2002, 2003) observed that the use of CAS, with its own peculiar syntax, may impact on students' use of conventional syntax. In her study of CAS and non-CAS students' written responses to examination questions it was clear that many CAS students used some CAS syntax (although not to excess) and that their solutions had more words, such as solve, than those of the non-CAS students. Ball (2003) suggests that we reconsider what it is that must be recorded in order both to achieve a solution and to communicate how the problem has been solved. Ball and Stacey (2003) propose the rubric RIPA (***R**easons–**I**nputs–**P**lan–(some)**A**nswers*) as a guide for teaching students how to record their solutions. When cognition is shared with a CAS it is no longer sufficient to instruct students to write down *their* working, since the CAS has done some of it for them. When the students in the CAS examination used words such as solve, they were communicating the plans of their solutions, rather than the detail of symbolic manipulation. Learning to communicate mathematics must have a different emphasis in the curriculum.

The above discussion tends us to the view that the benefits of CAS use in algebra can be large, but are dependent on the manner in which the CAS is used, and in particular on the relationships between the teacher, the student, and the ways CAS is used to construct knowledge in the classroom. The crucial relationship between the student and the tool is discussed in the next section.

7.3.2 Instrumentation of CAS

A number of different authors have noted that technological tools may be employed in qualitatively different ways by users and that this could be expected to influence their conceptual development. The types of usage outlined by Doerr and Zangor (1999, 2000) include: property investigation, computational, transformational, data collection and analysis, visualising, and checking. In comparison, Goos, Galbraith, Renshaw, and Geiger (2000, p. 312) have identified four qualitatively different types of student interaction with technology. In these the student can be subservient to the technology; the technology can be a replacement for pen and paper, a partner in explorations, or an extension of self integrated into mathematical working. In addition, Thomas (2001) has described how CAS may be used as a procedural or conceptual representation tool, with each use requiring qualitatively different perspectives on the mathematical objects. However, there is evidence emerging that many students initially use CAS only for checking working, or for procedural computations and have little understanding of conceptual use of CAS required in problem solving, for example.

7.3.2.1 Learning to use the tool as an instrument

The analyses of Artigue (1997) and Guin and Trouche (1999) in particular have made clear the distinction between what they call *tool* and *instrument* use. Based on the ideas of Rabardel (Rabardel, 1995; Vérillon & Rabardel, 1995) and Chevellard (see Artigue, 2002, for details of his anthropological approach) they outline the process of instrumental genesis, whereby a student, through use and application involving action schemes, transforms the CAS tool or artefact into an instrument, which can be used to achieve substantial intellectual goals. This process, which turns out to be unexpectedly complex (Artigue, 2002), involves the adaptation of the tool to a particular task, deciding what it might be useful for, how it might be applied, and development of the skills needed to use it for the task. In fact Rabardel speaks of combining two processes: instrumentation and instrumentalisation. In the former the subject adapts himself to the tool while in the latter he adapts the tool to himself. Trouche (2000, 2003) uses this idea, and also elaborates on the importance of instrumentation schemes of action, and the way in which the instrumentation process and the conceptualisation process are dependent on each other for the production of these schemes. He concludes that for instrumentation to occur, classroom organisation and instrumentation activity, both individually and collectively, must be clearly directed to particular conceptions (the teacher must ask "De quelles conceptions va-t-il favoriser l'émergence?" [Which emerging conceptions will it encourage?] (p. 261). This classroom direction could include collective discussion of different perspectives on concepts during which appropriate language for the concept is encouraged. One possible way to organise such a classroom with CAS calculators has been proposed by Guin and Trouche (2000), who suggest a session where there is a collective synthesis of work on a problem, led by a sherpa-student who manipulates their CAS calculator with its screen projected for all the class to see. The teacher then "institutionalised and decontextualised the mathematical knowledge he (or she) wanted to retain among student productions" (p. 13).

Another essential part of instrumentation described by Artigue (1997) is the analysis of the constraints imposed by CAS. Various categories of these constraints have been listed by Trouche (2000): internal constraints (e.g., limited accuracy), command constraints (e.g., correct syntax for commands), and organisation constraints (e.g., switching between exact and approximate modes). Drijvers (2002) likewise highlights the need to perceive as learning opportunities the apparent obstacles which arise due to the imbalance of technical and conceptual aspects during CAS instrumentation. He identifies twelve kinds of obstacles that students often encounter and suggests that they be made "the subject of classroom discussion in which the meaning of the techniques and the conceptions is developed" (p. 228).

Lagrange (1999b, 2000, 2001) stresses the important role of schemes and *techniques*, or mathematical activity between tasks and theories, in instrumental genesis for algebra (i.e., in the process of instrumentation). He describes (1999b) how instrumental genesis involves both the external use of the artefact along with internal instrument utilisation schemes, and gives examples of such schemes, along with their decisional, pragmatic, and interpretive dimensions. Commenting on the symbiotic nature of schemes and techniques he notes (p. 197) that "techniques without schemes are ineffective because they are not likely to evolve and cannot produce knowledge" and hence classroom discussion on techniques is essential to help students develop suitable schemes.

7.3.2.2 Classroom issues arising in instrumentation of CAS

Teachers using CAS in mathematics classrooms need to address the issue of instrumentation, and how it may be attained. A number of issues arise around this process. For example, Drijvers and van Herwaarden (2000) considered the instrumentation of CAS calculators in the algebra learning of 14 and 15 year-old students, focussing on conceptions of parameter in systems of equations. They suggested to students an *Isolate-Substitute-Solve* (ISS) instrumentation scheme (isolate a variable in one equation, substitute it into a second and then solve that equation), but found that students had many unforseen problems with it. The CAS techniques required were based on mathematical conceptions that students often did not possess such as *equations can be solved for different letters* and a *formula can be considered as an object rather than a process*. They concluded that both the technical sequence of actions performed on the machine and the conceptual aspects of instrumentation need explicit attention and that integration of the CAS with by-hand substitution and isolation techniques would have led to better results. This agrees with research of Trouche and Guin (1996) on understanding of the concept of limit. They point to "a significant difference between procedures used with (or without) calculators" (p. 328). Without CAS the students worked on the algebraic forms and tried to reduce them by factorisation to known forms, whereas with CAS they zoomed in on the graphic representation, trying to deduce the behaviour of the function at infinity. Since both activities have value in learning, they recommend integrating calculators and by-hand work.

Not all students will attain the level of instrumentation described by Guin and Trouche (1999) and Lagrange (1999a, 1999b), or indeed will see the need to. Students who use CAS have to work out its role in their learning. Many of the possibilities of CAS rely on students forming a partnership with it in their everyday mathematical work. Through their individual interactions they have to learn to decide what CAS is useful for and what will be better done by hand. When controlling the machine they have to be aware of possibilities and constraints, of possible differences between mathematical and CAS functioning, of symbolic notations, and internal algorithms. Then there is the issue of monitoring the operation of the CAS (e.g., the syntax and semantics of the input/output, the algebraic expectation, etc.), and the difficulties of navigating between screens and between menu operations. Driver (2001) reported that many students found CAS more difficult to use than graphics calculators. Aware of these possible difficulties King, Hillel, and Artigue (2001) raise questions about the instrumentation process. These include the need to know what mathematical knowledge is necessary to become an efficient user of CAS in reasonable control of the instrument, and how to manage the steep learning curve that CAS use requires. Guin and Trouche (2000) identify a sufficient mathematical background as a factor in instrumentation, and distinguish between those who have it and those who, dependent on the CAS, have difficulty reaching an initial instrumentation stage. They describe the key cross-over point to be when students accept the symbolic register as taking priority over the graphical register.

One of the roles of mathematical understanding in forming a partnership with CAS in learning is that it is required in order to formulate input (especially symbolic input) to the CAS, as well as to check and interpret output in its various representational formats (Pierce & Stacey, 2001a, 2001b, 2002; Hong, Thomas, & Kiernan, 2000). Tall (2000, p. 43) highlights one example of this, explaining how CAS use "a variety of representations for numbers, including integers, rationals, finite decimals, radicals such as $\sqrt{2}$, $\sqrt[10]{7}$, special mathematical numbers such as π, e." He maintains that this is a source of difficulty since many students do not have a coherent view of the number line. It is not clear to us the extent to which the need for a precise notational input and for checking the reasonableness of input and output in a CAS environment will either impede or facilitate mathematical understanding. However, we would hope that this would occur through the mechanisms of cognitive challenge, immediate receipt of negative CAS feedback, and gradual increase of task complexity.

A second, wider factor in the development of CAS as an instrument is the influence of CAS on social interactions in the classroom. Here aspects such as the influence of the teacher privileging on learning (Kendal, 2002; Kendal & Stacey, 1999, 2000), the role of CAS in classroom discussion, classroom organisation, and peer and teacher interactions (Guin & Trouche, 1999) require careful consideration (as discussed earlier in this chapter). Goos, Galbraith, Renshaw, and Geiger (2000, p. 318) have expressed the opinion that "Perhaps the most significant challenge for teachers lies in orchestrating collaborative inquiry so as to share control of the technology with students." To develop the potential of the CAS learning environment, they identify a hierarchy of roles for the interaction between the teacher and technology, using the metaphors of technology as *Master*, *Servant*, *Partner*, and *Extension of Self*. A technology-rich classroom can be difficult for the teacher to manage.

One opportunity for teachers to stimulate interesting discussion in the learning environment is by creating cognitive conflict. Unlike by-hand work, where students generally are usually not surprised by unexpected results, CAS can easily be used to provide algebraic phenomena that can puzzle students and even teachers (e.g., unusual notations, different internal algebra, etc.). CAS can also be employed to *scaffold* students' mathematical activity, with the potential to compensate for weaknesses in previous learning, such as is often seen in the algebra of beginning calculus students.

The realisation by teachers of the potential of CAS for student learning requires guidance through the process of instrumentation, learning to use the new technology, and a re-focusing of emphasis in the algebra curriculum. To make use of a CAS calculator, students and teachers will need to learn new technical skills and syntax. This is not always simple. For example, Zehavi, Wassertal, and Mann (2001) comment that even different shades of meaning in the words used as part of a mathematics vocabulary and a CAS vocabulary may cause some confusion. The obstacles noted by Drijvers (2002) (discussed above) also need to be specifically addressed through targeted teaching. To gain the benefits afforded by CAS, students also need to learn how to explore (e.g., to study a family of functions by varying parameters one at a time and in combination, noting the features of each of the available representations), using CAS like a laboratory (Kissane, 1995, 2001b). However, students need to learn to approach such exploration in a strategic rather than a random manner. Pierce (2002) provides a framework for Effective Use of CAS that is designed to give some guidance for curriculum planning and monitoring students' progress in learning to use the technical facility of CAS.

7.3.3 CAS and by-hand synergy

7.3.3.1 Integrating by-hand and CAS activity

A relatively new and significant research focus is the manner in which students integrate by-hand and CAS approaches to problems. Such integration is not straightforward and Artigue (1997), considering the integration of CAS in secondary school mathematical learning, describes relevant learning processes and didactical phenomena linked to the CAS transposition of mathematical knowledge. For example she cites the number of different actions available at any point, the number of different commands on the CAS, and the problem of differentiating the error messages as central in establishing integration. Lagrange (2000) stresses the role of *techniques,* the mathematical activity between tasks and theories, in the instrumentation process, arguing that new technological instruments, such as CAS, introduce new techniques which have to be linked to and coordinated with the usual by-hand techniques. Waits and Demana (1998) describe several features of this coordination. Firstly, they speak of some processes which should be carried out by hand before CAS is used (such as integration of functions) and others which may be carried out on the CAS immediately (e.g., partial fraction decomposition). Secondly, they believe that some by-hand symbolic skills are still required when using CAS in order to be able to interpret the reasonableness of CAS results. Pierce and Stacey (2001b) take the position that often it is *mental* rather than *by-hand* algebra that is critical. This is an area that needs further research and wider debate. The fact that many students have not developed the facility to coordinate CAS and by-hand activity may go some way to explaining why Weigand and Weller (2001) in their research project on the use of CAS for understanding quadratic and trigonometric functions, after noting that a student "switched from the [CAS] ...to pencil and paper, and tried to figure it out manually", comment that "This [integrated] working style is quite rare" (p. 99). Another possible reason identified by Artigue (2002) is that in many classrooms by-hand techniques may have an elevated status compared with instrumented techniques (i.e., using the CAS), which may lack mathematical status. Although CAS use need not necessarily decrease students' by-hand skills (Heid, 1988), students are likely to be more in control of their notation systems in the by-hand environment than they are in a CAS one (Heid, 2002b), and hence will need to learn the requirements of communication in the new system, and how it relates to by-hand working. This will almost certainly require teachers to work on their presentation of instrumented CAS techniques in a didactic contract (Artigue, 2002).

This raises issues of teachers' content knowledge and their ability to resequence or reconstruct teaching trajectories, since CAS and by-hand environments encourage different techniques for addressing tasks (e.g., methods may differ in syntax, variable use, equivalence of expressions and functions, and the meaning of operations and processes). Trouche and Guin (1996, p. 329) describe one aspect of this resequencing as teachers having "to integrate graphic calculators... and to organise, when it is possible, backward and forward motions between calculators, theoretical results and calculus by hand." Hence it is necessary to integrate CAS and by-hand activity in concert with an individual student's mathematical thinking and understanding. Drijvers (2002) lists the difficult transfer between CAS technique and pen and paper as one of the obstacles of using CAS, but which should be seen as a learning opportunity. In an earlier paper, Drijvers (2001, p. 226) describes the need to pay close attention to the by-hand methods, saying that "The equilibrium between paper-and-pencil work and machine [CAS] work during the instrumentation process may be quite delicate, and we probably did not pay enough attention to the 'traditional' approach." Monaghan, Sun, and Tall (1994, p. 282), considering the limit concept, highlighted the possibility of a student selectively moving between CAS and by-hand methods in order to gain the benefits from each.

Selectively therefore, the student may focus on the production of the limit object (using the computer) or on the limit process (by a paper-and-pencil or computer calculation). Therefore the student can see the two complementary facets of the limit procept as concept and process, in whichever order is desired.

These results suggest that it is important to carry out some procedures by hand, but the question is which? Monaghan (1997) presents the view that it should be those containing a principle that will become important in later development, or that is important for cognitive development. However, as Stacey, Kendal, and Pierce (2002) note, it is not yet clear which procedures these will turn out to be, and there is a role both for research and curriculum development to assist in answering this question (see above).

A number of research questions arise concerning how to achieve a synergy between by-hand and CAS methods.

- What are the commonalities and differences between methods integrating CAS and by-hand methods, and how can we assist students to discern them?
- Do students have difficulty relating a by-hand method to a corresponding CAS method? If so, why, and to what extent?
- Given a task do students tend to use one method or the other rather than automatically establishing a relationship between them?

One key difference in working with CAS is that it stresses the need for an overall strategy which the user must be very precise about, whereas with by-hand working, the user may lose sight of the overall solution strategy through having to concentrate on the detail of individual procedures. This need may encourage mental formulation of an algebraic expectation of CAS output which in turn will enhance algebraic understanding. A second key consideration in the relationship between CAS and by-hand working is raised by Drijvers and van Herwaarden (2000). They ask whether computer algebra will support the gradual formalisation of students' informal strategies. Taking this up, Heid (2002b) suggests studying the extent to which working in a formal CAS system will influence students' ability to reason and interpret in the mathematician's formal system.

One area in which CAS use may facilitate improvement is in problem-solving ability. Zehavi and Mann (1999) report on students' building of algebraic models from word problems. They found that CAS use enhanced reflection on problem solutions since it freed up time to think about the correctness of a solution. CAS also helped students create a network of relations between conceptual and procedural knowledge that enhanced their understanding. In addition, CAS can enable more realistic problems to be addressed and allow students to focus and reflect on the formulation of a mathematical model, along with interpretation and validation of the solution to the problem (Lesh & Doerr, 2002). Lesh and Doerr's concept of the importance of model-eliciting activities in the construction of students' mental models raises questions about the role and influence of CAS in the formation of the models arising from such activity.

7.4 Other Future Research

The above discussion leaves us with the potential for a full and exciting future with CAS, since there are many questions as yet unanswered, some of which have been suggested above. To these questions we can add versions of the broad questions proposed by King, Hillel, and Artigue (2001). Will epistemological obstacles change with CAS? Will CAS change mathematics? Should we be changing the curriculum to integrate CAS or writing a new one? Based on a meta-study of 146 CAS papers, Lagrange (2001) addressed approaches to these and other research questions, reflecting on the integration of CAS into student practices. He provides a useful multi-dimensional framework to consider research in the integration of CAS in mathematics learning. In particular, his framework highlights epistemological, semiotic, cognitive, and instrumental dimensions as necessary for such research. For algebra learning these dimensions emphasise the value of considering: the changes that CAS might bring to algebraic practices; changes in knowledge; possible obstacles, representation and semiotic aspects; new cognitive tools, connecting knowledge; how constraints of CAS may shape the action of the learner; the classroom setting and its organisation; and the time factor required when using CAS.

Kaput (2000) looks towards the future and discusses the possibly profound consequences of the development of devices offering classroom connectivity with CAS. New activity structures may be developed which give students the ability to design and pass structured mathematical objects (e.g., functions) and representations (e.g., graphs) among themselves and to the teacher and hence promote different learning. Other developments of technology, such as highly user-friendly versions of CAS such as Symbolic Math Guides for the TI series of calculators, or other cut-down systems which will be closely linked to particular curriculum topics and stages, will also provide new possibilities for improving learning for students, even for younger children. The benefits and disadvantages of using these rather than the very open tools is likely to be a new and important area of future research.

In this chapter we have outlined some of the recent research results relevant to the use of CAS in the learning of algebra. We have particularly highlighted issues surrounding curriculum, assessment, teaching, and student learning. As we have indicated a number of times, our partial synthesis of the studies and issues has not left us with a clear distillation of either where CAS use currently sits, or where the road ahead leads. However, we hope that what we may have done is to have placed on the roadside a sign or two which may indicate routes that could prove productive.

7.5 References

Aldon, G. (1996). *Derive* for 16-18 year old students. *International Derive Journal, 3*(3), 13-20.

Arcavi, A. (1994). Symbol sense: Informal sense-making in formal mathematics. *For the Learning of Mathematics, 14*(3), 24-35.

Artigue, M. (1997). Le logiciel 'DERIVE' comme revelateur de phenomenes didactiques lies a l'utilisation d'environments informatiques pour l'apprentissage [The symbolic manipulator Derive as a revealer of didactic phenomena in the use of calculating environments for learning]. *Educational Studies in Mathematics, 33*, 133-169.

Artigue, M. (2002). Learning mathematics in a CAS environment: The genesis of a reflection about instrumentation and the dialectics between technical and conceptual work. *International Journal of Computers for Mathematical Learning, 7*, 245-274.

Ball, L. (2001). Solving equations: Will a more general approach be possible with CAS? In H. Chick, K. Stacey, J. Vincent, & J. Vincent (Eds.), *The future of the teaching and learning of algebra* (Proceedings of the 12th ICMI Study Conference, pp. 48-52). Melbourne, Australia: The University of Melbourne.

Ball, L. (2002). So, what do I write down? I've solved the whole problem using CAS. In J. Böhm (Ed.), *VISIT-ME 2002* (Proceedings of the Vienna International Symposium on Integrating Technology in Mathematics Education) [On CD]. Available: bk teachware series: Support in Learning (No.SR-31, ISBN 3-901769-49-8)

Ball, L. (2003). Communication of mathematical thinking in examinations: A comparison of CAS and non-CAS student written responses. *2003 Symposium on Computer Algebra in Mathematics Education* [On-line]. Available: http://ltsn.mathstore.ac.uk/came

Ball, L., & Stacey, K. (2003). What should students record when solving problems with CAS? Reasons, information, the plan and some answers. In J. T. Fey, A. Cuoco, C. Kieran, L. Mullin, & R. M. Zbiek (Eds.), *Computer Algebra Systems in Secondary School Mathematics Education* (pp. 289-303). Reston, VA: NCTM.

Bergsten, C. (1996). Learning by computing? The case of calculus and a CAS. In L. Lum (Ed.), *Proceedings of the 7th annual international conference on Technology in Collegiate Mathematics* (pp. 36-39). Reading, MA: Addison Wesley.

Berry, J., Johnson, P., Maull, W. & Monaghan, J. (1999). Routine questions and examination performance. In O. Zaslavsky (Ed.), *Proceedings of the 23rd annual conference of the International Group for the Psychology of Mathematics Education* (Vol. 2, pp. 105-112). Haifa, Israel: Program Committee.

Björk, L.-E. & Brolin, H. (1998). Which traditional algebra and calculus skills are still important? In Z. Usiskin (Ed.), *Developments in school mathematics education around the world* (Proceedings of the Fourth UCSMP International Conference on Mathematics Education, Chicago (Vol. 4, pp 287-295). Reston, VA: NCTM.

Boers-van Oosterum, M. A. M. (1990). Understanding of variables and their uses acquired by students in traditional and computer-intensive algebra (Doctoral dissertation, University of Maryland College Park, 1990). *Dissertation Abstracts International, 51*, 1538A.

Bolam, R. (1982). *School-focused In-Service Training*. Heinemann.

Brousseau, G. (1986). Fondements et méthodes de la didactique des mathématiques [Foundations and methods of mathematical didactics]. *Recherches en Didactique des Mathematiques, 7*, 33-115.

Brown, R., & Nielsen, B. (2001). What algebra is required in "high stakes" system wide assessment? A comparison of three systems. In H. Chick, K. Stacey, J. Vincent, & J. Vincent (Eds.), *The future of the teaching and learning of algebra* (Proceedings of the 12th ICMI Study Conference, pp. 128-135). Melbourne, Australia: The University of Melbourne.

Cnop, I. (2001). New insight in mathematics by live CAS documents. In H. Chick, K. Stacey, J. Vincent, & J. Vincent (Eds.), *The future of the teaching and learning of algebra* (Proceedings of the 12th ICMI Study Conference, pp. 187-191). Melbourne, Australia: The University of Melbourne.

Crowley, L. (2000). *Cognitive Structures in College Algebra.* Unpublished doctoral dissertation, University of Warwick.

De Alwis, T. (2000). Computer algebra systems as visualisation and experimentation tools in mathematics education. In M. O. J. Thomas (Ed.), *Other Conference Papers of TIME 2000 an International Conference on the Technology in Mathematics Education* (pp. A36-A44). Auckland, NZ: The University of Auckland & AUT.

Delos Santos, A. G., & Thomas, M. O. J. (2002). Teaching derivative with graphic calculators: The role of a representational perspective. In W-C. Yang, S-C. Chu, T. de Alwis, & F. M. Bhatti (Eds.), *Proceedings of the 7th Asian Technology Conference in Mathematics 2002* (pp. 349-358). Melaka, Malaysia: Program Committee.

Doerr, H. M., & Zangor, R. (1999). Creating a tool: An analysis of the role of the graphing calculator in a pre-calculus classroom. In O. Zaslavsky (Ed.), *Proceedings of the 23rd annual conference of the International Group for the Psychology of Mathematics Education* (Vol. 2, pp. 265-272). Haifa, Israel: Program Committee.

Doerr, H. M., & Zangor, R. (2000). Creating meaning for and with the graphing calculator. *Educational Studies in Mathematics, 41*, 143-163.

Drijvers, P. (2000). Students encountering obstacles using a CAS, *International Journal of Computers for Mathematical Learning, 5*, 189-209.

Drijvers, P. (2001). The concept of parameter in a computer algebra environment. In H. Chick, K. Stacey, J. Vincent, & J. Vincent (Eds.), *The future of the teaching and learning of algebra* (Proceedings of the 12th ICMI Study Conference, pp. 221-227). Melbourne, Australia: The University of Melbourne.

Drijvers, P. (2002). Learning mathematics in a computer algebra environment: Obstacles are opportunities. *Zentralblatt für Didaktik der Mathematik, 34*(5), 221-228.

Drijvers, P. (2003). *Learning algebra in a computer algebra environment.* Utrecht, The Netherlands: CD-ß Press

Drijvers, P., & van Herwaarden, O. (2000). Instrumentation of ICT-tools: the case of algebra in a computer algebra environment. *The International Journal of Computer Algebra in Mathematics Education, 7*(4), 255-275.

Driver, D. (2001). The trial of algebraic calculators in senior mathematics by distance education. In H. Chick, K. Stacey, J. Vincent, & J. Vincent (Eds.), *The future of the teaching and learning of algebra* (Proceedings of the 12th ICMI Study Conference, pp. 228-236). Melbourne, Australia: The University of Melbourne.

Edwards, M. T. (2001). *The electronic "other": A study of calculator-based symbolic manipulation utilities with secondary school mathematics students.* Unpublished doctoral dissertation, The Ohio State University, Columbus, Ohio.

Flynn, P. (2002). Adapting "problems to prove" for CAS-permitted examinations'. In J. Böhm (Ed.), *VISIT-ME 2002* (Proceedings of the Vienna International Symposium on Integrating Technology in Mathematics Education) [On CD]. Available: bk teachware series: Support in Learning (No.SR-31, ISBN 3-901769-49-8)

Flynn, P. (2003). Using assessment principles to evaluate CAS-permitted examinations. *2003 Symposium on Computer Algebra in Mathematics Education* [On-line]. Available: http://ltsn.mathstore.ac.uk/came

Flynn, P., Berenson, L., & Stacey, K. (2002). Pushing the pen or pushing the button: A catalyst for debate over future goals for mathematical proficiency in the CAS-age. *Australian Senior Mathematics Journal, 16*(2), 7-19.

Goos, M., Galbraith, P., Renshaw, P. & Geiger, V. (2000). Reshaping teacher and students roles in technology-enriched classrooms, *Mathematics Education Research Journal, 12*(3), 303-320.

Graham, A., & Thomas, M. O. J. (2000). Building a versatile understanding of algebraic variables with a graphic calculator. *Educational Studies in Mathematics, 41*(3), 265-282.

Guin, D., & Trouche, L. (1999). The complex process of converting tools into mathematical instruments: The case of calculators, *International Journal of Computers for Mathematical Learning, 3*, 195-227.

Guin, D., & Trouche, L. (2000). Thinking of new devices to make viable symbolic calculators in the classroom. In T. Nakahara & M. Koyama (*Proceedings of the 24th annual conference of the International Group for the Psychology of Mathematics Education* (Vol. 3, 9-16). Hiroshima, Japan: Program Committee.

Hart, D. K. (1992). Building concept images: Supercalculators and students use of multiple representations in calculus (Doctoral dissertation, Oregon State University, 1991). *Dissertation Abstracts International, 52*, 4254A.

Heid, M. K. (1988). Resequencing skills and concepts in applied calculus using the computer as a tool. *Journal for Research in Mathematics Education, 19*, 3-25.

Heid, M. K. (1992). *Final report: Computer-intensive curriculum for secondary school algebra.* Final report for NSF project number MDR 8751499. University Park, PA: The Pennsylvania State University, Department of Curriculum and Instruction.

Heid, M. K. (1995). The interplay of mathematical understandings, facility with a computer algebra program, and the learning of mathematics. *Proceedings of the 17th annual meeting of the North American Chapter of PME* (pp. 221-225). Columbus, OH: Program Committee.

Heid, M. K. (2002a). Computer algebra systems in secondary mathematics classrooms: The time to act is now! *Mathematics Teacher, 95*(9), 662-667.

Heid, M. K. (2002b). How theories about the learning and knowing of mathematics can inform the use of CAS in school, mathematics: One perspective. *International Journal of Computer Algebra in Mathematics Education, 9*(2), 95-108.

Heid, K., Choate, J., Sheets, J., & Zbiek R.M. (Eds.) (1995). *Algebra in a technological world.* Reston, VA: NCTM.

Heid, M. K., & Edwards, M. T. (2001). Computer algebra system: Revolution or retrofit for today's mathematics classroom? *Theory into Practice, 40*(2), 128-141.

Herget, W., Heugl, H., Kutzler, B., & Lehmann, E. (2001). Indispensable manual calculation skills in a CAS environment, In J. Engelbrecht (Ed.), *Communications of the Third Southern Hemisphere Symposium. on Undergraduate Mathematics Teaching* (pp. 56-63). Kruger National Park, South Africa: Program Committee.

Hong, Y. Y, & Thomas, M. O. J. (2001a). Representational fluency and the Newton–Raphson method. *Proceedings of the International Conference of Computers in Education* (pp. 591-599). Seoul, Korea: Program Committee.

Hong, Y. Y., & Thomas, M. O. J. (2001b). Conceptual understanding of the Newton–Raphson method. *Science and Mathematics Education Papers (SAME papers)* (pp. 91-112). Waikato University, NZ: CSTER, University of Waikato.

Hong, Y. Y., Thomas, M. O. J., & Kiernan, C. (2000). Super-calculators and university entrance calculus examinations, In M. Chinnappan & M. O. J. Thomas (Eds.), *Mathematics Education Research Journal. Special Issue on Technology in Mathematics Learning and Teaching, 12*(3), 321-336.

Hong, Y. Y., Thomas, M. O. J., & Kwon, O. (2000). Understanding linear algebraic equations via super-calculator representations. In T. Nakahara & M. Koyama (*Proceedings of the 24th conference of the International Group for the Psychology of Mathematics Education* (Vol. 3, 57-64). Hiroshima, Japan: Program Committee.

Hunter, M., Marshall, P., Monaghan, J., Roper, T., & Wain, G. (1993). Using computer algebra systems with younger students, In J. Monaghan & T. Etchells (Eds.), *Computer algebra systems in the classroom* (pp. 68-82). Leeds, UK: The University of Leeds.

Kaput, J. J. (2000). Implications of the shift from isolated, expensive technology to connected, inexpensive, diverse and ubiquitous technologies. In M. O. J. Thomas (Ed.), *Proceedings of TIME 2000 an International Conference on the Technology in Mathematics Education* (pp. 1-24). Auckland, NZ: The University of Auckland & AUT.

Keller, B. A. (1994). Symbol sense and its development in two CAS environments. In L. Lum (Ed.), *Proceedings of the 6th annual conference on Technology in Collegiate Mathematics*, Reading, MA: Addison-Wesley.

Keller, B. A. & Russell, C. A. (1997). Effects of the TI-92 on calculus students solving symbolic problems. *The International Journal of Computer Algebra in Mathematics Education, 4*(1), 77-97.

Keller, B. A., Russell, C. A., & Thompson, H. (1999). A large-scale study clarifying the roles of the TI-92 and instructional format on student success in calculus. *The International Journal of Computer Algebra in Mathematics Education, 6*(3), 191-207.

Kendal, M. (2002). *Teaching and learning introductory differential calculus with a computer algebra system.* Unpublished doctoral thesis, The University of Melbourne, Australia. Available: http://thesis.lib.unimelb.edu.au/

Kendal, M., & Stacey, K. (1999). Varieties of teacher privileging for teaching calculus with computer algebra systems, *The International Journal of Computer Algebra in Mathematics Education, 6*(4), 233-247.

Kendal, M., & Stacey, K. (2000). Acquiring concepts of derivative: Teaching and learning with multiple representations and CAS. In T. Nakahara & M. Koyama (*Proceedings of the 24th annual conference of the International Group for the Psychology of Mathematics Education* (Vol. 3, 127-134). Hiroshima, Japan: Program Committee.

Kendal, M., & Stacey, K. (2002). Teachers in transition: Moving towards CAS-supported classrooms. *Zentralblatt für Didaktik der Mathematik, 34*(5), 196-203.

Kendal, M., & Stacey, K. (2003). Tracing learning of three representations with the differentiation competency framework, *Mathematics Education Research Journal, 15*(1), 22-41.

Kidron, I. (2001). Teaching Euler's algebraic methods in a calculus laboratory. In H. Chick, K. Stacey, J. Vincent, & J. Vincent (Eds.), *The future of the teaching and learning of algebra* (Proceedings of the 12th ICMI Study Conference, pp. 368-375). Melbourne, Australia: The University of Melbourne.

Kieran, C. (1992). The Learning and Teaching of School Algebra, In D.A. Grouws (Ed.), *Handbook of research on mathematics teaching and learning* (pp. 390-419). New York: Macmillan.

King, K., Hillel, J., & Artigue, M. (2001). Computer algebra systems in the learning and teaching of linear algebra: Some examples. In D. Holton (Eds.), *The teaching and learning of mathematics at university level: An ICMI Study* (pp. 349–356). Dordrecht, The Netherlands: Kluwer Academic.

Kissane, B. (1995). Technology in secondary school mathematics—the graphics calculator as personal mathematical assistant. In R. Hunting, G. Fitzsimmons, P. Clarkson, & A. Bishop (Eds.), *International Commission on Mathematics Instruction Conference on Regional Collaboration* (pp. 383-392). Melbourne, Australia: Monash University.

Kissane, B. (1999). The algebraic calculator and mathematics education. In W-C. Yang, D. Wang, S-C. Chu, & G. Fitz-Gerald, (Eds.), Proceedings of 4th Asian Technology Conference on Mathematics (pp. 123-132). Guangzhou, China: Program Committee.

Kissane, B. (2001a). The algebra curriculum and personal technology: Exploring the links. In A. Rogerson (Ed.), *Proceedings of the International Conference: New Ideas in Mathematics Education* (pp. 127-132). Palm Cove, Queensland, Australia: Program Committee.

Kissane, B. (2001b). Algebra and technology: Emerging issues. In B. Lee (Ed.), *Mathematics: Shaping Australia*. (Electronic Proceedings of the biennial conference of the Australian Association of Mathematics Teachers Canberra, January 2001. http://wwwstaff.murdoch.edu.au/~kissane/papers/AAMT2001.pdf

Kutzler, B. (1994). *Derive*—The future of teaching mathematics. *International Derive Journal, 1*(1), 37-48.

Lachambre, B., & Abboud-Blanchard, M. (1996). Training of mathematics teachers. *International DERIVE Journal, 3*(3), 109-124.

Lagrange, J.-b. (1999a). Complex calculators in the classroom: Theoretical and practical reflections on teaching pre-calculus. *International Journal of Computers for Mathematical Learning, 4*(1), 51-81.

Lagrange, J.-b. (1999b). Learning pre-calculus with complex calculators: Mediation and instrumental genesis. In O. Zaslavsky (Ed.), *Proceedings of the 23rd conference of the International Group for the Psychology of Mathematics Education* (Vol. 3, pp. 193-200). Haifa, Israel: Program Committee.

Lagrange, J.-b. (2000). L'Intégration d'instruments informatiques dans l'enseignement: Une approche par les techniques [The integration of calculators into teaching: An approach using techniques]. *Educational Studies in Mathematics, 43*(1), 1-30.

Lagrange, J.-b. (2001). A multi-dimensional study of the use of IC technologies: The case of computer algebra systems. In J. Novotná (Ed.), *Proceedings of the second conference of the European Society for Research in Mathematics Education* (pp. 170-182). Mariánské Lázne, Czech Republic: Charles University.

Leigh-Lancaster, D., & Stephens, M. (2001). Responding to the impact of increasing accessibility of CAS on national and other school systems at the senior secondary level. In H. Chick, K. Stacey, J. Vincent, & J. Vincent (Eds.), *The future of the teaching and learning of algebra* (Proceedings of the 12th ICMI Study Conference, pp. 400-404). Melbourne, Australia: The University of Melbourne.

Leigh-Lancaster, D. (2003). The Victorian Curriculum and Assessment Authority mathematical methods computer algebra pilot study and examinations. *2003 Symposium on Computer Algebra in Mathematics Education.* http://ltsn.mathstore. ac.uk/came

Leinbach, C. (2001). Using a CAS in the mathematics classroom: Two examples for exploring cubic curves. *The International Journal of Computer Algebra in Mathematics Education, 8*(2), 131-147.

Lesh, R. (2000). What mathematical abilities are most needed for success beyond school in a technology based age of information? In M. O. J. Thomas (Ed.), *Proceedings of TIME 2000 an International Conference on the Technology in Mathematics Education* (pp. 73-83). Auckland, NZ: The University of Auckland & AUT.

Lumb, S., Monaghan, J., & Mulligan, S. (2000). Issues arising when teachers make extensive use of computer algebra. *International Journal of Computer Algebra in Mathematics Education, 7*(4), 223-240.

Mayes, R. (1997). Current state of research into CAS in mathematics education. In J. Berry, M. Kronfellner, J. Monaghan, & B. Kutzler (Eds.), *The state of computer algebra in mathematics education* (pp. 171-189). Lund, Sweden: Chartwell-Bratt.

Meel, D. E. (1998). Honors students' calculus understandings: Comparing calculus and Mathematica and traditional calculus students. *CBMS Issues in Mathematics Education, 7*, 163-215.

Monaghan, J. (1997). Teaching and learning in a computer algebra environment. *The International Journal of Computer Algebra in Mathematics Education, 4*(3), 207-220.

Monaghan, J. (2000). Some issues surrounding the use of algebraic calculators in traditional examinations. *International Journal of Mathematical Education in Science and Technology, 31*(3), 381-392.

Monaghan, J., Sun, S. & Tall, D. O. (1994). Construction of the limit concept with a computer algebra system. In J. P. da Ponte & J. F. Matos (Eds.), *Proceedings of the 18th annual conference of the International Group for the Psychology of Mathematics Education* (Vol. 3, pp. 279-286). Lisbon, Portugal: Program Committee.

Nathan, M., & Koedinger, K. (2000). Teachers' and researchers' beliefs about the development of algebraic reasoning. *Journal for Research in Mathematics Education, 31*(2), 168-190.

O'Callaghan, B. R. (1998). Computer-intensive algebra and students' conceptual knowledge of functions. *Journal for Research in Mathematics Education, 29*(1), 21-40.

Palmiter, J. R. (1991). Effects of computer algebra systems on concept and skill acquisition in calculus. *Journal for Research in Mathematics Education, 22*(2), 151-156.

Pierce, R. (2002). *An exploration of algebraic insight and effective use of computer algebra systems*. Unpublished doctoral thesis, The University of Melbourne, Australia.

Pierce, R., & Stacey, K. (2001a). Observations on students' responses to learning in a CAS environment. *Mathematics Education Research Journal, 13*(1), 28-46.

Pierce, R., & Stacey, K. (2001b). A framework for algebraic insight. In J. Bobis, B. Perry, & M. Mitchelmore (Eds.), *Proceedings of the 24th conference of the Mathematics Education Research Group of Australasia* (Vol. 2, pp. 418-425). Sydney, Australia: MERGA.

Pierce R., & Stacey, K. (2002). Monitoring effective use of computer algebra systems. In B. Barton, K. Irwin, M. Pfannkuch, & M. O. J. Thomas (Eds.), *Mathematics Education in the South Pacific* (Proceedings of the 25th annual conference of the Mathematics Education Research Groups of Australasia, Auckland, Vol 2, pp 575-582.). Sydney: MERGA.

Rabardel, P. (1995). *Les hommes et les technologies, approche cognitive des instruments contemporains* [People and technology, a cognitive approach using contemporary instruments]. Paris: Armand Colin.

Recio, T. (1997). Didactical relevance of meaningless mathematics. *The International Journal of Computer Algebra in Mathematics Education, 5(1*), 15-27.

Repo, S. (1994). Understanding and reflexive abstraction: Learning the concept of derivative in the computer environment. *International DERIVE Journal, 1*(1), 97-113.

Slavit, D. (1996). Graphing calculators in a "hybrid" algebra II classroom. *For the Learning of Mathematics, 16*, 9-14.

Stacey, K., Asp, G. & McCrae B. (2000). Goals for CAS-active senior mathematics curriculum. In M. O. J. Thomas (Ed.), *Proceedings of TIME 2000 an International Conference on the Technology in Mathematics Education* (pp. 246-254). Auckland, NZ: The University of Auckland & AUT.

Stacey, K., Kendal, M., & Pierce, R. (2002). Teaching CAS in a time of transition. *The International Journal of Computer Algebra in Mathematics Education, 9*(2), 113-124.

Tall D. O. (2000). Technology and versatile thinking in mathematics. In M. O. J. Thomas (Ed.), *Proceedings of TIME 2000 an International Conference on the Technology in Mathematics Education* (pp. 33-50). Auckland, NZ: The University of Auckland & AUT.

Thomas, M. O. J. (2001). Building a conceptual algebra curriculum: The role of technological tools. In H. Chick, K. Stacey, J. Vincent, & J. Vincent (Eds.), *The future of the teaching and learning of algebra* (Proceedings of the 12th ICMI Study Conference, pp. 582-589). Melbourne, Australia: The University of Melbourne.

Trouche, L., & les 37 élèves d'une classe de Terminale S. (1998). *Expérimenter et prouver: Faire des mathématiques au lycée avec des calculatrices* [Experiment and proof: Doing mathematics in school with calculators]. IREM, France: Université Montpellier II.

Trouche. L. (2000). La parabole du gaucher et de la casserole à bec verseur: ètude des processus d'apprentissage dans un environnement de calculatrices symboliques [The left-handed parabola and the saucepan lip: A study of a learning module using a symbolic calculator environment]. *Educational Studies in Mathematics, 41*(3), 239-264.

Trouche, L. (2003). *Managing the complexity of human/machine interaction in a computer based learning environment: Guiding student's process command through instrumental orchestrations* [On line]. http://www.mathstore.ac.uk/came/events/reims/index. html.

Trouche, L., & Guin, D. (1996). Seeing is reality: How graphic calculators may influence the conceptualisation of limits. In F. Biddulph & K. Carr (Eds.), *Proceedings of the 20th annual conference of the International Group for the Psychology of Mathematics Education* (Vol. 4, pp. 323-333). Rotorua, NZ: Program Committee.

Vérillon, P., & Rabardel, P. (1995). Cognition and artifacts: A contribution to the study of thought in relation to instrumented activity. *European Journal of Psychology of Education, 10*(1), 77-101.

Waits, F., & Demana, F. (1998). *The role of hand-held computer symbolic algebra in mathematics education in the twenty-first century: A call for action*! A paper prepared for the introduction of NCTM Standards 2000 [On line]. Available: http://mathforum.org/technology/papers/waits/waits.html

Weigand, H-G., & Weller, H. (2001). Changes of working styles in a computer algebra environment - The case of functions. *International Journal of Computers for Mathematical Learning, 6*, 87-111.

Wilson, M. R., & Krapfl, C. M. (1994). The impact of graphics calculators on students' understanding of function. *Journal of Computers in Mathematics and Science Teaching, 13*, 252-265.

Wood, D.J., Bruner, J.S. & Ross, G. (1976). The role of tutoring in problem solving. *Journal of Psychology and Psychiatry, 17*, 89-100.

Wurnig, O., & Townend, S. (1997). Coursework, portfolios and learning with understanding. In J. Berry, M. Kronfellner, J. Monaghan, & B. Kutzler (Eds.), *The state of computer algebra in mathematics education* (pp. 76-83). Lund, Sweden: Chartwell-Bratt.

Zbiek, R. M. (1995). Her math, their math: an in-service teacher's growing understanding of mathematics and technology and her secondary students' algebra experience. *Proceedings of the 17th annual meeting of the North American Chapter of PME* (pp. 214-220). Columbus, OH: Program Committee.

Zehavi, N. (1996). New teaching practices using a CAS. In M. C. Borba, T. A. Souza, B. Hudson, & J. Fey (Eds.), The role of technology in the mathematics classroom. *Proceedings of Working Group 16, ICME-8*, Seville, Spain. UNESP, Sao Paulo, Brazil: Program Committee.

Zehavi, N., & Mann, G. (1999). Teaching mathematical modeling with a computer algebra system. In O. Zaslavsky (Ed.), *Proceedings of the 23rd annual conference of the International Group for the Psychology of Mathematics Education* (Vol. 4, pp. 345-352). Haifa, Israel: Program Committee.

Zehavi, N., Wassertel, L., & Mann, G. (2001). From a word problem to a family of word problems. In H. Chick, K. Stacey, J. Vincent, & J. Vincent (Eds.), *The future of the teaching and learning of algebra* (Proceedings of the 12th ICMI Study Conference, pp. 690-696). Melbourne, Australia: The University of Melbourne.

The Working Group on Algebra History in Mathematics Education

Leaders: Teresa Rojano and Luis Puig

Working Group Members:
Stephen Campbell, Aurora Gallardo, and Israel Kleiner.

The Working Group on Algebra History in Mathematics Education (L to R): Aurora Gallardo, Luis Puig, Israel Kleiner, Teresa Rojano, Stephen Campbell.

In the Discussion Document, written for the 12th ICMI Study Conference Proceedings, Teresa Rojano and Luis Puig identified five key components of algebra history. During the conference the members of the Working Group on Algebra History in Mathematics Education discussed these components in detail and identified a sixth component. These six key components are: the history of symbolism (i.e., the history of ways of representing quantities and operations in calculations), the history of methods for solving problems, the history of methods for solving equations, the history of the interactions of algebra with other mathematical domains (such as geometry), the development of the ideas of algebraic structures, and the history of the concept of number (i.e., the development of the algebraic concept of number).

The discussions of the Working Group on Algebra History were informal due to the small number of participants. Initially, discussion was stimulated by the Discussion Document and two additional manuscripts written by Luis Puig and Teresa Rojano.

- Luis Puig: *Characteristics of the algebraic and components of the history of algebra: A statement to start the discussion.*
- Teresa Rojano: *The case of pre-symbolic algebra and the operation of the unknown.*

The papers that the members wrote for the Conference Proceedings informed later discussions. These papers reflect their personal interests and expertise in researching the history of algebra. The authors and the titles of these papers are:

- Stephen Campbell: *Number theory and the transition from arithmetic to algebra: Connecting history and psychology* (pp. 147-154).
- Aurora Gallardo: *George Peacock and a historical approach to school algebra* (pp. 273-280).
- Israel Kleiner: *A historically focused course in abstract algebra* (725-731).

The members of the Working Group also shared interesting references about algebra history that enriched the discussion about the various key components and uses of history in mathematics education (e.g., research, preparation of curriculum material, and teachers' training). Two references were particularly useful, Fauvel and van Maanen (2000) and Katz (2000).

Each member of the Working Group undertook the responsibility to prepare and make an oral presentation that was delivered at the end of the conference. During the conference, they discussed each others ideas and their presentations related to particular aspects of the history of algebra and their uses. The members may be contacted from their e-mail addresses listed at the back of the book. The titles of their briefs are listed:

- Stephen Campbell: *The concept of number and intersection with other mathematical domains and the use of history in teaching.*
- Aurora Gallardo: *The concept of number and the use of history in research.*
- Israel Kleiner: *Emergence of the idea of algebraic structures and intersection of mathematical domains and the use of history in teaching.*
- Luis Puig: *Methods of solving problems and symbolism and the use of history in teacher training and in research.*
- Teresa Rojano: *Methods of solving equations and symbolism and the use of history in research.*

The extensive and thoughtful work produced by the members of the Working Group, prior to and during the Conference, is gratefully acknowledged. Their contributions to the Conference Proceedings, Working Group discussions, and the briefs they prepared during the conference provided important insights that assisted the authors during the writing process. Thanks are extended to Teresa Rojano and Luis Puig for their leadership of the Working Group on Algebra History in Mathematics Education and for their authorship of this chapter.

Chapter 8

The History of Algebra in Mathematics Education

Luis Puig and Teresa Rojano
Universidad de Valencia, Spain, and Centro de Investigación y Estudios Avanzados del Instituto Politécnico Nacional, Mexico

Abstract: In this chapter, we analyse key issues in algebra history from which some lessons can be extracted for the future of the teaching and learning of algebra. A comparative analysis of two types of pre-Vietan languages (before 16th century), and of the corresponding methods to solve problems, leads to conjecture the presence of didactic obstacles of an epistemological origin in the transition from arithmetic to algebraic thinking. This illustrates the value of historic and critical analysis for basic research design in mathematics education. Analysing the interrelationship between different evolution stages of the sign system of symbolic algebra and vernacular language supports the inference that a particular sign system represents a significant step in the evolution of algebra symbolism when it permits calculations at a syntactic level. Such analyses provide elements to identify features of *the algebraic* in the translation processes from natural language to the algebraic code. In particular, these elements can be used as a basis to study pupils' strategies when they solve word problems, and to conceive didactical routes for the teaching of solving methods of these problems. The examples discussed emphasise the importance of speaking of manifestations of *the algebraic* in the specific, in contrast to other perspectives that emphasise the nature of algebraic thinking in the general.

Key words: History of algebra, early algebraic language, didactic obstacles, the algebraic, Cartesian method, language stratum, mathematical sign system, symbolic algebra, word problems

8.1 Introduction

Two or three decades ago it was common to find articles that discussed the necessity or usefulness of knowing about the history of mathematics in order to teach or learn it or to research its teaching and learning. Hans Freudenthal, for example, called his famous article *Should a mathematics teacher know something about the history of mathematics?* (Freudenthal, 1981), but since then reasons for supporting the importance of using history in mathematics education have been extensively provided from a variety of perspectives. ICMI established a group more than 20 years ago to study the relationship between the history and pedagogy of mathematics, and during the last few years important publications have appeared that summarise the work done outside and inside this group, especially the ICMI Study *History in Mathematics Education* (Fauvel & van Maanen, 2000), but also Katz (2000), and Jahnke, Knoche, and Otte (1996). These books save us from the need to review recent work in this chapter, because an extensive account of everything is included in them. The content of this chapter will therefore focus on examining some key issues to study in the history of algebraic ideas to be used in mathematics education, rather than on why to do it or on establishing what has been done.

The idea is to look at the future of the teaching and learning of algebra in terms of the lessons that can be extracted from a historical perspective. In turn, our current knowledge of the difficulties that teachers and students face when learning (and teaching) algebra should tell us what aspects of the history of algebra are worth studying in depth. For instance, the present debate on the teaching of the manipulative aspects of algebra has led us to report on the history of algebraic symbolism in some depth (see Section 8.3) in order to find arguments to support decisions concerning the curriculum of algebra. One of the arguments that might be founded on a historical analysis is that the conceptual development of some algebraic notions is strongly related to sources of meaning arising from the syntactic manipulation of symbols.

The history of symbolism in algebra can be regarded as the history of the development of a system of signs that makes it possible to carry out calculations at a syntactic level to find the solution of a word problem without having to refer to the semantic level of the problem statement. In this sense, the evolution of algebraic symbolism is strongly related to the history of algebraic methods for solving problems. Relevant interrelations between these two components of the history of algebra will be discussed (see Section 8.2), and also, especially, the interrelation of the characteristics of a particular sign system with concepts and methods (Section 8.4), and the use of these analyses in research (Section 8.5).

From a historical point of view, functions are not formally a part of algebra. Nevertheless, they are a part of the teaching of algebra in many countries (see Chapter 13 of this book). Furthermore, a promising approach to the teaching of algebra is the functional one. In connection with this functional approach to the teaching of algebra it is important, therefore, to study the relationship in history between algebraic ideas and the idea of variation, the mathematics of change, variables (that vary), and functions. However, this is a task that we shall not tackle here.

8.2 Algebraic Problem Solving

8.2.1 The Cartesian Method as a paradigm of algebraic problem solving

What lies at the heart of algebraic problem solving is the expression of problems in the language of algebra by means of equations. In order to be able to compare the ways of writing equations that represent word problems in different historical texts so that the comparison brings out what is pertinent for teaching, a good strategy is to take as a reference what is done in the Cartesian Method, which is the algebraic method *par excellence* and may be considered as the canon of the methods traditionally taught in school systems.

Stacey and MacGregor (2001) have pointed out that a major reason for the difficulty that students have in using algebraic methods for solving problems is not understanding its basic logic—that is, the logic that underlies the Cartesian Method. There is a students' compulsion to calculate, based on their prior experiences with arithmetic problem solving. This tendency to operate backwards rather than forwards (Kieran, 1992) prevents students from finding sense in the actions of analysing the statement of the problem and translating it into equations which express, in algebraic language, the relations among quantities; actions of analysing and translating that are the main features of the Cartesian Method. Teaching models that take into account these tendencies are presented in Filloy, Rojano, and Rubio (2001). They state that in order to give sense to the Cartesian Method users should recognise the algebraic expressions, used in the solution of the problem, as expressions that involve unknowns. Competent use of expressions with unknowns is achieved when it makes sense to perform operations between the unknown and the data of the problem. In steps prior to competent use of the Cartesian Method, the pragmatics of the more concrete sign systems leads to using the letters as variables, passing through a stage in which the letters are only used as names and representations of generalised numbers, and a subsequent stage in which they are only used for representing what is unknown in the problem. These last two stages, both clearly distinct, are predecessors of the use of letters as unknowns and using algebraic expressions as relations between magnitudes, in particular as functional relations.

Furthermore, competent use of the Cartesian Method is linked with the creation of families of problems that are represented in the mathematical sign system (MSS) of algebra as canonical forms. This implies an evolution of the use of symbolisation in which, finally, the competent user can give meaning to a symbolic representation of the problem that arises from the particular concrete examples given in teaching. Students will make sense of the Cartesian Method when they become finally aware that by applying it they can solve families of problems, defined by the same scheme of solution. The integrated conception of the method needs the confidence of the user that the general application of its steps will necessarily lead to the solution of these families of problems.

In this section we will present an in-depth analysis of algebraic problem solving in history bearing in mind that this results from research on its teaching and learning. We will have to examine the characteristics of the Cartesian Method (Section 8.2.1), and the search for canonical forms that represent families of problems and its methods of resolution (Section 8.2.2). In the next section we deal with the history of symbolisation. First of all we will examine the formulation of the method proposed by Descartes (1596-1650, France). Indeed, the reason for calling this method *Cartesian* is that one part of Descartes' (1701-original posthumous publication, 1996) *Regulæ ad directionem ingenii (Rules for the direction of the mind)*[1] can be interpreted as an examination of the nature of the work of translating the verbal statement of an arithmetic-algebraic problem into the mathematical sign system (MSS) of algebra and its solution in that MSS.

Broken down into ideal steps, that is, those that would be taken by a competent user, the first step of the Cartesian Method consists in an analytical reading of the statement of the problem, which reduces it to a list of quantities and relations between quantities. The second step consists in choosing a quantity that will be represented by a letter (or several quantities that will be represented by different letters), and the third step consists in representing other quantities by means of algebraic expressions that describe the (arithmetic) relation that these quantities have with others that have already been represented by a letter or an algebraic expression. With the MSS of current school algebra this is done by maintaining the representation of each quantity by a different letter and taking care that each letter should represent a different quantity and combining the letters with signs for operations and with delimiters, while also observing certain rules of syntax that express the order in which the operations represented in the expression are performed. Descartes (1701) indicates that one makes an abstraction of the fact that some terms are known and others unknown. Treating known and unknown in the same way is precisely one of the fundamental features of the method's algebraic character, and Descartes himself pointed out that the basic nature of his method consisted in this [totum huius loci artificium consistet in eo, quod ignota pro cognitis supponendo possimus facilem & directam quærendi viam nobis proponere, etiam in difficultatibus quantumcumque intricatis (Descartes, 1701, pp. 61-62)]. The fourth step consists in establishing an equation (or as many equations as the number of different letters that it was decided to introduce in the second step) by equating two of the expressions written in the third step that represent the same quantity. In Descartes' rule XIX what gives meaning to the construction of the equation is the expression of a quantity in two different ways [Per hanc ratiocinandi methodum quarenda sunt tot magnitudines duobus modis differentibus expressa (Descartes, 1701, p. 66)].

This concludes the part of the method described in the *Regulæ* that corresponds to the translation of the statement of the problem into the MSS of algebra. The continuation of the method, which describes the solving of the equation, must be sought in the *Geometry* that Descartes published as an Appendix to the *Discourse on Method*, which is where he actually develops what he himself calls "his algebra".

In a letter to Mersenne (1588-1648, France) written in April 1637, which appears on pages 294-301 of the sixth volume of Cousin's edition (1826), Descartes says that he gives the rules of his algebra on page 372 of his *Geometry*. What Descartes begins to do on that page—to use the terminology of Freudenthal's phenomenology for a moment—is to take the equations themselves not as a means for organising phenomena but, in a movement of vertical mathematisation, as a field of objects subjected to phenomenological exploration which need a new means of organisation for that purpose. Starting from the idea that, if a is a root of an equation, $x - a$ divides into the corresponding polynomial, Descartes explores the number of roots of equations, the effect that replacing x with $y - a$ has on the roots, et cetera.

Cardano (1501-1576, Italy) had already studied the number of roots in some cases in the first chapter of his *Ars Magna or the Rules of Algebra* (see Cardano, 1963/1968, translated by T. Richard Witmer). The effect that changing one or more terms of one member of the equation has on the roots (which for Cardano meant changing to another canonical form) is presented in the seventh chapter. Viète (1540-1603, France) dedicated the book *De emendatione æquationum* to this issue but Descartes says that he begins his algebra precisely where Viète left it in that book[2].

In fact, in his *Geometry*, Descartes explains the method in a section called "Comment il faut venir aux Equations qui servent a resoudre les problemes", in which he also emphasises the similar treatment of known and unknown, and the writing of an equation based on the expression of a quantity in two different ways, "until we find it possible to express a single quantity in two ways. This will constitute an equation, since the terms of one of these two expressions are together equal to the terms of the other" (Descartes, 1925, p. 300). However, Descartes goes on from what can be found in the *Regulæ* (Descartes, 1701, 1996) with the development of the method, explaining that once all the equations have been constructed (either as many as there are letters, or fewer, in which case the problem is indeterminate), the equations must be transformed.

Here, Descartes does not expound the rules for the transformation of algebraic expressions. He assumes that they are known, but what he does say is the form that the canonical equation must have, indicating that the transformations must be done in such a way as eventually to obtain an equation:

> so as to obtain a value for each of the unknown lines; and so we must combine them until there remains a single unknown line which is equal to some known line, or whose square, cube, fourth power [lit. square of square], fifth power

[sursolide], sixth power [lit. square of cube], etc., is equal to the sum or difference of two or more quantities, one of which is known, while the others consist of mean proportionals between unity and this square, or cube, or fourth power [lit. square of square], etc., multiplied by other known lines. I may express this as follows:

$$z = b$$
$$\text{or} \quad z^2 = -az + bb,$$
$$\text{or} \quad z^3 = +az^2 + bbz - c^3,$$
$$\text{or} \quad z^4 = az^3 - c^3z + d^4, \text{ et cetera.}$$

That is, z, which I take for the unknown quantity, is equal to b; or, the square of z is equal to the square of b diminished by a multiplied by z ... (Descartes, 1925, p. 9; in square brackets we have added the names that Descartes uses for the species and that Smith does not retain in his translation).

Thus the method continues by transforming the written algebraic expressions and the resulting equations in order to reduce them to a canonical form. This implies that it has previously been determined which expressions and which equations will be considered canonical, and that one has a catalogue of all the possible canonical forms and procedures for solving each of them.

We have just shown the ones that Descartes presents specifically, but we could say that they all come down to a single canonical form, which Descartes presents broken down by degrees, for the form is the same in all cases. The breakdown by degrees is justified by the fact that the solving procedure is different for each degree (or else does not exist, depending on the degree). The form of the canonical equation, written in the most general form, is:

$$x^n = a_{n-1}x^{n-1} \pm a_{n-2}x^{n-2} \pm \ldots \pm a_2x^2 \pm a_1x \pm a_0$$

Thus Descartes makes the power of the highest degree without a coefficient (so that there is only one unknown and no known quantity on the left side of the equation) equal to the rest of the polynomial. As there are still unknown quantities (the other powers of the unknown) in the rest of the polynomial, he says that this is a known quantity (the monomial of degree zero) and quantities "consisting of certain proportional means between unity and this square or cube, etc.", an expression in which the idea that leads to the establishment of "degrees" is present, that is, the fact that $1 : x : x^2$, et cetera.

Thus the algebraic expressions that are considered canonical are the polynomials. This is so because the reiteration of the four elementary arithmetic operations, when these operations are performed on unknown quantities, leads to a situation in which all the multiplications (and divisions) produce a quantity ***multiplied by itself a certain number of times*** and multiplied by a specific number, that is, they produce a monomial, and the reiteration of additions (and subtractions), which can only be performed between monomials of the same degree (and this fact is crucial), produces an addition (and subtraction) of monomials.

Since the MSS of current school algebra and real numbers has been available, this has meant that the rules that make it possible to reduce any equation to a canonical form are the rules of literal calculus and transposition of terms and that there is only one canonical expression

$$a_nx^n + a_{n-1}x^{n-1} + a_{n-2}x^{n-2} + \ldots + a_2x^2 + a_1x + a_0$$

and one canonical equation

$$a_nx^n + a_{n-1}x^{n-1} + a_{n-2}x^{n-2} + \ldots + a_2x^2 + a_1x + a_0 = 0.$$

We have just seen that in Descartes' case the canonical forms are almost these. There are two differences. In the first place, Descartes does not establish as a canonical form a polynomial equal to zero, but the monomial of highest degree equal to the result of adding or subtracting the others. Thus Descartes' canonical forms are still linked to an arithmetical meaning of the equal sign and of equations, as they are a kind of formula which expresses a power of the unknown as the result of a series of arithmetical operations (even if these operations are performed also on the unknown). Making the polynomial equal to zero is something that Descartes does not do until 71 pages later, when he is discussing what he calls "his algebra", in which the equal sign acquires a full algebraic meaning, but this is done in a higher level of abstraction, when equations are taken as objects of mathematical organisation. Students' tendencies to rely on arithmetical meanings have to be dealt with through similar processes of abstraction.

In the second place, the current canonical form presents all the monomials joined by the plus sign, whereas in Descartes this is not the case. This is because the letters that represent coefficients, or unknown quantities, in Descartes' MSS always represent positive numbers, and the monomials are joined by the operations of addition and subtraction, which he conceives as two different operations. In the current canonical form, however, the only operation that appears is addition and the coefficients are any real numbers, because subtraction is no longer conceived as an operation with an entity of its own. This is the case even though Descartes admits the existence of negative roots ("false", in his terminology) and may write a monomial preceded by a minus sign although he is not subtracting from any other monomial, as in $z^2 = -az + bb$, in as much as the letters represent known numbers (lines) they can only be "true" numbers, that is, positive. It is symptomatic that when Descartes explains this equation by translating it into natural language, he changes the order so that it will make sense, and he writes: "le quarré de *z* est esgal au quarré de *b* moins *a* multiplié par *z*" [the square of *z* is equal to the square of *b* diminished by *a* multiplied by *z*]. Therefore the transition to the current canonical form requires an evolution of the concept of number.

Moreover, although the canonical forms in Descartes' text are written as polynomials, the monomials are still named by their species, with names that combine the basic species of square and cube, in a multiplicative form. However, Descartes breaks away from the geometrical connection of the names of the species by showing at the start of his *Geometry* that the product of one line and another line can be represented as a line and not as a surface, so that the species "square" and "cube" cease to be heterogeneous.

8.2.2 A history of problem and equation solving

We have interwoven this analysis of what is involved in the use of the Cartesian Method and the MSS of current school algebra with various observations taken from an examination of Descartes' writings and therefore from a study of the history of algebraic ideas. We have done so because we can say that the MSS of current algebra is already practically constituted in Descartes. The investigation of the history of algebraic ideas can now be carried out from the perspective given by this analysis.

Polynomials are, in fact, the conclusion in this history of all the forms that have been considered canonical at one time or another, but first it was necessary that the idea of the search for canonical forms should appear. For this idea to be able to appear it is necessary that problem solving should not be considered with the sole aim of obtaining the result of the specific problem in question, but that the solving process should include, to use Polya's terminology (Polya, 1887-1985, Hungary, Switzerland, USA), a fourth "looking back" phase with an epistemic character (Puig, 1996b), in which the solving procedure is analysed and problems are generated that can be solved with the same procedure or with variants or generalisations of that solving procedure. But it is also necessary that one should have an MSS in which the analysis of the solution can be made by relinquishing the specific numbers with which the calculations are performed. This requires that in some way one should be able to represent the numbers with which one calculates and the calculations that are performed with them as expressions.

The idea of searching for canonical forms then appears because of the need to reduce the number of expressions (equations) that are produced as a result of the translation of problems into equations that one already knows how to solve.

This idea of reducing to equations that one already knows how to solve leads to two projects as well as the project of identifying what will be called a canonical form: on the one hand, having a catalogue of the equations that one already knows how to solve, and, on the other, developing a calculus with expressions that enables one to transform any equation into one that can be solved.

This project takes a shape that for us is increasingly algebraic when the catalogue of expressions ceases to be constituted by accumulating solved problems, the corresponding expressions, and the techniques and procedures (or algorithms) for solving each of them, and ends up as a catalogue of all possible canonical forms[3].

However, the search for all possible canonical forms requires, on the one hand, the availability of an MSS in which expressions are represented precisely enough to make it possible to carry out a search for possibilities. This does not mean that the MSS has to be "symbolic" in the sense of the distinction between "rhetorical", "syncopated" and "symbolic" made by Nesselmann (1842), which we shall analyse in Section 8.3. This is testified by the fact that in the *Concise book of the calculation of al-jabr and al-muqâbala,* al-Khwârizmî (780-850, Khwarizm, now Uzbekistan) establishes such a catalogue of canonical forms in an MSS that consists only of natural language, in this case Arabic, and various geometrical figures, which are inserted in the text as representations (*sûra*, "figure", but also "representation" or even "photograph"), always preceded by the words "this is the representation" or "this is the figure". On the other hand, it modifies the project of constructing a catalogue of what one already knows how to solve and converts it into a project of knowing how to solve *all* the canonical forms.

This new project requires establishing sets of canonical forms that are complete in some sense. Thus, al-Khwârizmî establishes all the possibilities for what, for us, are linear and quadratic equations. His complete set[4] must contain all possible combinations of his three types of numbers: *mâl*, root and simple numbers[5]. The "types of numbers that appear in the calculations" of al-Khwârizmî correspond to Diophantus' *eidei* (a term used by Greek philosophers to mean species, type or form). However, Diophantus (mid 3rd Century, Greece) does not establish a complete set of normal forms or propose the possible combinations of *eidei*, nor, therefore, does he establish a calculation to reduce expressions to a normal form. The operations that Diophantus defines at the start of his *Arithmetic* (A.D. 250) and that are similar to *al-jabr* and *al-muqâbala* do not set out to reduce to a normal form, but simply to an equality of *eidei* (Klein, 1968, pp. 134-135). On the other hand, Diophantus' *eidei* cannot be identified with the powers of the unknown, but rather correspond to the Euclidean idea of something that is "given in form", one of the forms in which a geometric figure may have been given[6]. In fact, Diophantus defines the expressions *dynamis*, *cubos*, *dynamodynamis*, *dynamocubos*, et cetera, for specific numbers.

The continuation of the project is achieved by increasing the degree to the third, which also constitutes a naturally complete set of canonical forms. For this it is sufficient that the "types of numbers that are used in the calculations" should be conceived as Aristotelian magnitudes, as is done by ᶜUmar al-Khayyâm (1048-1131, Persia) in his *Treatise of algebra and al-muqâbala* (Rashed & Vahebzadeh, 1999).

The stumbling block in degrees greater than the fourth ultimately led to a modification of the project. Since an algorithm for a solution by means of radicals of the canonical forms after degree 4 could not be found, the question was transformed into a different question about whether the algorithm existed and it was specified in terms of the conditions of the existence of an algorithm. This was the nineteeth century work of Abel (1802-1829, Norway) and Galois (1811-1832, France), but with it the history of another algebra begins, abstract modern algebra, which can be summed up "in 100 words or less":

> Prior to the 19th century algebra meant essentially the study of polynomial equations. In the 20th century algebra became the study of abstract, axiomatic systems such as groups, rings, and fields. The transition from the so-called classical algebra of polynomial equations to the so-called modern algebra of axiom systems occurred in the 19th century. Modern algebra came into existence principally because mathematicians were unable to solve classical problems by classical (pre-19th century) means. They invented the concepts of group, ring, and field to help them solve such problems. (Kleiner, 1998, p. 105)

8.2.3 The algebraic

Mahoney (1971) and Høyrup (1994) discuss the characteristics of what we call "the algebraic" (i.e., referring to the abstract concept of being algebraic). The aspects of the history of algebraic ideas that we have examined in this section enable us to reformulate these characteristics of "the algebraic" as follows:

- The use of a system of signs to solve problems which allows us to express the content of the statement of the problem relevant to its solution (its "structure"), separated from what is not relevant to its solution. The history of symbolism in algebra can be regarded indeed as the history of the development of a system of signs that allows us to calculate on the syntactic level to find the solution of the problem without having to refer to the semantic level of the problem statement.
- The systematic search (usually combinatorial) for types of structures expressed by different expressions within this system of signs.
- The development of sets of rules to calculate on the syntactic level to reduce any expression to one of the types of structures.
- The search for rules (mainly algorithmic) to solve all types of structures.
- The absence of ontological commitment of the system of signs, that allows them to stand for any type of mathematical object.
- The analytical character of the use of the system of signs to reduce the statement of the problem to a canonical form.

These features of the algebraic make it possible to examine various interlinked components in the history of algebraic ideas:

1. The history of symbolism.
2. The history of algebraic problem solving.
3. The history of equation solving.
4. The history of the interactions of algebra with other mathematical domains.
5. The history of the emergence of the idea of algebraic structures.
6. The history of the concept of number.

In this section we have explored components (2) and (3). Campbell (2001) discusses (4), Kleiner (1998) discusses (5), and Gallardo (2001) and Campbell (2001) discuss (6). Furinghetti and Somaglia (2001) use the method of analysis, the history of which is part of component (2), as a common thread in the history of algebra to reflect on various critical aspects of the teaching and learning of algebra in the context of teacher training, in particular concerning "the symbolism and how to give meaning to its manipulation". In the following section we shall tackle some aspects of (1).

8.3 Algebraic Language: A History of Symbolisation

One perspective for analysing the history of algebra is the one that takes as a reference three stages in the evolution of its language: rhetorical, syncopated, and symbolic. This distinction was established by Nesselmann in the middle of the nineteenth century in his book *Die algebra der griechen* (Nesselmann, 1842), in which he says that the distinction derives from considering how the "formal representation of algebraic equations and operations" is realised (Nesselmann, 1842, p. 301). He applies the description of rhetorical to algebra in which the calculation is expressed completely, and in detail, by means of words of ordinary language. In this stage he places, for example, Al-Khwârizmî's algebra, in which problems and their solutions are expressed entirely in words. Syncopated algebra is algebra in which the exposition is also of a rhetorical nature, "but for certain frequently recurring concepts and operations it uses consistent abbreviations instead of complete words" (Nesselmann, 1842, p. 302). In this stage Nesselmann places "Diophantus and the later Europeans up to the middle of the seventeen century, although in his writings Viète had already sown the seed of modern algebra, which nevertheless only germinated some time after him" (Nesselmann, 1842, p. 302). The third stage is what Nesselmann calls 'symbolic algebra', in which all the possible forms and operations are represented in a sign system "independent of oral expression, which makes any rhetorical representation useless". From the time of this first characterisation of symbolic algebra by Nesselmann, therefore, the fundamental thing is not the mere fact of the existence of letters to represent quantities or of signs foreign to ordinary language to represent operations but the fact that one can operate with this sign system without having to resort to translating it into ordinary language. In Nesselmann's own words:

> We can perform an algebraic calculation from start to finish in a wholly understandable way without using a single written word, and, at least in comparatively simple calculations, we really only place a conjunction here and there between formulae so as to point directly to the connection between a particular formula and those that precede and follow it, in order to spare the reader the need to search and reread. (Nesselmann, 1842, p. 302)

If we accept Nesselmann's characterisation of the symbolic, the study of the history of this component of algebra will be guided by consideration of the extent to which, at a given moment or in a given text, the MSS makes it possible not only to represent the structure of the problem but also to calculate on the level of expressions without resorting to the level of content.

However, if we examine the *Concise book of the calculation of al-jabr and al-muqâbala* which al-Khwârizmî wrote to satisfy the wishes of the Caliph al-Ma'mûn (786-833, Bagdad, Iraq) to make known the technique of *al-jabr*, a technique of which the Caliph had heard but which had been lost, we can see that the representation of what is needed for the solution of the problems is done by means of two different tools.

One tool concerns the "types of numbers that appear in the calculations". These types of numbers are treasures (*mâl*, possession of money or treasure), roots (*jidr*) and simple numbers (*ᶜadad mufrad*), which are often a certain number of *dirhams* (the Arab currency). The conceptualisation is monetary, therefore, and the equations that al-Khwârizmî writes rhetorically in the Arabic language therefore have to do with a treasure, its root and a number of *dirhams*. However, these monetary expressions serve to represent any second-degree problem, whether it has to do with numbers, commercial transactions, geometric relations, or anything else.

The other tool concerns the unknown quantities. Al-Khwârizmî also uses another term, *shay'* (thing), when he has to translate the statement of a problem into an equation. He uses it to designate an unknown quantity, so that by having a name for the unknown quantity he can express arithmetic operations with the unknown rhetorically. The thing and the root have often been identified with our x, and the treasure with our x^2. Yet this identification is not present in al-Khwârizmî's text and only appears in later mediaeval Arab algebraists such as al-Karâji (about 970-1030, Persia) or ᶜUmar al-Khayyâm. Moreover, the two tools of representation are still differentiated in *Liber Abbaci*[7] by Leonardo Pisano (sometimes known as Fibonacci, about 1175-1240, Italy). In fact, Leonardo introduces the thing with the Latin term *res* in Chapter 12 of Boncompagni (1857, p. 191) when he defines the *Regula Recta* which needs a name for the unknown precisely so as to be able to calculate from it[8]. In contrast, al-Khwârizmî's algebra does not appear until 200 pages later, in Chapter 15, Boncompagni (1857, p. 406), where the names of the types of numbers are translated into Latin as *quadratus* (of which he later says that it is called *census*, retaining the monetary meaning of *mâl*[9], and this is the name that he uses until page 459, where the book ends), *radix* and *numerus simples*.

In al-Khwârizmî, *shay'* (thing) can represent one of the parts into which a number has been divided, for example in order to solve the following problem:

> I divided ten into two parts; then I multiplied each part by itself and added it gives fifty-eight dirhams. [What was each part?]

The solution of this problem begins "You make one of the parts thing and the other ten less thing" (Rosen, 1831, p. 28 of the text in Arabic[10]). In this case, when one continues with the construction of the equation, the thing is multiplied by itself and produces a treasure, and therefore the thing is identified with a root. But the thing can also represent a treasure that is mentioned in the statement of the problem. Such is the case in the next problem:

> Let there be a treasure, a third of which and three dirhams is taken away and then what remains is multiplied by itself and it gives the treasure. (Rosen, 1831, p. 40 of the text in Arabic) [Find the treasure.]

The unknown in this problem is the treasure (which is the result of multiplying something by itself), and in the course of the solution al-Khwârizmî identifies this treasure with the thing:

> Therefore multiply two thirds of thing, *that is, of the treasure*, less three dirhams by itself. (Rosen, 1831, p. 40 of the text in Arabic, our italics)

However when the calculations are done, as the thing is multiplied by itself it becomes a root, and the result of this multiplication becomes a treasure:

> Two thirds [of thing] multiplied by two thirds [of thing] gives four ninths of treasure and three subtractive dirhams by two thirds of thing gives two roots. Again, three subtractive dirhams by two thirds of thing gives two roots and minus three by minus three gives nine dirhams. Therefore they are four ninths of treasure and nine dirhams less four roots, equal to one root. (Rosen, 1831, pp. 40-41 of the text in Arabic)

As we see, thing, on the one hand, and treasure, root and simple numbers (*dirhams*), on the other, are not representing things of the same nature: thing serves to represent an unknown quantity so as to be able to calculate with it; treasure, root and simple numbers represent *types* or *species* of numbers.

In al-Khwârizmî's text there is only one name, "thing", for the unknown quantities, or more precisely there are no proper names for the unknown quantities, since "thing" functions as a common name: "a thing" is an unknown quantity. A competent user of al-Khwârizmî's sign system names an unknown quantity as "thing" and has to be careful in referring to other unknown quantities by compound expressions because a different name for them is not available in this system. He uses in fact a common name as a proper name. Students taught to name the unknown of a word problem with an x, frequently see the x as a common name meaning "unknown" and not a proper name referring to a specific unknown quantity, labelling then any unknown quantity with an x. The meaning they give to x does not correspond to its meaning in the current MSS of algebra, but to the meaning of a less abstract MSS. In what follows we are going to further examine ways of representing the unknown in history that are seen from the current MSS of algebra as less abstract, in order to have a tool kit for examining students' behaviours.

In al-Khwârizmî's text, "thing" is the name given to any unknown quantity to start the construction of the equation, and in the course of the calculations with thing that lead to the equation the quantities are named by their type or species, which is not an absolute property of the quantity but a property relative to the calculations that are being performed. That is, there are two different categories of things to be represented and of representations that are interlinked: unknown quantities and types of numbers.

In Babylonian algebra these two categories do not exist. Unknown quantities are represented by the sumerograms that signify "long" and "wide" in what Høyrup (2002) calls "a functionally abstract representation by means of mensurable segments". In the analysis that follows, we shall not deal with the peculiarities of Babylonian symbolisation, which Radford (2000) does from a socio-cultural perspective.

In Indian algebra this distinction is present, and the names that are used for each are not identified with one another. Thus, in Chapter I of the *Vija-Ganita* or *Avyacta-Ganita*, Chapter IV, which deals with arithmetic operations with unknown quantities, Bhâskara (1114-1185, India) begins by writing the following:

> "So much as" and the colours "black, blue, yellow and red" and others besides these, have been selected by venerable teachers for names of values of unknown quantities, for the purpose of reckoning therewith. (Colebrooke, 1817, p. 139)

And in Chapter VI, which deals with "Analysis by a Multiliteral Equation", he again introduces colours in abundance to represent unknown quantities, adding that letters can also be used:

> This is analysis by equation comprising several colours. In this, the unknown quantities are numerous, two, three or more. For which *yâvat-tâvat* and the several colours are to be put to represent the values. They have been settled by the ancient teachers of the science: viz. "so much as" (*yâvat-tâvat*), black (*calaca*), blue (*nîlaca*), yellow (*pîtaca*), red (*lôhitaca*), green (*haritaca*), white (*swêtaca*), variegated (*chitraca*), tawny (*capilaca*), tan-coloured (*pingala*), grey (*d'hûmraca*), pink (*pâtalaca*), white (*savalaca*), black (*syâmalaca*), another black (*mêchaca*), and so forth. Or letters are to be employed; that is the literal characters *c*, &c. as names of the unknown, to prevent the confounding of them. (Colebrooke, 1817, pp. 228-229)

As for the names of the species of numbers, they are *rûpa*, which means "form" or "species" (Colebrooke, 1817, p. 139), for absolute numbers, and *varga* and *ghana* for square and cube, respectively, and Bhâskara states certain rules of multiplication that enable him to write algebraic expressions with more than one unknown quantity represented:

When absolute number and colour (or letter) are multiplied one by the other, the product will be colour (or letter). When two, three or more homogeneous quantities are multiplied together, the product will be the square, cube or other [power] of the quantity. But if unlike quantities be multiplied, the result is their (*bhâvita*) 'to be' product or factum. (Colebrooke, 1817, p. 140)

In this case, "homogeneous" means that they are of the same colour, that is, that they represent the same quantity. The product of powers of different colours is represented by an expression that amounts to saying that the product has not been performed, it is an "indicated product". Furthermore, the names for unknown quantities are not used only for the first power of the unknown quantity. In fact, in order to represent the square of an unknown quantity what is written is not the term *varga* on its own, as happens with *mâl* or *dynamis* or *census*, but *varga* accompanied by the name of the unknown quantity, that is, *yâvat varga* or *câlaca varga*, et cetera. Also, Bhâskara uses complete words or the first syllable or, in the case of *varga* and *ghana* or *bhâvita*, sometimes the beginning of the word. Bhâskara writes the algebraic expression $x^3 + 3x^2y + 3xy^2 + y^3$, for example, as:

ya gh 1 *ya v. ca bh* 3 *ca v. ya bh* 3 *ca gh* 1 (Colebrooke, 1817, p. 248)

In this expression, *ya* is the abbreviation of *yâvat* and represents an unknown quantity, *ca* is the abbreviation of *câlaca* and represents another unknown quantity, *v* is the beginning of *varga*, square, *gh* is the beginning of *ghana*, cube, and *bh* is the beginning of *bhâvita*. In the formation of monomials, therefore, we find numbers (always present, even if it is the number 1), the names of the unknown quantities, the names of the species of numbers and the name of the indicated product.

Thus, Bhâskara's sign system has no difficulty in representing different unknown quantities by different signs, precisely by keeping the representations of quantities and types of numbers differentiated.

However, once thing is identified with root and the species of numbers are not limited to al-Khwârizmî's three, the distinction between the two categories becomes blurred. Thus, after defining the object of algebra as "absolute number and magnitudes that are measurable in as much as they are unknown but refer to a known thing by which they can be determined" and relating magnitudes to their Aristotelian definition, ᶜUmar al-Khayyâm is able to write, speaking now of a tradition:

Among algebraists it is the custom in their art to name the unknown that one wishes to determine "thing", its product by itself *mâl* [treasure], its product by its *mâl*, *ka'b* [cube], the product of its *mâl* by its likeness *mâl mâl*, the product of its *ka'b* by its *mâl* [as] *mâl ka'b*, the product of its *ka'b* by its likeness *ka'b ka'b*, and so on as far as you wish. From Euclid's book of the *Elements* it is known that these degrees are all proportional, meaning that the ratio of unity to the *jidr* [root] is equal to the ratio of the *jidr* to the *mâl* and it is equal to the ratio of the

> *mâl* to the *ka'b*. The ratio of the ^{c}*adad* [number] to the various *jidr* is therefore equal to the ratio of the *jidr* to the *mâl*, equal to the ratio of the *mâl* to the *ka'b*, and equal to the ratio of the *ka'b* to the *mâl mâl*, and this as far as you wish (Rashed & Vahebzadeh, 1999, pp. 120-122).

Leonardo Pisano, in turn, deals with what for him is algebra, that is, the technique developed by the Arabs from al-Khwârizmî's book, at the end of the *Liber Abbaci*, from page 406 to page 459, which is the last page in Boncompagni's edition. The first thing he does is to explain what the types of numbers are, the six normal forms, the algorithms for each form and the proofs of the algorithms, in the section *Incipit pars tertia de solutione quarumdam questionum secundum Modum algebre et almuchabale, scilicet ad proportionem et restaurationem* (Boncompagni, 1857, p. 406). [Here begins Part Three on the Solution of certain problems according to the method of algebra and almuchabala, namely proportion and restoration (Sigler, 2002, p. 554)]. It is in this section that he introduces the names *radix*, *quadratus* and *numerus simples* for the types of numbers (*propietates*, in his translation), and throughout this introduction to the *algebre et almuchabale* Leonardo, like al-Khwârizmî, uses *census*, *radix*, and *numerus* (or *denarius*, or *dragme*, also like al-Khwârizmî's *dirham*), without any appearance of the term *res*, which, as we have said, Leonardo introduced in another context (that of the *Regula Recta*) 200 pages earlier. The title of the following section is *Expliciunt introductiones algebre et almuchabale. Incipiunt questiones eiusdem* [Here ends the introduction to algebra and almuchabala. Here begin the problems on algebra and almuchabala (Sigler, 2002, p. 558)], and in it, just as al-Khwârizmî uses *shay'* in his problems, Leonardo uses *res*, but unlike al-Khwârizmî he explicitly identifies thing with root. He does so when he discusses the first problem in the section, the statement of which is:

> Si vis dividere 10 in duas partes, que insimul multiplicate faciant quartam multiplicationis maioris partis in se; [If you wish to divide 10 into two parts such that their product is a quarter of the product of the greater part by itself]

The instruction given by Leonardo is not to call the greater part thing, as al-Khwârizmî would have done, but rather to represent the greater part by the root (pone pro maiori parte radicem quam appellabis rem), and he explains that he calls the root thing and then goes on to say:

> remanebunt pro minori parte 10, minus re; que multiplicata in re, venient 10 res minus censu; *et ex multiplicata re in se provenit census; quia cum multiplicatur radix in se, provenit quadratus ipsius radicis*; ergo decem rei, minus censu, equantur quarte parti census. (Boncompagni, 1857, p. 410)

> [there will remain for the smaller part 10 minus the thing and it, multiplied by the thing, yields 10 things minus the census; and the multiplication of the thing

by itself yields the census, because when the root is multiplied by itself the square of the root results; therefore ten things minus the census are equal to a fourth of the census] (Sigler, 2002, p. 558)

In this solution, he does not limit himself to performing calculations with thing but, in order to do so, explains that multiplying thing by itself gives *census* because when root is multiplied by itself it gives the square of that root. In what follows he introduces thing directly and interchanges root and thing without further explanation.

The history of the development of symbolic language for algebra was marked both by this identification of root and thing and by the need for both categories (species of numbers and unknown quantities) to be represented. Indeed, according to Cajori (1928), in the symbolisation of the powers of the unknown in algebra, two general plans can be distinguished: one that develops abbreviations from the names thing, root, *censo*, et cetera, the "Abbreviate Plan", and the "Index Plan", in which one limits oneself "to simply indicating by a numeral the power of the unknown quantity" (Cajori, 1928, p. 339). This observation of Cajori's becomes even more pertinent if we examine the consequences that each of the two plans has for the representation of each of the two categories: unknown quantities and species of numbers.

Let's begin with the Abbreviate Plan. What is actually represented in it is species of numbers, and thing is only represented insofar as it is identified with one of the species, root. This has two consequences: since what is represented by the symbol is not a quantity (which is of a specific species) but only the *species* that the quantity in question is, in this symbolisation it is not possible to distinguish different quantities with different signs. Moreover, as the signs by which the species are represented are abbreviations of the names of the species, the rules of calculation, specifically the rules for multiplying expressions, cannot be derived from the signs themselves but have to be established in multiplication tables.

It is also worth pointing out that the lack of effectiveness of this symbolism for calculation on the level of expression has different features in each of the two series of names for species that have been developed in history. In fact, when the species were generalised beyond cubic numbers the names of the species were constructed from those of the second and third powers, but this was done in two different ways, one "additive" and the other "multiplicative". The additive way was used by Diophantus, and also by Abû Kâmil (about 850-930, Egypt), al-Karâji, as-Samaw'al (1130-1180, Iraq), ᶜUmar al-Khayyâm, Sharaf al-Dîn al-Tûsî (1135-1213, Persia) and most of the Arab mathematicians, including those of the Arab West (al-Andalus and the Maghreb) such as ibn al-Bannâ (1256-1321, Morocco) or al-Qalasâdi (1412-1486, Spain), and Leonardo Pisano and Viète in the Christian West. In it, *dynamocubos*, or *mâl ka'b*, or *census cubus*, or *quadrato-cubus* represent the fifth power, and juxtaposing the two words forms the new name. The multiplicative way was used by Sinân ibn al-Fath[11](s.X) among the Eastern Arabs, Bhâskara in India, and Luca Pacioli (1445-1517, Italy), Cardano, Tartaglia (1499-1557, Italy), Pedro Nunes (1502-1578, Portugal), Pérez de Moya (1513-1592, Spain) and, in general, most of the algebraists of the Christian West. In it, *mâl ka'b*, *census cubi* or *censo de cubo* represents the sixth power, and the new name is often formed with the genitive.

In the case of the additive combination the species multiplication table is simple, since the name of the product of two species is the juxtaposition of the names of the factors, and therefore multiplication of species can be converted into a rule of syntax. The additive combination also generates names for all the species, as any number can be obtained as a sum of twos and threes, but the name of each species is not unique. This makes it necessary to observe equality between names, as Leonardo does in the quote below, for example. It is also convenient to establish one or other of the synonyms as canonical, as Sesiano (1999) says the Arabs used to do[12].

> et est multiplicare per cubum cubi, sicut multiplicare per censum census census [and multiplying by the cube of the cube is like multiplying by censo censo censo]. (Boncompagni, 1857, p. 447)

In the case of the multiplicative combination, the species multiplication table cannot give rise to a syntactic operation between names because, for example, the product of *censo* of *cubo* and *censo* is *censo* of *censo* of *censo*, and this name cannot be derived from the two previous names but can only be obtained by resorting to the meaning of each name in the series of species. Moreover, the multiplicative combination does not generate names for all the species, as not all numbers can be expressed as products of twos and threes. Thus, Sinân ibn al-Fath had to introduce a special name, *madâd*, for the fifth power, the first "that is not a cube or a square", and similarly, in the Christian West, the names *primero relato*[13] or *sursolidum* and *segundo relato* or *bisursolidum* were used for the fifth and seventh powers, and the subsequent names for the higher powers that are not cubes or squares. However, although in this case there is this impossibility of generating syntactic operativity in the production of signs for species and in multiplication, which is one of the most frequent operations, the fact that compound names are formed with the genitive creates the possibility of the nesting or embedding of expressions (Høyrup, 2000), a possibility that is not present in the additive combination. In other words, the additive combination permits the syntactic operativity of multiplication, since the name *censo cubo* is formed in a way that is similar to the multiplication $x^2 \cdot x^3$, whereas the multiplicative combination opens up the possibility of embedding expressions, since the name *censo* of *cubo* is formed in a way that is similar to the power of a power $(x^3)^2$.

In the Abbreviate Plan, in as much as the "abbreviated" text is no different from the text originally written in the vernacular except for a few words that are abbreviated, the fact that some words are replaced by their first syllable, their first letter or some other sign can hardly add syntactic operativity to what was already present in the text written in the vernacular, and the features that we have just described are present similarly both when the names of species are written in full, *mâl mâl* or *censo* of *censo,* and when they are written abbreviated, as is done by Diophantus, al-Qalasâdi, Cardano or Pérez de Moya.

The case of the Western Arab algebraists such as al-Qalasâdi is not exactly the same, as it seems fairly likely that the abbreviations come from the use of a dust board on which the calculations were performed, and not from abbreviating a written text. Indeed, Abdeljaouad (2002) emphasises that al-Qalasâdi explicitly associates algebraic symbolism with the use of a dust board for calculating, *lawha*, on which the operations have to be performed, in *at-Tabsira al-wâdhiha fi masâ'il al-'adad allâ'iha* (c. 1413):

> Write the operation on one side of the *lawha* and above the thing (*shay'*) place the sign *shin* [the first letter of the word *shay'*] or three dots [there are three dots above the letter *shin*], above *mâl* place the sign *mim* [the first letter of the word *mâl*], above *ka'b* the sign *kaf* [the first letter of the word *ka'b*], and do not put anything above the number because the absence of a sign is also a sign. (Abdeljaouad, 2002, p. 14, the comments in brackets are ours)

As a consequence of this origin, algebraic expressions in Western Arabic texts do not generally appear as part of a text written in the vernacular with words being replaced by their abbreviations, but rather they tend to appear accompanying the vernacular text, introduced by the words "this is its image (figure or representation)". According to Djebbar (1985), Abdeljaouad calls these symbols "symbols of illustration", distinguishing them, within the symbols characteristic of syncopated algebra, from abbreviations such as those used by Diophantus, for example, which he calls "symbols of substitution". The operativity of the symbolic expressions of the Arabs of the West therefore lies in the arrangement of signs in tables on the dust board and in the actions of writing and rubbing out on it, rather than in syntactic operations with the signs themselves.

The transition from syncopated algebra to symbolic algebra begins with Viète, for whom the *logistica speciosa*, the analytical art to which he wished to give this name rather than that of algebra, was calculation with species or *formae rerum*, forms of things. But in order to represent this calculation by species, Viète developed symbolic expressions in which what is represented by letters is not the species but the known or unknown quantities. For example, Viète writes the equation $x^2 - 2bx = c$ as follows:

$$A \text{ quad} - B \text{ in } A \text{ 2, æquetur } Z \text{ plano.}$$

In Viète's sign system[14], therefore, a monomial contains a letter to represent a quantity and a name of a species, which is written in vernacular language or sometimes abbreviated. In his sign system, Viète differentiates between the signs that he uses to represent quantities and those that he uses to represent species. The latter enables him (as we have seen happens in Bhâskara's sign system, but not in the other cases) to represent various different unknown quantities by different signs (i.e., biunivocally) which are useful in the Cartesian Method, a crucial characteristic of the sign system. Nevertheless, the sign system is not suitable for calculation on the syntactic level, since it is still necessary to make use of species multiplication tables, as species are represented by their names (in this case, formed in accordance with the additive combination) or abbreviations of their names.

Symbolisation made calculation on the syntactic level possible when Viète's letters ceased to be accompanied by the names of species. But for this it was necessary to adopt the symbolisation developed in the Index Plan.

Indeed, what characterises the Index Plan is the fact that, instead of being represented by their name or an abbreviation of it, species are represented by a number that expresses their position in the succession, so that the multiplication of species can be converted into a syntactic rule by identifying it with the sum of the numbers that represent them[15]. This is already present in Chuquet's *Triparty* written in French in 1484[16] (Chuquet, 1445-1488, France). However, it was scarcely known until the end of the nineteenth century when Aristide Marre published it (Paradís, 1993).

However, in Chuquet's sign system (and this also happened in Bombelli's sign system, which was to be better known and more influential), as numbers are another way of representing what was represented in the Abbreviate Plan by abbreviations, that is, species, the only thing that is represented is species. Thus, Chuquet writes 3^2 for our $3x^2$ or 3^1 for our $3x$, so that his system is efficient for calculation on the syntactic level, but it cannot represent more than one unknown quantity.

It was necessary to combine Viète's letters for representing unknown (and known) quantities and Chuquet's and Bombelli's (Bombelli, 1526-1572, Italy) numbers for representing species so that the two categories might be represented in a clearly differentiated and efficient way for syntactic calculation and the sign system of symbolic algebra[17] might be fixed, which happened with Descartes and Euler (1707-1783, Switzerland).

8.4 Algebraic Language: Pre-Vietic Moments in the History of its Evolution

Many historians consider Viète's *The Analytic Art* (Witmer, 1983) as the work that inaugurates the symbolic stage of algebra (Klein, 1968). Those texts in which there is an explicit, systematic use of algebraic syntax fall into this category of symbolic. This possibility of classifying a historic algebraic text by using Nesselmann's categories has proved useful precisely when speaking of how symbolic the language in that text is and how its condition of rhetorical, syncopated or symbolic is related to the nature of the methods that it applies.

As we have seen, although this classification of Nesselmann's does not correspond to a chronological order, it does define a virtual axis of evolution of the language of algebra. On this axis it is possible to identify significant moments of the history of algebra, in which one can see advances not only in the language itself but also in concepts and methods.

In what follows we shall examine the effect that the nature of the sign system has on concepts and methods. Indeed, seen from this perspective of language and methods, in various texts of the pre-Vietic stages one can perceive, on the one hand, clearly differentiated languages, and, on the other, characteristics that are common to them. We are referring to the *abacus* books, for example, written between the thirteenth and sixteenth centuries and characterised by being devoted to solving practical problems expressed in ordinary language (Italian), using methods of oriental mathematics (Egmond, 1980). We are also referring to *De Numeris Datis* authored by Jordanus de Nemore (about 1225-1260, Germany), considered the first book of advanced algebra (Hughes, 1981), written in Latin and devoted to solving systems of equations that can be reduced to a quadratic equation. The difference in languages is very clear, for whereas the *abacus* books are completely rhetorical, *De Numeris Datis* incorporates the use of literals to denote unknowns and constants. We are not going to analyse here the special way in which letters function in Nemore's book, which is not one of the ones that we examined in Section 8.2. Such an analysis (see Puig, 1994) was useful for analysing and understanding the behaviour of certain students (Puig, 1996a), who reproduced Nemore's sign system when prompted to write "in algebra" the solution they had made of a problem by arithmetic means.

On the other hand, the contribution of *De Numeris Datis* goes beyond the fact that it uses letters to represent quantities, since "general numbers" appear in its statements and arguments. In other words, the processes for solving systems of equations are not performed on particular numbers, but rather it is precisely the use of literals that conveys the sense of the generality of the method, which one supposes should not depend on the particular numerical characteristics of coefficients. In this book, the sequence of propositions or statements of problems and their solutions constitutes a general method in itself, consisting in reducing each proposition to a *canonical form* which makes it possible to find the value of the unknown from what is known (the data). The propositions are linked together in the sequence in such a way that each canonical form found is added to a repertoire that is applied to the solution of new problems (propositions).

There is a contrast between *De Numeris Datis* and the *abacus* books which consists in the fact that, whereas in *De Numeris Datis* one sees the generality of a method through the expression of data (general numbers) by means of literals, in the *abacus* books the process for solving each problem is closely related to its numerical characteristics, that is, to the numerical specificity of the data. The result is that problems which, from a modern viewpoint, could be tackled with the same method are solved in the *abacus* books by means of very different procedures. We shall now give examples of an *abacus* text and a *De Numeris Datis* text to illustrate the contrast between the two points in pre-symbolic algebra to which these works correspond.

Problem 1 is from the *abacus* book *Trattato di Fioretti* by Mazzinghi, M. A. di (ca 1350, Italy), edited by Arrighi, G. (Mazzinghi, 1967). In the solution to Problem 1 (shown in Figures 8.1, 8.2, 8.3), we include the rhetorical version of the solving process in old Italian alongside a translation into modern symbolism. The end of the first part of the solution (Figure 8.1) shows a difference with modern algebra. In the current manipulative algebra, justification of the commutation of numeric coefficients of the unknown in each particular case is unnecessary. Nevertheless, in abacus problems (as in the one presented here), equivalence of expressions such as $19(2y)$ and $2(19)y$ is justified in words as follows: "if 19 multiplied by the double of the second part makes 228, in the same way, the double of 19 (that is, 38) multiplied by the second part, will make 228". That is, in abacus problems, a rule is phrased specifically for each particular case. This is a characteristic of this type of texts: general rules exist only in practice; they are evoked and expressed every time and for every particular case they are applied.

Original version in old Italian	**Translation into modern symbolism**
Fa' di 19, 3 parti nella proportionalità chontinua che, multiplichato la prima chontro alla altre 2 e lla sechonda parte multiplichato all'altre 2 e lla terza parte multiplichante all'altre 2, e quelle 3 somme agunte insieme faccino 228. Adimandasi qualj sono le dette parti.	Find three numbers, x, y, z such that: $x + y + z = 19$ As $x/y = y/z$ $x(y + z) + y(x + z) + z(x + y) = 228$
Conciosiachosachè Maestro Antonio sottile scriva e' chasi e' qualj non ànno asolutione alle quantità numeralj, nientedimeno io mi sforçerò di porre chasi e' qualj aranno asolutione a quantità che facilmente si potranno provare, cioè che con facilità acciò che lo 'nteletto possa essere chapacie di tutto.	
Ora al nostro chaso, qui è da sapere che se si multiplicha 19 nel doppio della seconda parte sarà 228, overo se si multiplicha lo doppio del 19, cioè 38, nella seconda parte farà 228. E che questo sia vero il voglio chiarire. Noi abbiamo proposto che si divida 19 in 3 parti nella proportionalità chontinua che, muliplichata la prima per l'altre 2 e la seconda per l'altre 2 e l'altra, cioè la terça, per l'altre 2 et agunte le dette multiplichatione, insieme faccino 228.	Solution: You have to know that $(19)(2y) = 228$. Which is equal to $2(19)y = 228$; $38y = 228$. This has to be explained.

Figure 8.1. Solution of Problem 1 from *Trattato di Fioretti* (first part).

The end of the second part of the solution (Figure 8.2) results in a long expression. When translated into algebra symbolism, the transformation of

$$xy + yy + yx + yz + yy + zy$$

in

$$xy + xy + yy + yy + zy + zy$$

might be interpreted as a permutation of terms containing unknown quantities. However, analysing the rhetoric version in abacus, it can be noticed that the intention is to make explicit that the first set of six terms is really a set of three terms, each of them appearing twice. This last one, in turn, is rephrased as "[the multiplication of] the second part by the addition of the other all three" which in modern symbolism can be expressed as:

$$xy + xy + yy + yy + zy + zy = y(x + y + z) + y(x + y + z)$$

where the addition $x + y + z$ is given (19).

Then the following rule is applied: *multiplying one quantity by another one, twice, is equal to multiplying the first quantity by the double of the second one*. Here once again, the wording of this rule in the original text refers to the particular case in question: "when multiplying the second part by the addition of the other all three, twice, is like multiplying the second part by the double of the addition of the all three". In algebra symbolism it can be expressed as:

$$y(x + y + z) + y(x + y + z) = y(2(x + y + z)).$$

In turn, the equality $y(2(x + y + z)) = 2y(x + y + z)$ is expressed as: "multiplying the second part by the double of the addition of the all three is as much as multiplying the double of the second part by the addition of the all three". In this way, this "rule of doubles" is applied in the previous steps until the chain of equalities $2y(19) = 2y(x + y + z) = 228$ is reached in the first part of the solution to Problem 1 (see Figure 8.1). Finally, the value of one of the unknowns is found from $12 = 2y$ in the third part (see Figure 8.3). From this point on, what is used to find the value of the other two unknown quantities is the Babylonian method to solve quadratics, which involves operations on known quantities (see Figure 8.3).

In this abacus problem the same general rule (multiplication of one quantity by the double of another is the same as the multiplication of the double of that quantity by the second one) is reworded for specific numbers, in every step it is applied. This is also a characteristic of the abacus books, in which can be observed the application of the same solving method to a large family of similar problems, reworded specifically for each particular case, every time it is applied, without any abbreviation process in any problem in the long list. This characteristic contrasts with that of the methods used to solve problems in *De Numeris Datis*, in which the use of letters to symbolise numbers (general numbers) permits the application of canonical forms to the solution of new problems. One clear manifestation of the generality of the methods developed in this text is that at the end of each problem solved with general numbers, an example with specific numbers is presented. All this is illustrated below with problems 1.1 and 1.2 from the *De Numeris Datis* (see Figures 8.4, 8.5, 8.6, & 8.7).

Onde è manifesto, che quando le quantità sono nella proportionalità chontinua, che lla multiplichatione della prima nella terza fa quanto la multiplichatione della seconda in sé; e questo assai volte l'abbiamo mostro.	As $x/y = y/z$ then $xz = yy$.
Onde quando e' si multiplicha la prima per l'altre 2, è quanto a multiplichare la prima per la seconda e quanto la seconda in sè imperochè multiplichando la prima nell'altre 2 e' si multiplicha la prima nella seconda et la prima nella terça; et multiplichando la prima nella terça è chome a multiplichare la seconda in sè.	Therefore $x(y + z) = xy + yy$ since $x(y + z) = xy + xz$.
Adunque a multiplichare la prima nell'altre 2 è chome a multiplichare la prima nella sechonda e chome a multiplichare la seconda in sè.	Therefore $x(y + z) = xy + yy$
E a multiplichare la seconda per l'altre 2 è quanto a multiplichare la seconda per la prima et la seconda per la terça.	also $y(x + z) = yx + yz$
E a multiplichare la terça parte per l'altre 2 è chome a multiplichare la seconda in sè et chome a multiplicharc la terça per la seconda.	and $z(x + y) = yy + zy$
Adunque queste 3 multiplichationi, che sono 6, èchome a multiplichare la prima per la seconda 2 volte et la seconda in sè 2 volte et la terça per la seconda 2 volte. Onsz, se bene aprj l'ochio, egli è chome a multiplichare la seconda parte nella somma di tute e 3 le parti, 2 volte. E a multiplichare la seconda parte nella somma di tutte e 3 due volte è chome a multiplichare la seconda parte nel doppio della somma di tutte a 3, overo quanto a multiplichare lo doppio della seconda parte nella somma di tutte et 3.	These 3 multiplications are in fact six terms and can be written as follows: $xy + yy + yx + yz + yy + zy =$ $xy + xy + yy + yy + zy + zy$ $= xy + yy + zy + xy + yy + zy$ $= y(x + y + z) + y(x + y + z)$ $= y(2(x + y + z))$ $= 2y(x + y + z)$

Figure 8.2. Solution of Problem 1 from *Trattato di Fioretti* (second part).

Figure 8.4 shows Proposition 1 from Book One of *De Numeris Datis*. We have included the basic definitions used, Hughes's English translation of the proposition, and in Figure 8.5 a translation of the solving process into modern symbolism. Neither Hughes's English translation nor the translation into modern symbolism are literal. A more literal translation can be seen in Puig (1994).

In Hughes's (1981) interpretation, step [1] corresponds to the construction of the equation, that is, to the formulation of the problem in terms of what is known (a and b) and what has to be found (x and y); steps [2] to [4] are transformations applied to [1] to arrive at the canonical form [5]; and [6] is the numerical example of [5]. In

Proposition 2 from the same book, the use of general numbers is evident, with the assignment of literals to quantities, whether known or unknown. Figures 8.6 and 8.7 give this in translation and in modern symbolism.

Onde, adunque, multiplichando el doppio della seconda parte per 19, che è somma di tutte et 3, farà 228.	$2y(19) = 2y(x + y + z)$ $= 228$
Adunque partendo 228 in 19 ne verrà el doppio della seconda parte. E partito 228 in 19 vienne 12, e il 12 è 2 chotanti della seconda parte.	$228 \div 19 = 2y$ $228 \div 19 = 12$ $12 = 2y$
Adunque la seconda parte è 6.	$y = 6$
E, questo trovato, l'altre 2 rimarrano 13; dove dirai: perchè le parti sono in chontinua proportione, che tanto fa a multiplichare la prima parte per la terça quanto la seconda in sè; onde abisognia di dividere 13 in 2 parti che lla prima multiplichata per l'altra faccia 36, cioè il quadrato di 6.	Found y, $x + z = 13$, from which we say: because of $x/y = y/z$, we have $xz = yy$. Therefore, we want to find x, z, such that : a) $13 = x + z$ b) $xz = 36 = 6^2$
E a questo si dà questa reghola: dimeça 13, è la metà 6½, multiplichata in sè, fanno 42¼, trane 36 rimanghono 6 e ¼, pigliane la radice che è 2½ e agiugnilo alla metà di 13, cioè a 6½, fanno 9. E tanto è la maggiore parte.	To the above corresponds the rule: $13 \div 2 = 6\frac{1}{2}$ $(6\frac{1}{2})(6\frac{1}{2}) = 42\frac{1}{4}$ $42\frac{1}{4} - 36 = 6\frac{1}{4}$ $(6\frac{1}{4})^{1/2} = 2\frac{1}{2}$ $2\frac{1}{2} + 6\frac{1}{2} = 9$ $x = 9$
e, per la altra, traj 2½ della metà du 13, cioè di 6½, rimanghono 4. E tanto è la minore.	To find z: $6\frac{1}{2} - 2\frac{1}{2} = 4$ $z = 4$
Onde è diviso 13 in 2 parte che, multiplichata l'una per l'altra, fanno 36. Adunque abbiamo fatto, di 19, 3 parti in chontinua proportione che, la prima multiplichata per l'altre 2 e lla seconda multiplichata per l'altre 2 et la terça multiplichata per l'altre 2 e, quelle multiplichationi agunte insieme fanno 228.	$13 = 9 + 4$ $(9)(4) = 36$, from which, $19 = 9 + 6 + 4$ $9/6 = 6/4$ $9(6 + 4) + 6(9 + 4) + 4(9 + 6) = 228$
E la minore è 4, la meçano 6, la maggione è 9.	lesser: 4, middle: 6, larger: 9

Figure 8.3. Solution of Problem 1 from *Trattato di Fioretti* (third part).

Definitions
1. *A number is given* whose quantity is known;

2. *A number is given in relation to another* where the ratio of the one to the other is given;

3. *A ratio is given* whose denomination [i.e., name] is known.

I.1. If a given number is separated into two parts whose difference is known, then each of the parts can be found.

Since the lesser part and the difference equal the larger, the lesser with another equal to itself together with the difference make the given number. Subtracting therefore the difference from the total, what remains is twice the lesser. Halving this yields the smaller and, consequently, the greater part. (Hughes, 1981, p. 127)

Figure 8.4. Proposition 1 from Book One of *De Numeris Datis*. *(Reprinted with permission from Jordanus de Nemore.* De numeris datis: A critical edition and translation. *Barnabas Hughes (Ed/Trans), copyright 1981 by University of California Press. All rights reserved.)*

Let the lesser and larger parts be called x and y respectively, and the sum and the difference of the parts a and b.

Since $y + (x - y) = x$ [1]

then $y + y + (x - y) = a$ [2]

Therefore $a - (x - y) = 2y$ [3]

and $\frac{1}{2}(a - (x - y)) = y$ [4]

Canonical form $\frac{1}{2}(a - b)$ (Hughes, 1981, p.189) [5]

For example, separate 10 into two parts whose difference is 2.
If that is subtracted from 10, 8 remains, whose half is 4.
This is the smaller number and the other is 6. (Hughes, 1981, p. 127) [6]

Figure 8.5. Modern translation of Proposition 1 from Book One of *De Numeris Datis* *(Reprinted with permission from Jordanus de Nemore.* De numeris datis: A critical edition and translation. *Barnabas Hughes (Ed/Trans), copyright 1981 by University of California Press. All rights reserved.)*

I.2. If a given number is separated into as many parts as desired whose successive differences are known, then each of the parts can be found.
Given is the number *a* which is divided into *w*, *x*, *y*, and *z* the least of the parts.

Since the successive differences of all these are given, each difference can be expressed in terms of the difference of each number with *z*. Therefore let *f* be the difference of *w* and *z*, and the sum of *g* and *h* be the sum of the differences of *x* and *z* with *y* and *z*. Now because *z* makes each of those equal to each of these, it is obvious that thrice *z* with the sum of *f*, *g*, and *h* equals those three. Therefore four times *z* with the sum of *f*, *g*, and *h* equals *a*. Subtracting this sum from *a* leaves four times *z*, which is now known; hence *z* is found. By addition of differences the other parts are found. (Hughes, 1981, pp. 127-128)
For example, separate 40 into four parts whose successive differences are 4, 3, and 2. Therefore the difference of the first and last is 9, of the second and last is 5, and of the third and last is 2. Their sum is 16. This subtracted from 40 leaves 24 whose fourth is 6, the least of the four parts. By adding this to 9, 5, and 2, the other three parts are found, namely 8, 11, and 15. (Hughes, 1981, p. 128)

Figure 8.6. Translation of Proposition 2 from Book One of *De Numeris Datis (Reprinted with permission from Jordanus de Nemore.* De numeris datis: A critical edition and translation. *Barnabas Hughes (Ed/Trans), copyright 1981 by University of California Press. All rights reserved.)*

Find w, x, y & z, such that
$w + x + y + z = a$
$w - x = b$
$x - y = c$
$y - z = d$.
Let z be the lesser part.
Since $w - x = b, x - y = c$ and $y - z = d$,
then b, c and d can be expressed in terms of the difference between z and each of the other numbers:
$f = w - z$, and
$g + h = (x - z) + (y - z)$.
Since z is such that $f + z = w$, $g + z = x$ and $h + z = y$,
It follows that
$3z + f + g + h = w + x + y$.
Therefore:
$4z + f + g + h = a$
$a - (f + g + h) = 4z$
Now $4z$ is known, hence z can be found.
Canonical form: $z = 1/4\ (a - (b + 2c + 3d))$ (Hughes, 1981, p. 189)

Figure 8.7. Translation into modern symbolism of Proposition 2 from Book One of *De Numeris Datis. (Reprinted with permission from Jordanus de Nemore.* De numeris datis: A critical edition and translation. *Barnabas Hughes (Ed/Trans), copyright 1981 by University of California Press. All rights reserved.)*

As one follows the developments of the solutions in the examples just given, in the *Trattato di Fioretti* (*abbaco*) and in *De Numeris Datis*, one observes that these mediaeval works correspond to two clearly differentiated levels of language, but one also observes that a characteristic that they have in common is the fact that in them there is no systematic treatment of the operations performed on the terms (of the equation) that involve unknown quantities. That is, there is no operation on the unknown. An indication of operation between literals that appears in *De Numeris Datis* is the juxtaposition of characters to indicate a sum of magnitudes, but this symbolisation of an operation does not go beyond the level of the expression, that is, it is not translated into syntax rules applied to these new symbols.

This "non-operation on the unknown", which might seem to be related to the pre-symbolic character of the two texts that we are discussing, has led to the formulation, in the field of mathematics education, of conjectures as to the presence of "didactic cuts" in the processes of transition from arithmetic to algebraic thought. For example, it has been conjectured that one of these "cuts" is located precisely at the moment when, for the first time, students face the need to operate on what is unknown in the solution of linear equations with terms containing x on both sides of the equals sign. In the following section, a brief description of the main outcomes from the clinical study "From arithmetic to algebraic thought" (Filloy & Rojano, 1989) related to this conjecture illustrates how historical analysis can be used in research design in mathematics education.

8.5 Using the History of Algebra in Education Research: A Didactic Cut of Epistemological Origin

A historical analysis of the evolution of algebraic language was decisive for the formulation of the conjecture about a didactic cut and the identification of the value of observing the point at which learners have to operate on what is unknown for the first time. In the study "From arithmetic to algebraic thought" (Filloy & Rojano, 1989), the notion of a rupture or cut in the evolution of understanding, used by Bachelard (1947), and the corresponding notion of an epistemological obstacle serve as central elements that link the domains of history and education.

In the clinical study to which we have referred, one of the most eloquent manifestations of the rupture (didactic cut) mentioned above is the typical spontaneous response of students (in a clinical interview) to solving equations such as $2x + 3 = 5x$ (i.e., of the form $Ax + B = Cx$). Children of 12-13 years of age who try to solve equations of this sort for the first time tend to assign an arbitrary value to the unknown on the right hand side (for example, 2) and solve an equation of the form $2x + 3 = 10$. In this way, they reduce the new type of equation to an equation of the type $Ax + B = C$, which they know how to solve with arithmetic tools (undoing the operations on the givens). When these students are asked to find the value of x in the equation $x + 5 = x + x$, they respond "this x (one on the right hand side) has a value of 5, and the other two (one on the right and the other on the left hand side) can have any value (the same value for both)".

This type of response corresponds to what Filloy and Rojano (1984 and 1989) call "the polysemy of x". The term "polysemy of x" refers to the interpretation of literal symbols when, in the same (algebraic) statement, the same symbol is assigned meanings belonging to different semantic fields. In one case, the symbol x is interpreted as a specific unknown ($x = 5$), that is, the corresponding semantic field is that of equations with numerical solutions. In the other case, the same symbol is interpreted as a general number (the two instances of x can have any value, but the same value); here the corresponding semantic field is that of algebraic identities. The study thus shows that, at the point of the cut, when students have to tackle tasks in which it is necessary to operate on the unknown they do not spontaneously transfer operativity with numbers to algebraic objects such as unknowns. Moreover, it becomes clear that the new type of equation gives rise to (faulty) readings of literal symbols in an equation in which different semantic fields of symbolic algebra are mixed together.

The attempt to devise theoretical explanations for the kind of spontaneous responses that children give, and for the codes with which they express themselves at moments of transition towards algebraic thinking has given rise to a research agenda that adds a semiotic perspective to the historical perspective. With a semiotic perspective it is possible, among other things: to speak of the different stages of development of algebraic language in terms of mathematical sign systems (MSSs) of algebra; to formulate criteria by which it is possible to say how abstract a particular MSS is with respect to another; to incorporate the analysis of intermediate sign systems; to make a theoretical reflection on processes of translating the text of a problem into algebraic code; to refer to the algebraic nature of solving processes for problems and equations in terms of MSSs; and to use this for devising schemes of analysis both for significant works in the history of algebra and for observed phenomena related to learning and the use of algebraic language.

The foregoing is an example of how a historical analysis of works of pre-Vietic algebraists has made it possible to design an experimental setting for observing the phenomena of the transition from arithmetic to algebra on the ontological level, and of how at the end of the study, in response to the need to propose theories to explain the phenomena observed, once again we turn to history.

8.6 Final Remarks

As it was mentioned in the introduction of this chapter, the idea was to analyse key issues in algebra history from which some lessons could be extracted for the future of the teaching and learning of algebra. Sections 8.4 and 8.5 present an example of how comparison of two types of pre-Vietan languages (MSS) and of the corresponding methods to solve problems serves as a basis to formulate conjectures about the nature of difficulties that novices encounter in their transit to algebraic thinking. In this case, for instance, the value of the historic and critical analysis rests on the possibility of getting to the bottom of the epistemological origin of the didactic obstacles (cuts) that are present in the transition. Therefore mathematics education benefits by drawing in basic research on the processes of the acquisition of algebraic language.

But the matter of the different types of MSS that appear alongside the history of algebra symbolism is more complex than what is revealed in the examples above. Section 8.3, *Algebraic language: a history of symbolism*, aimed to make this clear, by approaching the theme of the interrelationship between different evolution stages of the MSS of symbolic algebra and the vernacular language. From this analysis, it can be inferred (among other things) that the different ways of designating unknown quantities and their powers (for instance with the "Abbreviate Plan" or with the "Index Plan") are strongly related to the possibility of syntactic operation on such algebraic objects. In turn, this implies that a MSS represents a significant step in the evolution of algebra symbolism, when it permits calculations to be performed at a syntactic level.

Another issue that arises from the reflection on the relationship between natural language and pre-symbolic stages of algebra, specifically with regards to the designation of unknown quantities, is that the work of Viète is crucial to the Cartesian method, because the language used in this text allows different unknown quantities to be represented with different signs (biunivocally). All these analysis and reflections have implications for the didactics of algebra concerning, on the one hand, the interaction between natural language and the MSS of algebra in the translation processes from a problem text to the algebraic code, and on the other hand, to the possibility of identifying features of "the algebraic" in such processes (see Section 8.2.3, *The algebraic*). The former analyses provide basic elements to study and to characterise pupils' strategies and productions when they solve arithmetic-algebraic word problems, and to conceive didactical routes for the teaching of solving methods of such problems.

The cases approached in this chapter give account of the value of the historic-critical analysis for the field of research and the didactics of algebra itself, especially the analysis of key issues of the evolution of the symbolic language of algebra. The examples discussed emphasise the importance of speaking of manifestations of "the algebraic" in the specific, in contrast to other perspectives that intend to define the nature of algebraic thinking in the general.

At this point, it is suitable to restate the idea, expressed at the beginning, about drawing on results from empirical studies carried out with pupils who are learning algebra to guide ourselves in the identification of moments in the history of algebra that are worthy to be researched in depth, in the terms that have been exposed here.

8.7 References

Abdeljaouad, M. (2002). Le manuscrit mathématique de Jerba: Une pratique des symboles algébriques maghrébins en pleine maturité [The mathematical manuscript of Jerba: A practice of Maghrebian algebraic symbols in full maturity]. *Septième Colloque Maghrébin sur l'histoire des mathématiques arabes*. Marrakech, 30-31 Mai et 1er Juin 2002.

Bachelard, G. (1947). *La formation de l'esprit scientifique. Contribution à une Psychanalyse de la connaissance objective* [The formation of scientific thought. Contribution to a psychoanalysis of objective knowledge]. Paris: Librairie Philosophique J. Vrin.

Boncompagni, B. (Ed.) (1857). *Scritti di Leonardo Pisano matematico del secolo decimoterzo. I. Il liber abbaci di Leonardo Pisano* [Writings of Leonardo Pisano, mathematician of the XIII century. I. The *liber abbaci* of Leonardo Pisano]. Roma: Tipografia delle Scienze Matematiche e Fisiche.

Cajori, F. (1928-1929). *A history of mathematical notations*. Chicago, IL: Open Court Publishing Co. (Reprinted New York: Dover, 1993.)

Campbell, S. R. (2001). Number theory and the transition from arithmetic to algebra: Connecting history and psychology. In H. Chick, K. Stacey, J. Vincent, & J. Vincent (Eds.), *The future of the teaching and learning of algebra* (Proceedings of the 12th ICMI Study Conference, pp. 147-154). Melbourne, Australia: The University of Melbourne.

Cardano, G. (1968). (Trans. T.R. Witmer). *Ars magna or the rules of algebra*. Cambridge, MA: MIT Press [Reprinted New York: Dover, 1993].

Colebrooke, H. T. (Ed. & Trans.) (1817). *Algebra with arithmetic and mensuration from the sanscrit. Brahmegupta and Bháscara*. London: John Murray.

Descartes, R. (1701). *Opuscula posthuma physica et mathematica* [Posthumous works on physics and mathematics]. Amsterdam: Typographia P. & Blaev J.

Descartes, R. (1826). *Œuvres de Descartes* [Works of Descartes]. Paris: Victor Cousin (chez F. G. Levrault, libraire).

Descartes, R. (1925). *The geometry of René Descartes* (with a facsimile of the first edition) (D. E. Smith & M. L. Latham, Trans.). Chicago, IL: Open Court Publishing Co. (Reprinted New York: Dover, 1954.)

Descartes, R. (1996). *Regulæ ad directionem ingenii* [Rules for the direction of mind]. In *Œuvres de Descartes*. Tome X. Édition de Charles Adam et Paul Tannery. Paris: Librairie Philosophique J. Vrin.

Djebbar, A. (1985). *Enseignement et recherche mathématiques dans le Maghreb des XIIIème - XIVème siècles* [Mathematic teaching and research in the Maghreb during the XIII and XIV centuries]. D'Orsay, France: Université Paris-Sud.

Egmond, V. W. (1980). *Practical mathematics in the Italian renaissance*. [A catalog of Italian abacus manuscripts and printed books to 1600]. Firenze, Italy: Annali dell'Instituto e Museo di. Storia della Scienza, fascicolo 1, Instituto e Museo di Storia della Scienza.

Fauvel, J., & van Maanen, J. (Eds.) (2000). *History in mathematics education. The ICMI study*. Dordrecht, The Netherlands: Kluwer Academic.

Filloy, E., & Rojano, T. (1984). From an arithmetical to an algebraic thought. In J. M. Moser (Ed.), *Proceedings of the Sixth Annual Meeting of PME-NA* (pp. 51-56). Madison, WI: University of Wisconsin.

Filloy, E., & Rojano, T. (1989). Solving equations: The transition from arithmetic to algebra, *For the Learning of Mathematics, 9*(2), 19-25.

Filloy, E., Rojano, T., & Rubio, G. (2001). Propositions concerning the resolution of arithmetical-algebraic problems. In R. Sutherland, T. Rojano, A. Bell, & R. Lins (Eds.), *Perspectives on school algebra* (pp. 155-176). Dordrecht, The Netherlands: Kluwer.

Freudenthal, H. (1981). Should a mathematics teacher know something about the history of mathematics? *For the Learning of Mathematics 2(*1), 30-33.

Furinghetti, F., & Somaglia, A. (2001). The method of analysis as a common thread in the history of algebra: Reflections for teaching. In H. Chick, K. Stacey, J. Vincent, & J. Vincent (Eds.), *The future of the teaching and learning of algebra* (Proceedings of the 12th ICMI Study Conference. pp. 265-272). Melbourne, Australia: The University of Melbourne.

Gallardo, A. (2001). George Peacock and a historical approach to school algebra. In H. Chick, K. Stacey, J. Vincent, & J. Vincent (Eds.), *The future of the teaching and learning of algebra* (Proceedings of the 12th ICMI Study Conference, pp. 273-280). Melbourne, Australia: The University of Melbourne.

Høyrup, J. (1991). 'Oxford' and 'Cremona': On the relations between two versions of al-Khwârizmî's algebra. *Filosofi og videnskabsteori på Roskilde Universitetcenter. 3. Række: Preprint og Reprints* nr. 1. Cited in notes, p. 11.

Høyrup, J. (1994). The antecedents of algebra. *Filosofi og videnskabsteori på Roskilde Universitetcenter. 3. Række: Preprint og Reprints* 1994 nr. 1.

Høyrup, J. (2000). Embedding: Multi-purpose device for understanding mathematics and its development, or empty generalization? *Filosofi og videnskabsteori på Roskilde Universitetcenter. 3. Række: Preprint og Reprints* 2000 nr. 8.

Høyrup, J. (2002). *Lengths, widths, surfaces. A portrait of old Babylonian algebra and its kin.* New York: Springer Verlag.

Hughes, B. (Ed.) (1981). *Jordanus de Nemore. De numeris datis.* Berkeley, CA: University of California Press.

Hughes, B. (1986). Gerard of Cremona's translation of al-Khwârizmî's al-jabr: A critical edition. *Mediaeval Studies 48*, 211-263.

Hughes, B. (1989). *Robert of Chester's translation of al-Khwârizmî's al-jabr: A new critical edition.* Boethius, Band XIV. Stuttgart, Germany: Franz Steiner Verlag.

Jahnke, H. N., Knoche, N., & Otte, M. (Eds.) (1996). *History of mathematics and education: Ideas and experiences.* Göttingen, Sweden: Vandenhoek & Ruprecht.

Katz, V. (Ed.) (2000). *Using history to teach mathematics: An international perspective.* Washington, DC: Mathematical Association of America.

Kieran, Carolyn. (1992). The learning and teaching of school algebra. In Douglas Grouws (Ed.), *Handbook of research on mathematics teaching and learning* (pp. 390-419). New York: MacMillan.

Klein, J. (1968). *Greek mathematical thought and the origins of algebra.* Cambridge, MA: MIT Press. (Reprinted in New York: Dover, 1992.)

Kleiner, I. (1998). A historically focused course in abstract algebra. *Mathematical Magazine, 71*(2), 105-111.

Mahoney, M. S. (1971). Babylonian algebra: Form vs. content. *Studies in History and Philosophy of Science (1)*, 369-380.

Mazzinghi, M. A. di. (1967). *Trattato di Fioretti* [Fioretti's treatise]. (G. Arrighi, Ed.) Pisa, Italy: Domus Galileana.

Nesselman, G. H. F. (1842). *Versuch einer kritischen geschichte der algebra, 1. Teil. Die Algebra der Griechen* [Essay on a critical history of algebra. 1st Part. The algebra of Greeks]. Berlin: G. Reimer.

Paradís, J. (1993). La triparty en la science des nombres de Nicolas Chuquet [The triparty in the science of numbers by Nicolas Chuquet]. In E. Filloy, L. Puig, & T. Rojano (Eds.), *Memorias del tercer simposio internacional sobre investigación en educación matemática, historia de las ideas algebraicas* (pp. 31-63). México, DF: CINVESTAV/PNFAPM.

Pérez de Moya, J. (1776). *Arithmética práctica, y especulativa. Decimatercia impresión* [Practical and speculative arithmetic. 13th printing]. Madrid, Spain: En la Imprenta de D. Antonio de Sancha.

Puig, L. (1994). El *De numeris datis* de Jordanus Nemorarius como sistema matemático de signos [The *De numeris datis* by Jordanus Nemorarius as a mathematical sign system]. *Mathesis, 10*, 47-92.

Puig, L. (1996a). Pupils' prompted production of a medieval mathematical sign system. In L. Puig & Á. Gutiérrez (Eds.), *Proceedings of the 20th annual conference of the International Group for the Psychology of Mathematics Education* (Vol. 1, pp. 77-84). Valencia, Spain: Program Committee.

Puig, L. (1996b). *Elementos de resolución de problemas* [Elements of problem solving]. Granada, Spain: Comares.

Puig, L. (1998). Componentes de una historia del álgebra. El texto de al-Khwârizmî restaurado [Components of a history of algebra. Al-Khwârizmî's text restored]. In F. Hitt (Ed.). *Investigaciones en matemática educativa II* (pp. 109-131). México, DF: Grupo Editorial Iberoamérica.

Radford, L. G. (2000). The historical origins of algebraic thinking. In R. Sutherland, T. Rojano, A. Bell, & R. Lins (Eds.), *Perspectives on school algebra* (pp. 13-36). Dordrecht, The Netherlands: Kluwer Academic.

Rashed, R. (1984). *Entre arithmétique et algèbre. Recherches sur l'histoire des mathématiques arabes* [Between arithmetic and algebra. Researchs on the history of Arab mathematics]. Paris: Les Belles Lettres.

Rashed, R., & Vahebzadeh, B. (1999). *Al-Khayyâm mathématicien* [Al-Khayyâm mathematician]. Paris: Librairie Scientifique et Technique Albert Blanchard.

Rosen, F. (1831). *The algebra of Mohammed Ben Musa*. London: Oriental Translation Fund.

Sesiano, J. (1999). *Une introduction à l'histoire de l'algèbre. Résolution des équations des Mésopotamiens à la Renaissance* [An introduction to the history of algebra. Equation solving from Mesopotamia to the Renaissance]. Lausanne, Switzerland: Presses Polytechniques et Universitaires Romandes.

Sigler, L. E. (2002). *Fibonacci's liber abaci. A translation into modern English of Leonardo Pisano's book of calculation*. New York, Berlin, Heidelberg: Springer Verlag.

Stacey, K., & MacGregor, M. (2001). Learning the algebraic method of solving problems. *Journal of Mathematical Behavior*, 18(2), 149-167.

Tannery, P. (Ed.) (1893). *Diophanti Alexandrini opera omnia cum graecis commentariis* [Complete works of Diophantos of Alejandria with Greek comments]. (Reprinted 1974, Vols. 1-2). Stuttgart, Germany: B. G. Teubner.

Taisbak, C. M. 2003. *Euclid's data. The importance of being given*. Copenhagen, Denmark: Museum Tusculanum Press.

Witmer, T. R. (Ed., trans.) (1968). *Girolamo Cardano, The great art or the rules of algebra*. Cambridge, Mass., and London: M.I.T. Press. [Reprinted New York: Dover, 1993.]

Witmer, T. R. (Ed.) (1983). *François Viète. The analytic art*. Kent, OH: The Kent State University Press.

[1] The canonical edition of Descartes' works is the one by Charles Adam and Paul Tannery, *Œuvres de Descartes*, volume X of which contains the original Latin of the rules. These *Regulæ ad directionem ingenii* were not published in Descartes' lifetime and appeared in print for the first time in a collection of previously unpublished texts in Holland in 1701, with the title *Opuscula posthuma physica et mathematica*. The first French translation is contained in volume eleven of the 1826 edition by Victor Cousin, *Œuvres de Descartes*.

[2] *De emendatione æquationum* is included in the edition by Witmer called *The Analytical Art* but does not, in fact, contain only that book by Viète. For Cardano's *Ars Magna* there is also an edition and English translation by Witmer. In both cases Witmer not only translates the Latin into English but also translates Cardano's and Viète's sign systems into modern algebra, so that his editions are not very useful for studying those sign systems.

[3] Babylonian algebra does not satisfy this criterion, even though 1) there are catalogues of techniques and of problems that they can solve; 2) they use the sumerograms that signify "long" and "wide" to represent quantities that have nothing to do with geometrical figures; 3) the solving procedures are analytic; and 4) configurations are reduced to others that they know how to solve (cf. Høyrup, 2002). But Diophantus' *Arithmetic* also does not satisfy it.

[4] Although discussions about priority are not important from the viewpoint of didactic research, it is worth pointing out that we do not know of any text prior to al-Khwârizmî's *Concise book of the calculation of al-jabr and al-muqâbala* in which a complete set of canonical forms is established. In this respect, what constitutes a radical novelty in al-Khwârizmî's book is not the procedures that he explains for solving each of the canonical forms, since the procedures can be found in earlier texts which in some cases are extremely ancient. Instead the novelty lies in the fact that he begins by establishing a complete set of possibilities and expounds algorithms for solving all the possibilities. In other words, before al-Khwârizmî it was known how to solve quadratic equations with standardised procedures, and perhaps it was even known how to solve any quadratic problem, but it was not known that it was known how to solve all quadratic problems.

[5] A detailed discussion of the monetary conceptualisation of al-Khwârizmî's "types of numbers" and the unsuitability of translating *mâl*, which literally means "treasure", "possession (of money)", as "square" can be seen in Puig (1998).

[6] See definitions of "having been given" at the start of Euclid's book *Data* (Taisbak, 2003).

[7] The Latin text of the *Liber Abbaci* was published by Boncompagni (1857). There is also an English translation by Sigler (2002).

[8] Leonardo Pisano's Regula Recta corresponds to Kieran's (1992) "forward operations" and his Regula Versa corresponds to the "backward operations". Leonardo introduces the term "thing" to have a name for the unknown in order to calculate forwards.

[9] The translation of *mâl* as *census* was the one that proved most popular in Christian Western mathematics, and it is the one that Gerard of Cremona used in his translation of al-Khwârizmî's book (Hughes, 1986). Robert of Chester (ca 1150, England), who also translated al-Khwârizmî's book into Latin, translated *mâl* as *substantia* (Hughes, 1989).

[10] We are quoting from the edition by Rosen (1831). The translation, however, is not his, and in composing it we have taken into consideration the observations made by Høyrup (1991) and consulted the Latin translation made by Gerard of Cremona (1114-1187, Italy), published by Hughes (1986), in order to stay closer to al-Khwârizmî's text.

[11] The text by Sinân ibn al-Fath in which he expounds his names for the powers can be read on page 21, note 11 of Rashed (1984): "the first number, the second root, the third *mâl*, the fourth cube (*maka'ab*), the fifth *mâl mâl*, the sixth *madâd*, the seventh *mâl* cube, and then there is the eighth proportion and the ninth and thus whatever you wish".

[12] "if the exponent is divisible by 3, that is, of the form $3m$, the power of the unknown is designated by *ka'b* [cube] repeated m times; if it is of the form $3m + 2$, *mâl* will precede *ka'b* repeated m times; if it is of the form $3m + 1$, two *mâl* will be followed by $m - 1$ *ka'b*" (Sesiano, 1999, p. 57).

[13] These are the names used by Pérez de Moya in his book *Arithmética Práctica, y Especulativa,* written in Spanish, first published in 1562 and reprinted repeatedly during a period of over 200 years. We are quoting from the thirteenth impression (Pérez de Moya, 1776). The oldest mention of a name of this type that has been conserved is in a letter by Psellus, a Byzantine of the eleventh century (1018-ca 1080), published by Tannery (1893, vol. 2, pp. 37-42), in which the fifth power is called "the first that cannot be expressed (*alogos*)".

[14] However, Viète or his publisher also continued to use the sign system in which what is abbreviated is the species, in *Francisci Vietae Fontenaeensis de aequationum recognitione et emendatione tractatus duo per Alexandrum Andersonum.* Paris, Laquehay, 1615. In this book, after writing the general equation A quad $-$ B in A 2, æquetur Z plano, Viète expounds a particular case of it in which B is 1, Z plano is 20 and A is $1N$ (i.e., in the manner of Diophantus, 1 *arithmos* or temporarily indeterminate number) and then, replacing these values, he writes $1Q - 2N$, æquabitur 20, where the letters Q and N represent the species. (This cannot be seen in the edition by Witmer, 1983, as he does not retain Viète's sign system in his translation.)

[15] It is worth pointing out that in order that the production of the syntactic rule should be simple it is necessary that the numbering of species should assign 1 to thing and not to "simple numbers", because if the first position is assigned to simple numbers, as is done by Sinân ibn al-Fath and Luca Pacioli, the product of two species is no longer the species that is at the sum of the positions. But assigning 1 to thing implies assigning position 0 to simple numbers, which requires 0 not to be just a mark of absence in the positional writing of numbers brought from India, but to be integrated into the succession of numbers and, consequently, to serve as an instrument for numbering positions.

[16] Among Arab algebraists of the East there is a precedent in as-Samaw'al's representation of polynomials by tables. In these tables the positions of the species are numbered, and the number (coefficient) of each species is written in the corresponding position. The positions include a zero position for "simple numbers" and two series of numbers, ascending for positive powers and descending for negative powers (which are called "parts of *mâl*", "parts of *cubo*", etc.). These tables clearly come from transferring the calculations done on the dust board onto paper, and therefore they function as "symbols of illustration" and their operativity lies in the actions that are performed on the dust board.

[17] The sign system of symbolic algebra does not consist only of the signs whose history we have traced here, which are its basic pieces, but also of signs for operations and equality, delimiters, rules of preference of operations and of syntax in general, whose history it is also pertinent to study.

The Working Group on Symbols and Language

Leaders: Jean-Philippe Drouhard and Desmond Fearnley-Sander

Working Group Members:
Bernadette Baker, Nadine Bednarz, Dave Hewitt, Brenda Menzel, Jarmila Novotná, Mabel Panizza, Cyril Quinlan, Anne Teppo, and Maria Trigueros.

The Working Group on Symbols and Language. Seated (L to R): Desmond Fearnley-Sander, Brenda Menzel, Nadine Bednarz, Bernadette Baker, Cyril Quinlan.
Standing (L to R): Anne Teppo, Dave Hewitt, Maria Trigueros, Jarmila Novotná, Mabel Panizza, Jean-Philippe Drouhard.

Prior to the Conference, each member of the Working Group on Symbols and Language prepared a paper for the 12th ICMI Study Conference Proceedings. These papers reflected members' expertise and prior experiences in teaching and researching the symbolic and language aspects of algebra. The authors (sometimes with co-authors) and the titles of their papers are listed:

- Bernadette Baker, C. Hemenway, & Maria Trigueros: *On transformations of basic functions* (pp. 41-47).
- Nadine Bednarz: *A problem-solving approach to algebra: Accounting for the reasonings and notations developed by students* (pp. 69-78).
- Jean-Philippe Drouhard: *Research in language aspects of algebra: A turning point?* (pp.238-242)
- Desmond Fearnley-Sander: *Algebra worlds* (pp. 243-251).

- Dave Hewitt: *On learning to adopt formal algebraic notation* (pp. 305-312).
- B. Menzel: *Language conceptions of algebra are idiosyncratic* (pp. 446-453).
- Jarmila Novotná & M. Kubínová: *The influence of symbolic algebraic descriptions in word problem assignments on grasping processes and on solving strategies* (pp. 494-500).
- Cyril Quinlan: *Importance of view regarding algebraic symbols* (pp. 507-514).
- Mabel Panizza: *Generalization and control in algebra.* This paper can be obtained from author (due to technical difficulties it was not included in the Conference Proceedings).
- Anne Teppo & W. Esty: *Mathematical contexts and the perception of meaning in algebraic symbols* (pp. 577-581).
- Maria Trigueros & S. Ursini: *Approaching the study of algebra through the concept of variable* (pp. 598-605).

During the first two group sessions all of the members met together for general discussion. Then, to focus the work of the group, four themes were selected:

- The gradual development of symbolisation processes in early algebra learning.
- Parallels between learning one's natural language as a young child and learning algebraic language.
- The consideration of possible changes in standard algebraic notation to facilitate learning and remove potential ambiguity.
- Language aspects from a semiotic/linguistic perspective.

Sub-groups met to discuss these themes. The results of the work within each sub-group were shared during final whole-group discussions and summarised as briefs in a power point presentation reported at the end of the conference. The authors (see e-mail addresses listed at the back of the book) and the titles of their briefs are listed:

- Nadine Bednarz, Jarmila Novotná: *Beginning algebra: First encounters with algebraic language.*
- Bernadette Baker, Desmond Fearnley-Sander, Maria Trigueros: *Algebraic notations: Variables, equations, functions.*
- Dave Hewitt, Brenda Menzel: *Language aspects of algebra: In classroom practice and in theories of learning.*
- Jean-Philippe Drouhard, Mabel Panizza, Anne Teppo: *Language aspects of algebra: From a semiotic/linguistic perspective.*

The members' extensive and thoughtful work before (e.g., papers in the Conference Proceedings) and during the conference (e.g., briefs and contributions to discussions) provided important insights that assisted Jean-Philippe Drouhard and Anne Teppo during the writing process.

Thanks are extended to Jarmila Novotná and her husband for preparing the power-point presentation for the final Group Presentation and to Jean-Philippe Drouhard and Anne Teppo who co-authored the chapter. Finally, Desmond Fearnley-Sander and Jean-Philippe Drouhard are congratulated and thanked for their leadership of the Working Group on Symbols and Language.

Chapter 9

Symbols and Language

Jean-Philippe Drouhard and Anne R. Teppo
IUFM & IREM de Nice, UMR ADEF, France, and Bozeman, Montana, USA

Abstract: Algebraic symbols are investigated from both a linguistic and a semiotic perspective. In the first part of the chapter, a theoretical framework is presented based on Frege's notions of *sense* and *denotation* and language aspects of the algebraic symbol system that affect how individuals *read* collections of symbols are discussed. The chapter then focuses on the interpretative nature of assigning meaning to symbols, including discussion of a hierarchical framework based on an interpretation of Deacon's work. Meaning-making is also related to the activity of symbolising. The notion of a language focus is related to teacher preparation, and the chapter concludes with a brief examination of possible future directions for research related to algebraic symbols and language. Examples from research are given throughout the chapter to illustrate instructional aspects of the topics discussed.

Key words: Algebraic language, linguistics, meaning-making, semiotics, symbols

9.1 Introduction

Abstract symbols reside within a complex system of rules and internal relationships that make it possible to both communicate and generate powerful mathematical ideas. In addition, these symbols serve as external *representations* of mental objects. Thus, any discussion of symbols and language must deal not only with syntactical issues, but also with the interpretative nature of algebraic communication and understanding.

9.1.1 Symbols and meaning

To set the stage for this chapter, we offer the following example to illustrate the range of mathematical understandings that may be *symbolised* by the same collection of *marks on paper*. Before reading further, consider the nature of the mathematical object that comes to your own mind when confronted with the identity $2(x + 3) = 2x + 6$.

We presented this identity as part of the Symbols and Language Working Group presentation to the other participants of the ICMI Algebra Conference. Our intention was to illustrate the notion of the *equivalence of combinations of operations*. From this perspective, the identity was not a statement about numbers, but rather, a statement about alternative sequences of operations. That is, the operations represented by the expression to the left of the equals sign could be replaced by the equivalent sequence of operations given to the right. The role of x was that of a placeholder, making it possible to shift the focus from talking about numbers to talking about sequences of operations. Here, the identity was being regarded as a useful tool for making algebraic transformations.

What was illuminating about this example was not the personal meaning that we had associated with it, but the range of mathematical entities that was invoked in the minds of others. One of the conference participants took issue with our interpretation. For this person, the symbols were seen as an illustration of a *true* statement. For him, the role of x as a placeholder was important in that it made it possible to express relationships true for all values of the variable. Here, the truth-value of the sentence was of paramount interest. Later, a university mathematics instructor, when presented with this identity, objected to the decontextualised nature of the sentence. She remarked that she would never give such a sentence to her students without first framing it with appropriate words, such as, "Show that, for any real number x, we have $2(x + 3) = 2x + 6$". Here, the identity was taken as an example of a true statement and a representative of the distributive law.

Kieran and Sfard (1999) report on yet a different interpretation. They developed a 30-day module for 7th grade students (11 years old) designed to introduce algebra using a graphical approach to functions. It was decided to orient the material in such a way that, after instruction, a student might provide the following explanation for why $3(x + 2)$ equalled $3x + 6$ and not $3x + 2$:

> Well, if you take the graph of the function $x + 2$ and multiply the y-value of each point by 3, you will get another linear graph. That second graph has a y-intercept of 6 and a slope of 3, so its expression is $3x + 6$ and not $3x + 2$. (p. 3)

From this perspective, such a student would tend to regard identities as statements of *equivalent functions*. Kieran and Sfard also provide a rule-based interpretation. A 12 year-old student who had just completed a more traditional introduction to algebra gave the following response during an interview. When asked to explain why the expressions $3(x + 2)$ and $3x + 6$ are equivalent, he remarked, "Yeah ... because they're rules ... They are there so that you can follow them, so that everybody'll do the same thing" (p. 1). The student appears to be focused on the mastery of a set of manipulation rules. To him, symbols may be just carriers of practical rules, developed to obtain results. Notice here, regardless of the meaning assigned to the symbols, working with the given expressions, equations, et cetera, involves manipulation of symbols under a set of rules to achieve the result.

These examples present five different interpretations of the same, or similar, collections of symbols. The cognitive objects to which the symbols and words refer are mental constructs that reflect different webs of meaning that, for each individual, might be said to be part of their personally constructed system of *algebra*. The examples from Kieran and Sfard (1999) illustrate how the contexts of past mathematical experiences play a role in shaping one's interpretation of a given situation. Sfard (2000) also emphasises the importance of context in the development of mathematical meaning. She maintains that the *street mathematics* used by Brazilian street vendors is truly different from similar mathematical calculations carried out by school children with pencil-and-paper.

> Clearly, for those who only know how to use numbers in everyday discourse, the number is not the same object as it is for the learner of mathematics in school. Although the same names appear in both discourses, these names lead to entirely different connotations and, thus, constitute different signs. ... The Brazilian vendor and the European pupil refer to completely different objects using the same number name. (p. 86)

Before we investigate the interpretative nature of *algebraic meaning*, we need to place our discussion of algebraic symbols and language within a broader framework. In the following sections we introduce and explain the vocabulary we will be using, examine several general characteristics of language, and discuss particular syntactical and semantic aspects of the algebraic language.

9.1.2 What is this algebraic language?

Reading more than one research paper about algebraic symbols and language may provoke an intense sense of confusion. Not that such papers are confusing by themselves: on the contrary, most address interesting problems, and present rich and deep theoretical frameworks. The problem is that their perspectives are different. However, this difference is not a mere question of translating from one theoretical vocabulary to another. The various frameworks are neither equivalent nor opposite; while certain parts match, other parts are mutually incompatible. In addition—and this is the real source of confusion—different papers use the same keywords (language, sign, symbol, meaning, sense, representation) to express quite different concepts. Therefore, we will begin by clarifying certain terms, and give a precise meaning to the words *language* and *symbol*, as we are applying them to the case of algebra.

Consider the occurrence of (algebraic) symbols and (algebraic) language in textbooks. Although used in students' handbooks, blackboards, teachers' and students' speech, et cetera, the incidence of symbols and language in textbooks is more homogeneous and, thus, easier to study (textbooks are good corpuses, as linguists say). These printed elements on a textbook's page can be separated into three categories: *natural language*, *symbolic writings*, and *compound representations*. Algebraic textbooks are full of sentences and phrases written in English (or any other idiom), such as "the number of matches is given by the following expression: ...". We call this the *natural language* component. We call expressions such as 4, $\frac{2+3}{6}$, $2n + 1$ or $2x - 3 = 7$ *symbolic writings*. Finally, there is almost no textbook without some sort of graph, drawing, or scheme as in Figure 9.1.

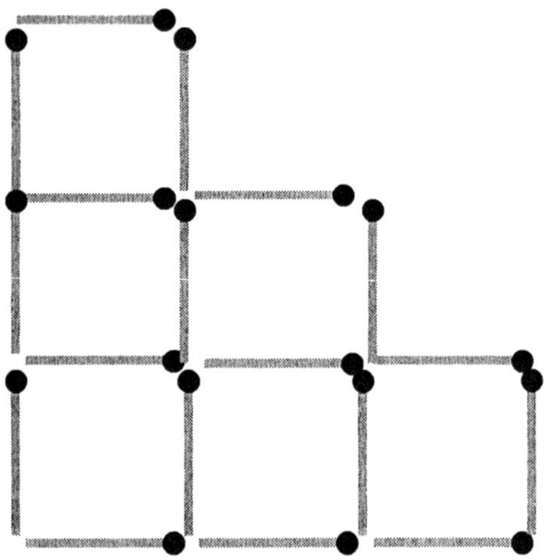

1	2	3	4	5
3	5	7	9	11

Figure 9.1. Typical diagrams in introductory algebra textbooks.

We designate the example in Figure 9.1 a *compound representation*, using the word *compound* to indicate that such entities can consist of both symbolic writings (in the second example, the 2 and the 11 are symbolic writings), drawings, and some natural language for labels et cetera.

We drew the above characterisations from Laborde (1982, 1990). Guillerault and Laborde (1982) also ascribe a similar three-component characteristic to all mathematical texts. What is called the "mathematical language" is composed of both natural language and a symbolic system, with the symbolic system further broken down into symbolic writings and compound representations. Following Laborde (using a free translation of Laborde's French terms), we characterise *algebraic language* as the set composed of natural language, algebraic symbolic writings, and algebraic compound representation (with the algebraic language being a subset of Laborde's "mathematical language").

9.1.3 Components of algebraic language

9.1.3.1 Definitions of language

Different linguists through different series of characteristics variously define languages as particular systems of signs. For example, Chomsky (1965) describes the generation of sentences by a formal "grammar". Martinet, quoted in Ducrot and Todorov (1972, p. 73), addresses "articulation", whereby sentences can be broken up into words and words into sounds. Although characteristics vary according to each specialist, the question of how linguists define a language is not of concern here. Rather, the relevant point is that we can describe the language components of the algebraic language (and the related teaching and learning problems) using linguistic levels of analysis, such as *syntax* (the organisation and transformations of symbols), *semantics* (the level of meaning), and *pragmatics* (the relation between signs and their users). Language can also be described by linguistic concepts such as *sense*, *ambiguity*, et cetera. An important distinction is that although languages are made up of words considered to be signs not all systems of signs are languages. Ducrot and Todorov (1972) define "signs" as entities that have, for a defined group of users, a particular form (the signifier) and a "meaning" (the signified). The general framework for studies on systems of signs is referred to as *semiotics*. We will return to these ideas in a later section.

The *natural language* component of algebra (English, or French, or Spanish, or any other idiom) is, clearly, a language. What is not as obvious is that the *symbolic writings* component is also a language, a point independently demonstrated by Kirshner (1987a, 1987b, 1989) and Drouhard (1989, 1992, 1995)—(see also Drouhard, Léonard, Maurel, Pécal, and Sackur, 1994).

To do this, the authors first chose a precise definition for what is a language, and then carefully proved that the symbolic writings system matched all the requirements of that definition. Thus, as with the natural language component of algebra, symbolic writings can be described using linguistic levels of analysis and linguistics concepts. In particular, at the syntactical level of analysis, the Chomskyan generative and transformational grammar is relevant (see the Kirshner and Drouhard references listed above). At the semantic level of analysis, the ideas of Gottlob Frege (1842-1925, Germany), as discussed in Section 9.2.1, are relevant (Arzarello, Bazzini, & Chiappini, 1994; Drouhard, 1995).

The *system of compound representations*, on the contrary, is *not* a language, since it does not fit with a particular characterisation of a language (grammar-ruled, or articulated, etc.). Although it is just a system of signs (or *semiotic system*, which is equivalent), compound representations are indeed complex entities, often made up of more elementary compound representations. They have meanings, and their *semantics* are rich and sophisticated. Compound representations are described and analysed by semiotics rather than by linguistics. Many authors (e.g., Davis & McGowen, 2001; Deacon, 1997) have shown the relevance of the ideas of Charles Sanders Peirce (1839-1914, USA) for thinking about the language of mathematics, and we discuss below how to apply them to algebraic signs.

We claim that signs (if they are *not* part of a language) must be studied with a semiotic framework (and not a linguistic one) whilst linguistic signs like x (which *are* part of a language) might be better studied first with a linguistic framework. Without such a clarification, observations about learning and teaching might well either remain at an informal level without an underlying theoretical framework, or remain confused through the application of unsuitable frameworks. This point is not just academic. Within the framework we have presented (and will discuss further in Section 9.4.1), we claim that Peirce's classification of signs (according to Davis & McGowen, 2001) is not useful for linguistic symbols (and therefore inappropriate for understanding students' difficulties with algebraic letters, for instance). In addition, the notion of *syntactic sense* is relevant just as a metaphor for understanding students' difficulties with graphs or tables.

9.2 Semantics of Symbolic Expressions: Frege

9.2.1 Meaning: Denotation and sense

A common statement that teachers present to beginning algebra students is that letters *represent* (or refer to, designate, etc.) numbers. This statement is efficient and easy to say, but not necessarily illuminating to students. In fact many students, when asked, mention the *letters* as the origin of their difficulties in algebra (Sackur, 1995). They complain that they have understood mathematics until letters appear. For many students, letters mean distress. As mathematics educators, we cannot put the blame for this distress on the students' laziness or stupidity. Such a massive learning problem must be based on non-apparent, but real difficulties. The question "What is a letter that *represents* a number?" may be reformulated as "What does a letter (an algebraic expression) *mean*?" To investigate this question from a mathematical perspective we turn to the work of the German logician Gottlob Frege (1848-1925) and, in particular, to that aspect of his work that addressed what exactly mathematical sentences are: "What are letters?" and "What is the meaning of a mathematical sentence?"

Frege, in his 1892 article entitled *Sense and Meaning* presented in 1891 at a conference (cited in McGuinness, 1984, pp. 157-177), examined meaning in terms of the (mathematical) *value* of expressions—meaning that is neither situational nor conceptual. In order to make sense of Frege's work we decided to use modern words rather than the literal translations. His breakthrough consisted in considering this meaning not as an elementary concept but instead as composed of two complementary phenomena: ***denotation*** (or *reference*, in German "*Bedeutung*") and *sense* (in German, "*Sinn*"). For example, *Melbourne* and *The capital of the state of Victoria* have the same denotation: the actual city. However, these two phrases don't have the same sense: the second sentence emphasises the official role of the city while the first stresses its name. Frege illustrated this general distinction with natural language phrases (the Morning Star, the Evening Star, and Venus); but then made a detailed analysis of the denotation and sense of mathematical expressions (or expressions of the mathematical symbolic language), and that is the interesting point for us in this chapter.

Frege considered four cases, each one being a generalisation of the previous one: arithmetic expressions (e.g., $2 + 3$), arithmetic statements (e.g., $2 + 3 = 5$), algebraic expressions (e.g., $x + 3$), and algebraic statements (e.g., $x + 3 = 7$). We will first analyse all four cases in terms of their denotations, and then discuss the sense attributed to algebraic expressions and statements.

Denotation of arithmetic expressions (e.g., $3 - 1$). Arithmetic expressions (i.e., expressions without letters) denote numbers. For example, 2, $1 + 1$, $3 - 1$, $6 \div 3$ denote the same number (two). "The different expressions," says Frege, "correspond to different conceptions and aspects, but nevertheless always to the same thing" (McGuinness 1984, p. 139). To form a more complete theory than that developed by Frege, we must also consider the case of well-written meaningless expressions, such as $\frac{1}{0}$. Such an expression has a denotation that we can arbitrarily note as U (undefined).

Denotation of arithmetic statements (e.g., $2\times4 + 3 = 11$). Frege introduced a little conceptual revolution regarding arithmetic statements (i.e., writings with = or < signs etc.) by considering that such writings do not refer to a statement of fact (such as "the double of 4 added to 3 is equal to 11") but instead, to a truth value (here: TRUE). In general, every arithmetic statement denotes a truth value (TRUE or FALSE), or U if at least one of the sub-expressions is undefined (for example $\frac{1}{0}$). Note that here Frege is only describing the denotation component of the arithmetic statements' meaning. Obviously, Frege considers the particular nature of the facts as an aspect of the meaning of a statement, but he assigns this aspect to the *sense* component rather that to the *denotation.*

Denotation of algebraic expressions (e.g., $2x + 3$). Following Frege's article *What is a Function?* (1904), in McGuinness (1984, pp. 285-292), we can ask what is the meaning of the familiar statement that algebraic expressions (e.g., $2x + 3$) represent indefinite or indeterminate numbers. What is an *indeterminate* number? Does it exist, as in something like *the set* of *indeterminate* numbers? Although *indeterminate* is not a mathematical property, like *linear*, it is used on an everyday basis by mathematicians. Frege says:

> Of course we may speak of indefiniteness here; but here the word "indefinite" is not an adjective of "number", but "indefinitely" is an adverb, e.g., of the verb "to indicate". We cannot say that *n* designates an indefinite number, but we *can* say that it indicates numbers indefinitely. And so it is always when letters are used in arithmetic, except for the few cases (π, *e*, *i*) where they occur as proper names; but then they designate definite, invariable numbers. (Frege, 1904, in McGuinness, 1984, p. 288)

This second breakthrough by Frege rejects the idea that expressions denote indeterminate numbers. Instead, Frege proposes (phrased in modern words) that an expression denotes a function (i.e., the function that associates, to every value of the letters, the value taken by the expression). This might be interpreted as the origin of the modern distinction between polynomials (elements of a polynomial ring R[*x*]) and polynomial functions, except that not all algebraic expressions (e.g., $b+\sqrt{1-a}$) are polynomial or even rational.

Interestingly, saying that expressions denote indeterminate numbers is not any different than calling a function by the name of its image (the good old function *f*(*x*)). Using Frege's perspective, the denotation of $2x + 3$ is a function, and the *indefinite number* is the generic number for the image of this function.

Of course, there are other ways of looking at $2x + 3$ without thinking of it as a function. For example, a 20th century mathematician is quite happy to say that $2x + 3$ is an element in a polynomial ring R[*x*]. We will examine how semiotics considers meaning as an interpretive act later in Section 9.4.2. Here, our discussion uses Frege's examples to illuminate his framework.

Denotation of algebraic statements (e.g., $2x + 3 = 11$). Consistent with his analysis of arithmetic expressions and statements, Frege considered that algebraic statements denoted functions (being algebraic, i.e., with letters) whose images are truth values (being statements). In modern formal terms, an algebraic statement *s* denotes a function δ_s from $(\mathbb{R}\cup\{\text{U}\})^n$ to {TRUE, FALSE, U}.

For example, $2x + 3 = 11$ denotes a function whose value is FALSE for all $\mathbb{R}$ except 4 (where it is TRUE) and it is U when *x* is U. The identity $2(x + 3) = 2x + 6$ denotes a function with a value TRUE for all real numbers (and U when *x* is U). Of course we presume here that $n = 1$ (no other variable than *x*). Teachers know well that many students have difficulties in interpreting $2x + 3 = 11$ in the context of two-variable equations (where $n = 2$).

9.2.2 Sense of algebraic writings

What is the difference in meaning between, say, $2x(x + 1)$ and $2x^2 + 2x$? The denotation of these two expressions is the same: the function $x \rightarrow 2x^2 + 2x$. Similarly, what is the difference in meaning between, $2x^2 + 2x = 0$ and $x = 0$ or $x = -1$", their denotation being the same? Both statements denote functions that are false for all values of $\mathbb{R}$ except 0 and 1 and U. This difference can be found in what Frege calls the *sense*. The sense, according to Frege, is in the way denotation is *given*; the sense of writings permits us to know how it is made. In the case of an arithmetical expression (say, 2×(3 + 4)) the sense indicates how to find its denotation: here, one has to double the sum of 3 and 4. The sense also gives us information on what can be done. For example, $2x^2 + 2x$ can be factored, while $2x(x + 1)$ can be expanded.

At this point the reader might feel that the description of what algebraic writings represent has reached a rather complicated level of abstraction and is of no use in everyday classroom practice. Nothing would be more wrong! With this kind of analysis in mind, let us observe the behaviour of an *expert*, such as a teacher or a more advanced student, dealing with a complex inequality, for example, something tedious like:

$$2 - \frac{\sqrt{x+3}}{x} > 0$$

Writing always takes its sense within a certain context; that is, it is the user that gives the writing its sense. For example, an expert, observing the inequality might make frequent shifts between its sense ("Well, I can find the solution by adding $\frac{\sqrt{x+3}}{x}$ to each side, squaring the result, ..." etc.) and (what we call) its denotation ("Uh, oh! I cannot have simultaneously [true] $x + 3 \geq 0$ and $x \neq 0$. So..." etc.). Observing experts and, by contrast, less able students, we find that the capability to deal with not only the sense but also the denotation of writings is the key to a correct attitude in algebra. Actually, it could be said that without denotation, algebra would be just what a lot of students believe it is—the mechanical application of meaningless rules to meaningless writings.

Algebraic activity—such as solving an equation, reducing an expression, or simplifying a sum of fractions—are all games that produce a sequence of changing sense, while, simultaneously, leaving the denotation the same. In each case, the very last line is a writing that is equivalent to the first line. However, the sense of this last line is that its denotation is obvious to determine: for example, the denotation of a statement such as $x = 1$ is very obvious.

Many authors in algebra education have worked on this notion, with the same vocabulary (Arzarello, Bazzini, & Chiappini, 1994) or more or less closely related concepts (e.g., Artigue, Abboud, Drouhard, & Lagrange, 1994; Boero, 1993; Nicaud, Bouhineau, & Gélis, 2001; Vergnaud, 1996). All stressed the importance of attending to the much underestimated sense component of the meaning. They note that students with poor capabilities to recognise this aspect of the meaning of an expression often make endless calculations because they do not know in what direction to go and when to stop.

The notion of a pedagogical emphasis on both components of mathematical meaning provides a useful framework for evaluating instructional approaches. For example, consider the common claim of some studies, particularly related to CAS (Computer Algebra Systems), that computer tools help students avoid tedious calculations, facilitating instead high-level reasoning (see for instance Lagrange, 2001, p. 376). Applying Frege's theories, such claims are false, if not harmful. Considering that CAS help students by letting them *think* instead of dully calculate, comes down to overemphasising the denotation component at the expense of the sense component of meaning. Students can only acquire the sense of writings after a rather long practice, not simply through having observed a CAS (or a teacher) solve some writings a couple of times. We suggest that an exaggerated reduction in algebraic calculation practice in the curriculum may lead to an effect opposite to the one desired. Although students have more time to solve rich and realistic problems using, possibly, high-level reasoning, in fact, the writings they produce and read, may be more meaningless than ever. This is because sense is neither more nor less important than the denotation component of the meaning.

9.2.3 Teaching denotation and sense

In teaching algebra, the *use* of a spreadsheet can be considered as a robust metaphor for denotation (and therefore for sense). The *dynamical* aspect of denotation (which is the functional relationship between the value of the variables and the value of the expression) appears here quite clearly. For example, in Figure 9.2 the content of a cell (say, **C2**) is based on the content of another cell (say, **A2**) via a so-called *formula* (here, =2*A2+1) so that the displayed content of **C2** depends *dynamically* on the content of **A2**.

	A	B	C	D	E
1					
2			=2*A2+1		
3					

Figure 9.2. Extract of a spreadsheet, with the 'display formulae' option.

The *formula* that expresses the relationship between the contents of the two cells is the strict equivalent of an algebraic expression like $2x + 1$ or, better said, a translation of this expression in the language of spreadsheets. Such translations are a particular case of what Duval (2000, pp. 1-63) calls "conversion" in the more general domain of semiotic systems: Conversion is the transformation of representation of an object by changing register. Here, both algebraic expressions and spreadsheet formulae are not mere semiotic systems but also languages (e.g., sentences like =2*A2+1 are generated by a grammar, see above).

Figures 9.3 and 9.4 display the dynamical aspect of denotation. Here, a change of the displayed value in **A2** (to change 2 into 4, for instance) involves a quasi-simultaneous change of the value that appears in **C2** (changing the 5 into a 9).

	A	B	C	D	E
1					
2	2		5		
3					

Figure 9.3. Insertion of **2** into **A2**.

	A	B	C	D	E
1					
2	4		9		
3					

Figure 9.4. Change from **2** to **4** into **A2**.

Moreover, as Figure 9.5 shows, if the column A contains a vertical series of numerical values, it is possible to *copy downwards* C2 in order to see the various values of C2 corresponding to the various values of A2. Then we obtain:

	A	B	C	D	E
1					
2	1		3		
3	2		5		
4	3		7		
5	4		9		

Figure 9.5. Formula 'copied downwards'.

Here, it is not the dynamical aspect of the function $\delta_{\text{"}2x+1\text{"}}$ which appears but rather (a part of) the *graph* of this function (where a function is considered as the set of ordered pairs). Spreadsheets can also be used to show the relationship between sense and denotation. Many exercises can be done (see Rojano & Sutherland, 2001) to determine if various formulae (for instance, =2*A2+1, =2*(A2+1) and =2*(A2+0.5)) are equivalent, all producing the same numbers. Students can also be asked to modify a formula (for instance, =2*(A2+1)) to produce the same numbers as those generated by a second (for instance, =2*A2+1).

Eventually, as Figures 9.6 and 9.7 show, the spreadsheet formulae language can be used to express Boolean functions. It is rather easy to write a formula that, when a condition is satisfied, gives true or, for the contrary, false:

	A	B	C	D	E
1					
2		=2*A2-4		=IF(B2=C2, true, false)	
3		=2*A3-4		=IF(B3=C3, true, false)	

Figure 9.6. Showing the 'display formula' option.

This can produce the following illustration of the denotation of an algebraic statement.

	A	B	C	D	E
1					
2	1	-2	0	false	
3	2	0	0	true	
4	3	2	0	false	
5	4	4	0	false	

Figure 9.7. Denotation of the equation $2x - 4 = 0$.

Exercises such as the above can facilitate the shift between a focus on the denotation of the statement $2x - 4 = 0$ (i.e., **true**, **false**) and its sense (two times **A3** minus **4** is actually **0 (zero)** when the value of **A3** is 2).

9.3 Teaching and Learning Problems

Algebraic thought is made overt through the three components of natural language, symbolic writings, and compound representations. Acquiring a mastery of these components, however, is not straightforward. In this section we consider language aspects that may affect how individuals *read* symbolic writings, and discuss some of the (mathematical) attributes of the language components that present obstacles to achieving algebraic fluency.

9.3.1 Syntax

Languages allow us to structure experiences in particular ways, enabling us "to see and say things which are not possible without access to such a tool" (Sutherland 2001, p. 570). Languages are compositional: they make it possible to combine a limited collection of sounds (symbols) to form a limitless number of words (expressions) and sentences (equations and inequalities). Syntactic rules indicate how these elements can be combined to form higher, meaningful units, such as phrases, clauses, and sentences (McArthur, 1992).

The language aspects of algebra can be highlighted by thinking about how they are used to structure certain mathematical situations and reason about them. Algebraic activity involves operating on the unknown, stating generalisations about patterns of operations, describing functional relationships, and making statements about equivalent combinations of operations. Basic elements, including numbers, variables, and operations can be combined to form other entities of increasing complexity, such as expressions, equations, functions, and theorems.

9.3.1.1 Structure sense

Mastering the syntax or the rules for manipulating symbolic writings may be problematic for many students. Linchevski and Livneh (1999) propose that it is necessary for students to possess a "structural sense" in order to successfully employ the conventional rules of symbol manipulation. They maintain that algebraic structure is difficult to comprehend if a student does not have an appropriately developed understanding of the structural properties associated with the arithmetic of the number system.

For example, students given the equation $4 + n - 2 + 5 = 11 + 3 + 5$ have been observed to simplify it as $4 + n - 7 = 19$, detaching one of the terms from its indicated operation. Using the ideas of Frege, it can be said that such students have not yet developed an appropriate set of rules for their construction of the *sense* of the original equation (the *way* in which the denotation is given). Developing a structural sense for algebra can be a challenge.

> We should not underestimate the cognitive obstacles involved. ... [Recent] studies consistently show that the justification of algebraic relationships as generalisations of numerical relationships is not easily accessible to the young student, and that the "classic" approach to make algebra meaningful to the students via the "arithmetic" connection is quite questionable. (Linchevski & Livneh, 1999, p. 192)

Linchevski and Livneh suggest that, to promote structure sense, "instruction should promote the search for decomposition and recomposition of expressions and guarantee that the mental gymnastics needed in manipulating expressions makes *sense* [our emphasis]" (p. 191). For example, constructing the calculation for 8×23 by decomposing the 23 to produce $8 \times 10 = 80$; $8 \times 3 = 24$; and $80 + 80 + 24 = 184$.

9.3.1.2 How symbol-strings are apprehended

In their written form, natural languages possess a linear structure related to the temporal order in which words are used to communicate. In contrast, many algebraic symbol strings are *read* in non-linear ways. In many cases, particular notational conventions or syntactical rules override the strictly left-to-right processing of symbolic writings, as in the following examples.

$$\text{Evaluate: } \int_0^1 x^2\,dx \qquad \text{Simplify: } 6 - 2(3 - 3x)^2$$

The order in which signs are read is determined by the ways in which individuals interpret these collections of symbols. Consider the following symbolic writing:

$$\frac{4-3x^2}{x+9}$$

If this expression is *said* aloud, one reading could be, "Three times x squared is subtracted from 4, and the result is divided by the sum of x and nine." Here, the writing is perceived operationally, as a set of directions to be *carried out*. On the other hand, it could be viewed more *globally* as one expression *over* another, denoting a fraction.

Note also that, in regard to *speaking* algebra, symbol strings can be considered to be ideograms (i.e., like Chinese characters) and share an essential feature with ideograms. To write a given word, say *man*, Cantonese Chinese, Mandarin Chinese and—to some extent—Japanese use the same written ideogram. However the corresponding spoken words are totally different ("yàhn", "rén", and "hito") since the three languages are different. In the same way, $(2x + 3)^2$ is uttered "two x plus three, squared" in English; "deux x plus trois, au carré" in French; and "dos equis más tres, al cuadrado" in Spanish.

Pirie and Martin (1997) have addressed the notion of "seeing" a symbolic writing as an entire entity rather than as a sequenced string of symbols. They describe a set of lessons given to less able students at a secondary school that were based on helping students to take a more global view of symbols in order to understand linear equations.

> It is just this need to see an equation as a whole, in one instant, rather than as an accumulation of items and operations processed over time, that is crucial for the complete understanding of linear equations. (p. 160)

By changing the way the collection of symbols was perceived, students were able to work intuitively with equations of the form $ax + b = cx + d$; a form that has been regarded in much of the literature as being on the far side of the "didactic cut" described by Filloy and Rojano (1984) and explained in this book in Chapter 8.

9.3.2 Sources of ambiguity

The symbolic writings of algebra are compact and powerful. This aspect of the language makes it possible to move fluently through layers of abstraction and compress complex mathematical thoughts into efficient symbol strings. At the same time, however, these characteristics make symbolic writings very opaque for the learner. There are deep ambiguities in symbol use that are advantageous to the expert, but difficult for the novice.

9.3.2.1 Process/object duality

Many collections of symbols can be interpreted either as expressing a *process* or as denoting a mathematical *object*. For example, the expression $2x - 6$ may be viewed procedurally as a set of directions for operating on the variable x. From another perspective, this expression can also be taken as denoting an object in its own right;

that which results from carrying out the particular operations. Gray and Tall (1994) have coined the term *procept* to describe such symbols that represent both a process and the object of that process.

Sfard and Linchevski (1994) explain the process/object duality of algebraic symbols in terms of hierarchical levels of mathematical understanding involving two modes of thinking—operational and structural. According to their notion of *reification*, moving from a focus on process to seeing that process as an object in its own right (e.g., as an expression or a function) involves a significant cognitive restructuring. Thus, the interpretation of a collection of symbols "depends on what one is *prepared* to notice and *able* to perceive" (p. 192).

The difference between the notion of a procept and Frege's framework is subtle. In our interpretation, this difference lies in how an individual's perception is characterised. Frege sees sense and denotation as representing two complementary phenomena that, together, comprise the meaning assigned to a symbolic writing. On the other hand, Gray and Tall use a procept to describe a separation in perception—the idea that symbolic writings are perceived *either* procedurally, or structurally. They characterise an individual who is fluent in algebra as one who can move easily back and forth between these two points of view, according to the dictates of the given mathematical context. Sfard takes the notion of perceptual focus one step further, placing the operational and structural properties of symbolic writings at different levels of cognitive abstraction. Rather than being mutually exclusive, all three characterisations of meaning are useful, helping to illuminate different aspects of a complex set of mental processes that may be occurring as students attempt to make sense out of collections of algebraic symbols.

9.3.2.2 Minimal symbol set

Algebraic language is potentially ambiguous. As mentioned above, collections of symbols can be seen as either representations of procedures or can be taken structurally to stand for mathematical objects. There are other sources of ambiguity associated with the language's use of a minimal symbol set. For example, the letter x, according to the context, *is* a variable (which means, according to Frege, that x denotes numbers in an indeterminate way), whether this denotation is for an unknown, a placeholder, or the argument of a function. Students also perceive yet another use for variable. When encountered in an algebraic sentence x signals "here's something to be calculated" (Novotná & Kubínová, 2001, p. 496). In this interpretation, x is regarded not as an unknown or generalised number, but as an indicator for a process that must be carried out in order to reach an answer.

As an additional example, consider the different kinds of algebraic sentence in which = appears. This symbol can be used to indicate equality of numbers ($5 + 3 = 8$), equivalence of expressions ($a - (-b) = a + b$), or to define a function ($f(x) = 2x + 7$). The nature of the algebraic relationship in each case is conceptually different because of the nature of the *objects* that are linked by the equals sign. Frege made a distinction here between arithmetic and algebraic statements. Students also bring with them from arithmetic the notion that = means *yields* from their encounters with exercises such as 4 + 7 = □.

Ambiguity can be controlled by paying attention to the mathematical context in which multi-use symbols appear. For example, Trigueros and Ursini (2001) recommend that students be presented with simultaneous situations involving the various uses of letters and encouraged to examine how the letters behave in each situation. They provide the following questions as an example of an activity for an introductory algebra course (see p. 603).

> Consider the expression $2x + 9 = 0$. What does the x represent? How many values can x take to make the expression true? Can you find the value of x?
>
> Consider the expression $2x + 9$. What does the x represent? How many values can x take to make the expression true? Is it possible to determine its value?
>
> Consider the expression y = 2x + 9. What does the x and y represent? How many values can x and y take to make the expression true? Is it possible to determine their values?

9.3.3 Pragmatics: Tools and use

Donald (1991) points out that the use of a physical notation system influences the kind of thought that is possible. Historically, the invention of *graphic representations* produced a shift away from the oral, narrative cultures of prehistoric man to the present-day *theoretic* culture, with a concurrent shift from the use of *internal* to *external* memory devices (cf. Havelock, 1986; Goody, 1987, 1994). Such external coding facilitates analytic thought, which Donald characterises as, among other things, "formal arguments, systematic taxonomies, induction, deduction, verification, differentiation, quantification, [and] idealisation ..." (pp. 273-274). Villarreal (2000, p. 2) makes a similar point. He quotes a student, trying to solve a mathematical question, who asks, "... may I think on paper?" This question highlights the notion of "paper as a place where thought can develop ... [where] student, pencil-and-paper [form] a *thinking collective*". Here, the medium becomes an "object that is mediating human thought".

Goldin and Kaput (1996) focus on the "information-carrying capabilities" of different media and how these influence features of the systems "instantiated within them" (p. 410-411). Thus, algebraic symbols manipulated with a pencil-and-paper medium facilitate a different kind of activity and thought than, say, the set of symbols with which one interacts in a spreadsheet environment.

For example, consider a spreadsheet exercise similar to those described in Section 9.2.3. The equivalence of the three formulae =2*A2+2, =2*(A2+1) and =4*A2–(2*A2–2), can be easily checked by using the *copy down* feature. Here, the particular medium places the emphasis on denotation, through a functional approach. Using pencil-and-paper, however, the same activity focuses students' attention on the sense of the expressions $2x + 2$, $2(x + 1)$, and $4x - (2x - 2)$. The physical actions of symbol manipulation highlight syntactical aspects of the algebraic language.

A useful perspective for thinking about the relationship between tools and use is to shift from a focus on mathematics as a collected body of knowledge and concentrate, instead, on the *activity* of *doing* mathematics. From an instructional stance, the emphasis is thus turned away from the end product and concentrated on the *processes* of meaning-making, which include, among other mathematical activity, social discourse and the act of symbolising.

Although the area of mathematical discourse is extremely important, it is not a focus of this chapter. We just briefly mention here a few aspects of natural language as they affect the way that *mathematics* is spoken. The activity of symbolising will be addressed, also briefly, in Section 9.5.

9.3.4 Natural language aspects

Mathematical language (with its three components) is integral to mathematics learning. The natural language component plays a key role. "Mathematics education begins and proceeds in language, it advances and stumbles because of language, and its outcomes are often assessed in language ..." (Durkin, 1991, p. 3). Words are used to gain access to previously learned concepts, to help direct the learner's attention to particular aspects of a given situation, and to label newly developing concepts. Such words can be technical, with a meaning specific to mathematics; sub technical, with a meaning depending on context; or general, everyday language. In this section, we move away from Frege's notions of sense and denotation, looking instead at how language is *used* to support mathematics learning.

9.3.4.1 Technical vocabulary

The ability to explain technical words in natural language may be problematic for students. In a study conducted in Australia, 13-year-old students were asked to define a list of 20 common mathematical terms. The average number of terms that the 646 students could define in their own words was four. Even when symbols, diagrams, or examples were also considered for an acceptable response, the average number of terms correctly defined was only eleven (Miller, 1993).

The results of the study may be interpreted as measuring students' lack of appropriate concept development. These results also highlight the important role that teachers play in such development, as they model the use of technical vocabulary. For example, Miller describes how a group of teachers from the researched classrooms reported that they always used the word "answer" to refer to the quotient in a division problem, explaining that they "knew that the students would not know the meaning of the word 'quotient'." Given this response, it is not surprising that only 7 percent of the 646 students were able to give an acceptable definition of *quotient* using words, diagrams, examples, or symbols.

A deficiency in a technical vocabulary may inhibit a student's ability to effectively use text-based materials and consequently have an impact on their algebra skills. A recent study of students enrolled in a USA college course covering high school-level algebra found that a measure of the students' algebra vocabulary was positively correlated with algebra achievement, but not with general reading vocabulary (Miles, 1999). The researcher cautions "one cannot assume that because an individual has a good general vocabulary that the individual has a vocabulary sufficient for learning and concept formation in a specific domain" (p. 45-46).

Assessing a student's ability to manipulate symbolic writings only partially characterises that student's algebraic development. It is also important to provide students opportunities to *say* mathematics, either aloud or in writing. As an example, recently, Kirkpatrick and Teppo considered one student's response to the following item taken from a final exam in a one-semester college course designed to introduce students to the *Language of Mathematics* (see Esty, 1999).

Write out in words the pronunciation of the given symbol sentence:

$$f(x + h) = (x + h)^2$$

Answer: The product of x plus h, times f equals the product of x plus h squared.

9.3.4.2 Expressing mathematical ideas

Careful attention needs to be paid to the *way* in which natural language is used to talk about mathematical situations and relations. Sfard (2000) draws attention to the crucial role of *naming* in the movement from procedural thinking to structural thinking.

> Introduction of nouns into those places in which, until now, people had only been talking about processes ... refocuses the discourse. ... This ontological shift from an operational to a structural focus is well felt, for example, in the transition from the expression *These things cost five dollars* to the expression *The cost of these things is five dollars*. (p. 68)

Such a subtle restatement (linguists call it "nominalisation") moves the discourse away from a focus on *getting an answer* to that of considering the answer as an object in its own right. In the first case cost is something obtained by a process, and the focus is on the things and the underlying, implicit process. In the second case the cost is a representative of an algebraic object—a variable representing an unknown. To shift the focus to *cost*, the noun, away from a particular number and the process giving rise to it (*cost* as a verb), directs the students' attention from a procedural orientation to one in which it is possible to explicitly examine the algebraic concepts implicit in the symbolic manipulations.

At the other extreme from the careful use of natural language to purposefully focus instructional discourse, is the use of "ostensive" language (Freudenthal, 1980).

Many students will use demonstrative words such as *it*, *this*, and *that*, along with finger pointing (linguists call that "deictics") to explain procedures. Such imprecise language may indicate a student's lack of an appropriate vocabulary or be a symptom of his or her lack of exposure to correct mathematical discourse. It is important for teachers to model the use of correct terminology and to insist that their students divest themselves of habits of imprecise mathematical speech.

9.3.5 Metalinguistic awareness

What one is prepared to notice and able to perceive when using algebraic language may be a function not only of one's operational or structural orientation, but also of one's *metalinguistic* ability. MacGregor and Price (1999) have attempted to adapt this notion from the field of literacy development as a useful lens through which to examine students' difficulties with learning algebra. They describe a metalinguistic ability in algebra in terms of three types of awareness.

Firstly, *symbol awareness* "includes knowing that numerals, letters, and other mathematical signs can be treated as symbols detached from real-world referents" (p. 452). This *coming to recognise that something can stand for something else* has also been described as *representational insight* in cognitive psychological studies investigating young children's early development of symbolic thought (Kinzel 2001, p. 114). Bednarz (2001) emphasises that students' conceptions of symbolism form "an indispensable foundation for algebraic reasoning activity" (p. 72). Quinlan (2001, pp. 513-514) provides examples of test questions designed to measure an aspect of symbol awareness.

Secondly, *syntax awareness* "includes recognition of well-formedness in algebraic expressions and the ability to make judgements about how syntactic structure controls both meaning and the making of inferences" (MacGregor & Price 1999, p. 452). Thirdly, awareness of *potential ambiguity* includes "the recognition that an expression may have more than one interpretation, depending on how structural relationships or referential terms are interpreted" (MacGregor & Price, 1999, p. 457).

MacGregor and Price point out that little research has been done to see if students' level of metalinguistic ability in language and/or in algebra affects their ability to learn algebra. In their own study, they observed that some students with good language metalinguistic skills did not do well on algebraic items based on "familiar school algebra tasks". They hypothesised that these students "were not sufficiently aware that the algebraic sign system has its own grammatical rules and conventions that are not intuitively obvious and have to be learned" (p. 462).

9.4 Meaning as an Interpretive Act

In previous sections, we considered characteristics of the mathematical aspects of algebraic language, using Frege's characterisation of meaning. We turn our attention now to the more cognitive aspects of language use and to other theoretical frameworks. We first discuss the notion of a *mathematical object* and the interpretative nature of meaning making. We present a brief outline of a semiotic framework (based on the work of Peirce) for thinking about how individuals assign meaning to systems of signs. We end with a discussion clarifying the terms *meaning* and u*nderstanding* in relation to algebraic activity.

9.4.1 Mathematical objects

Systems of symbolic language and representation are essential for mathematical thinking "because, unlike the other fields of knowledge (botany, geology, astronomy, physics), there is no other way of gaining access to the mathematical objects" (Duval 1999, p. 4). The questions we address in this section are "What exactly is it that these symbols stand for?" and "What are the mental objects that are the intended references in the minds of both the writer and the perceiver?" It is interesting to notice that Frege did not intend to answer these questions. He focused on the nature of the denoting process, not the nature of the denoted objects. This and later sections contribute additional perspectives with which to think about language issues.

Devlin (2000) describes the mathematician's abstract patterns as "skeletons of things in the world" (p. 77). Mathematics can be thought of as a "pair of conceptual spectacles that enable us to see what would otherwise be invisible" (p. 74). A particular kind of notation, in this case, a system of signs and symbols, is therefore necessary to "describe on paper a pattern that exists only in the human mind" (p. 77).

Sfard (2000) uses the metaphor of a world of "virtual reality" to characterise the nature of mathematical thought and discourse. In this world, objects under discussion cannot be directly mediated by perceptual experience. Rather, a system of "symbolic substitutes" must be used for communication. For example, the numeral "5" is a physical representation of the numerosity of a collection of objects—a way to make visible the mental construct of "number." Even such a "simple" concept as number is an interpretative act—remember Sfard's comments about the Brazilian street vendors.

The notion of a mathematical object is an illusive one. It is a mental construction that has been variously characterised as "schemes of operation and webs of meaning" by Thompson (in Sfard & Thompson, 1994, p. 8); as a "metaphor that shapes the abstract world in the image of tangible reality" by Sfard (in Sfard & Thompson, 1994, p. 24); and as a discursive object that comes "into existence exclusively by and within the discourse" by Dörfler (2000, p. 123). Since mathematical constructs are purely mental objects, their existence can only be inferred from observable behaviour, and, to complicate things further, through the use of "mathematical discourse" involving a system of language and symbols.

Traditionally, symbols were viewed as representing a "mind-independent intangible object" (Sfard 2000, p. 92). As such, symbols could be used to transfer an external reality—a kind of ready-made mathematics—from one individual to another. Symbols were thought of as pointing to something else—some externally given, pre-existing referent. The advent of semiotics and the work of Saussure, Peirce, and Vygotsky led to a "reconceptualisation of the issue of the construction of meaning in general, and of mathematical objects in particular" (Sfard 2000, p. 44). "Mathematical ideas, as located through notation, are not endowed with a universal meaning but rather derive their meaning through the way in which an individual attends to them" (Brown 1997, p. 15).

We found in Panizza (in her presentation to the Working Group on Symbols and Language, *Generalisation and Control in Algebra*) an example of this perspective. She discussed a typology for students' spontaneous generalisations in algebra, based on their origin, for a particular subject, in a particular moment (see Panizza, 2002). She found there is a *semiotic* spontaneous generalisation (based on the subject's analysis of the content of a particular semiotic representation). Two other types of spontaneous generalisations are: the *conceptual* (based on the content to which the statement refers) and the *logical* (based on an inadequate understanding of logical connectors or rules of reasoning). This typology is not a classification since a spontaneous generalisation, in a particular situation, may concern more than one of these types simultaneously.

This post-objectivist stance regards meaning as an interpretive act. What have been variously called signs, marks, or signifiers take their meaning from the ways in which individuals act with them, within particular contexts and discourses. Meaning, depending on your epistemological perspective, may be the result of an individual construction or the outcome of social interaction. The meaning attached to a symbol is considered to be affected by its perceived mathematical context as well as by the kinds of knowledge and past experience that the individual brings to the task at hand. In order to understand this meaning, it is necessary to take into account how that individual uses the symbol in a particular context. "We can only speak of signs as [having a particular meaning] for a particular person, at a particular time, in a given context" (Davis & McGowen, 2001, p. 17). Brown (1997) provides a concise summary of current epistemological perspectives on the development of mathematical meaning and the role played by symbols in such learning.

9.4.2 Deacon: Three semiotic levels

Several frameworks have been developed to deal with the complexity that arises when the meaning assigned to any symbol is taken as an interpretative act. These frameworks propose similar icon-index-symbol distinctions in the ways that individuals make sense of signs: what it is that an individual is able to focus on and the nature of the meaning derived from this focus (see, for example the work of Peirce, in Buchler, 1955, and Schütz, in Brown, 1997). Here we report on a framework based on Deacon's (1997) interpretations of Peirce's three semiotic levels, and related to mathematical understanding, as described by Davis and McGowen (2001). In this framework, a set of hierarchical levels is distinguished by the way in which individuals interpret "squiggles" or "marks." Deacon uses Peirce's terminology of "iconic," "indexical," and "symbolic" to differentiate the three types of interpretation.

At a lower level, a "mark" is interpreted as an *icon*, so called because the sign "brings to mind for that person something else, which it resembles" (p. 6). For example, when the drawing of a clock face 🕒 is taken to stand for a clock itself. An interesting point worth noting is that *there are very few, if any, icons* in the algebraic sign system (strictly defined as the symbolic language of equations, inequalities etc.). Actually, there are scarcely icons in *languages* (considered as particular semiotic systems). For example, in oral English there are very few spoken words that *resemble* the objects they denote, except maybe some animals' names, such as cuckoo that resembles the sound of the bird. Linguists say that signs have, in principle, no relationship with the objects they represent (or in other words, that this relationship is said to be *arbitrary*). In the case of algebraic symbolic language, maybe only the placeholder □ could be considered as an icon: in the expression 2 + □ = 5 the small empty square may suggest the missing quantity.

On the other hand we can find much more easily icons in the *compound representation* component of the mathematical language (see Section 9.1.3.1). Technically, this component is *not* a language, but is instead a mere semiotic system. Also in the drawings and schemes presented in algebra textbooks such as that displayed in Figure 9.8, the little strokes used to represent matches are *icons* of matches. Also, the dots "•" are icons of units of square numbers.

Note that, in general, it may be quite difficult to determine with certainty when a given sign is an icon, in particular when the represented thing is abstract. Is this O an icon of a circle? Does it resemble a circle? How can I know? Has anybody ever *seen* an actual mathematical circle?

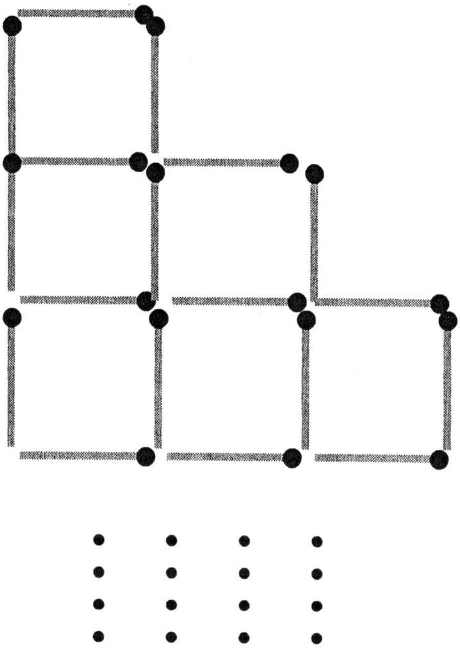

Figure 9.8. Examples of drawings and schemes presented in algebra textbooks.

At the next level, signs are interpreted *indexically*. Davis & McGowen (2001, p. 10) wrote "an index is a sign that refers to something by association" and described (p. 16) *indexical behaviour* as "a way of acting in a remembered context". For example, the equation "$3x^2 - 10x \quad 8 = 0$" may be interpreted indexically by linking the form of the equation and its solution process to similar situations. At the third level, signs become *symbols*. "Symbols, unlike icons and indexes do not stand alone: they form part of a connected symbolic system" (p. 12).

9.4.3 Meaning and understanding

In Section 9.2.1 we discussed the notion of *meaning* in terms of Frege's use of *sense*. We now shift from his focus on the process of assigning meaning to consider meaning in terms of the interpretative nature of mathematical objects. We also make a distinction between *meaning* and *understanding*.

Meaning refers to the type of mental entity that an individual associates with a particular symbol. In contrast, the term *algebraic understanding* characterises the way in which a student relates the sign and its meaning to a larger, connected set of relationships—that is, a way of representing, organising, and acting mathematically within a particular syntactic structure. The *rote* use of symbols, where it is possible to manipulate algebraic symbols in a purely rule-based, mechanistic manner, is different from manipulation based on understanding. According to Davis and McGowen (2001), learning and understanding can be said to take place at a symbolic level, while rote performance is indexical (note that meaning can be at an iconic, indexical, or symbolic level).

Understanding is inferred from particular types of behaviour. For example, students should be able to interpret symbolic writings *flexibly*, seeing, for instance, $2x + 6$ as an expression representing an unknown number, as a function, or as an equivalent set of operations to the expression $2(x + 3)$. (It is worth noting that Frege would consider that all three examples have the same functional denotation, here we expand the type of object that is denoted.) Sfard and Thompson (1994, p. 25) characterise such behaviour as "the ability to match an interpretation to the context in which the [expression] is used." It is the same with Gray and Tall's (1994) notion of "procept." Another indicator of understanding is a student's ability to use a particular symbol in ways that are new to him or her—Sfard's so-called *novel uses* (Sfard, 2000). For example, recognising that x^2 in the equation $ax^2 + bx + c = 0$ stands for the *operation of squaring*, and that the place holder x^2 is represented by b^2 in the particular equation $c^2 = a^2 + b^2 - 2ab \cos C$. (See Section 9.5.1.)

Explaining procedures is also taken to be an indication of understanding. Students should be able to "justify their decisions in ways that go beyond a mere recitation of the list of movements through which they went while transforming formal expressions" (Kieran & Sfard 1999, p. 2). For example, it could be inferred that a student understood the counting *structure* of the dot arrays in a sequence of square numbers if she explained her actions as, "To find the number of dots in the next square, I add twice the number of dots on the side of the old square to the sides of that square, and then add one more dot." (See Section 9.5.2.)

Radford (2002) examined a slightly different aspect of symbolic meaning-making in his research into students' abilities to render key information given in a story problem into an appropriate equation. His example uses the short story "*Kelly has 2 more candies than Manuel, Josée has 5 more candies than Manuel. All together they have 37 candies.*" Students were to designate Manuel's number of candies by x and to write and solve an equation for the story problem. Radford points out how the focus of the students' attention must shift during this activity. "The 'heroes', so to speak, of the [symbolic narrative] are no longer Kelly, Manuel, or Josée, but the numerical relationships between the amount of candies" (pp. 4-83).

Radford acknowledges the presence of a "limbo", which parallels Sfard's didactical dilemma (see Section 9.5.1), "in which students have neither fully left the original story, nor have fully entered the symbolic narrative" (pp. 4-86). Part of this dilemma is the movement that must be made from natural language, with its extensive descriptive vocabulary, to algebraic symbolism where objects are designated by a very limited set of characters. Additionally, Radford's example highlights the way that algebraic language is designed to see and think about the world in a particular way. It may be important at some point in their learning experiences to call students' attention to the fact that, when moving to a symbolic narrative, the focus shifts to a different (mathematical) set of objects.

9.4.3.1 Hierarchical levels

The meaning that an individual associates with a particular sign or collection of symbols can be interpreted within a hierarchical structure, with different levels characterised by the type of abstraction that an individual employs to construct a mental image from the physical sign. For example, the notion of *procept* distinguishes between a focus on actions carried out with or upon a collection of symbols and the abstraction of a mathematical *noun* to designate the (mental) result of such activity. Sfard (2000) postulates a type of mental reorganisation that might take place as a student internalises a particular symbol use. The meaning of the symbol, initially developed to signify a particular context, becomes abstracted to signify a class of similar situations or objects. The semiotic framework of icon, index, and symbol also describes a hierarchy of levels of meaning.

The fact that mathematical notation can be seen from many points of view, as being icons, indexes, or symbols, perceived as procepts, denoting functions, having a sense, and so on, creates instructional challenges. Students may be observed engaging in similar overt behaviours while operating within different levels of meaning and understanding. Kieran and Sfard (1999) caution that we must be careful that students don't fall "victim to the addictive power of algebraic manipulations and [forget] all about the meaning of the symbols" (p. 16). There is a tension here, however. Expert behaviour relies on the capability to forget the meaning; it is one of the strengths of algebraic manipulation to be able to work without context, with meaning simultaneously both forgotten but available. We would like to insist on this point: expert behaviour is neither being aware of the meaning of symbols all the time, nor forgetting it all the time. Instead, it relies on the capability to reach the meaning of the symbols on demand.

It is equally important to be aware of potential differences in interpretation between teacher and student, an "essential distinction in reference between the producer and interpreter of signs" (Davis & McGowen 2001, p. 26). Textbook use is a prime example of such a distinction. A student may be *reading* collections of symbols, at best indexically, whereas for the author the same collection of symbols evokes a rich symbolic association of meanings.

To be effective, instruction needs to be aligned with the students' levels of understanding. Baker, Hemenway, and Trigueros (2001), studied the success of college-level pre-calculus students (around 18 years old) in developing an understanding of the process of transformations of functions (in both algebraic and graphical form). They found that "students who don't have an object conception of function are not able to recognise transformations on functions even after being taught to use them as a starting point for the analysis of properties of general functions" (p. 46). Their study also addresses an aspect of language that is not treated in this chapter, but is discussed elsewhere in this book: forming links between graphical and symbolic representations (see Chapters 6 and 7 especially).

9.4.4 Algebraic language awareness

It is important that students be made aware of what it means to be symbolically literate. We propose to coin the word *algebracy* to describe this awareness. The notion of algebraic language awareness parallels that of "linguistic awareness," which is described as "the development of greater awareness among school children of the nature and purpose of language" (McArthur, 1992, p. 575). Such an awareness includes getting a sense of algebra as a way of mathematically structuring and manipulating experience, understanding why a symbolic language is used to represent and solve particular problems, and being able to recognise such problems described in symbolic form (Malara & Navara, 2001). Metalinguistic awareness (MacGregor & Price, 1999) is another important component of algebracy. Students need to develop a sense of how syntax provides a structure within which meanings can be applied in a relational manner to symbols. In order to acquire such an awareness, it is important that the teacher provide opportunities for the students to discuss the how and the why of the use of algebraic symbols.

The metaphor of transparency has been applied to the process of assigning meaning to algebraic symbols (Kaput, 1999). Symbols become transparent when students can *look through* them to focus on the mathematical concepts represented by particular symbolic writings. On the other hand, symbols remain opaque when students *look at* the physical marks on paper, perceived to be governed by sets of meaningless manipulation rules. We would like to add a third type of *seeing*, that of *studying the medium*. Continuing the metaphor, if the way of seeing is likened to meaning residing behind clear glass or upon a glass window, an important aspect of algebraic understanding is gained from understanding the *glass* itself, or an explicit study of the language aspects of algebra. "The ability to see the rich landscape of mathematical objects hidden behind formal expressions" (Kieran & Sfard 1999, p. 3) is facilitated by a capability of focusing *on* the window and on understanding *how* we see through this glass.

9.5 Symbolising as an Activity

It is not our intention in this chapter to address the teaching and learning aspects of algebraic language and notation. These aspects are treated in depth in other chapters in this book (see, for example, Chapters 4 & 5). Here we briefly examine the *activity* of symbolising, develop further the notion of algebraic language awareness, and discuss the implications of a focus on symbols and language on teacher preparation.

9.5.1 Developing meaning for symbols

Meaning-making and symbolising can be seen as inexorably connected to the learning process (Cobb, 2000). Symbolising is integral to mathematical activity. The assumed reflexive relationship between symbol use and mathematical meaning implies that a student's use of symbols involves some type of meaning, and that the development of meaning involves modifications in ways of symbolising. "Viewed in these terms, teaching and instructional design both involve attempts to support the development of students' ways of symbolising as part of the process of supporting the development of mathematical meaning" (p. 19).

Sfard (2000) points out the didactical dilemma that exists when students attempt to engage with new ideas before they have the words or symbols with which to carry out such discourse. She describes the circular nature of such a situation.

> It is the discursive activity, including its continuous production of symbols, that creates the need for mathematical objects; and these are mathematical objects (or rather the object-mediated use of symbols) that, in turn, influence the discourse and push it into new directions (p. 47). ... The act of introduction creates a "semantic space" yet to be filled with meaning. The signifier enters the language game before the rules of this game have been sufficiently specified and before the signifier has acquired the power of evoking a familiar experiential resonance (p. 58).

When encountering new symbols, students regard these in a "templates-driven" fashion. "The signifier is not yet conceived by its user as standing for something else. In fact, the user may not even perceive the sign as a self-sustaining entity" (p. 77). In this phase, students rely on their present understandings of the immediate context in which the signifier appears as a way of using the new symbol. "The very first use already [frames] the discourse and greatly [delineates] the set of possible linguistic applications" (p. 63). Students may be able to communicate using the new symbol at this point, but they are as yet unable to "reason why things work" (p. 76).

Sfard defines the transition from templates-driven to "object-mediated" symbol use as being a "rich and multifaceted event" (p. 81). The goal is to enable students to use the symbol as a representation for something else. "The transition from signifier-as-an-object-in-itself to signifier-as-a-representation-of-another-object is a quantum leap in a subject's consciousness" (p. 79). An indication of this change in perception is a student's ability to use the symbol in flexible ways. "The ability to create novel uses is often regarded as an ultimate criterion of meaningfulness and understanding" (p. 78).

For example, consider the following situation (taken from Teppo & Esty, 2001). The equation in a) can be solved in a straightforward fashion. All a student needs to do is make a one-to-one match between the symbol pattern of the Quadratic Theorem and that of the given equation. The symbols x and x^2 can be taken as *standing for themselves* in the given symbol strings. In contrast, the equation in part b) cannot be solved without a deeper knowledge of the role played by dummy variables.

> Given the Quadratic Theorem:
>
> If $a \neq 0$, then $ax^2 + bx + c = 0$ is equivalent to
>
> $$x = \frac{-b \pm \sqrt{(b^2 - 4ac)}}{2a}$$
>
> Use the theorem to solve the following:
>
> a) $5x^2 - 7x - 12 = 0$.
>
> b) In the Law of Cosines, solve for b: $c^2 = a^2 + b^2 - 2ab \cos C$.

The symbols x, a, b, and c in the Quadratic Theorem are used to represent a particular sequence of operations, and can stand for mathematical entities other than themselves. In b) the operation of squaring in the theorem needs to be the focus of the student's attention—the x^2 represents *squaring*, even if it is not "x" that is squared in the given equation. "To recognise that the Quadratic Theorem is relevant in [problem b)], it is necessary to regard squaring as an *object divorced from a particular symbolic representation*" (Teppo & Esty, 2001, p. 579).

9.5.2 Horizontal and vertical mathematising

Symbolising plays an active role in the processes of horizontal and vertical mathematising, distinctions introduced by Treffers in 1978 (Freudenthal, 1991). Horizontal mathematising "leads from the world of life to the world of symbols," while in the process of vertical mathematising "symbols are shaped, reshaped, and manipulated, mechanically, comprehendingly, reflectively" (Freudenthal, 1991, p. 41).

Rasmussen, Zandieh, King, and Teppo (in press) extend Treffers' idea, examining how the notions of horizontal and vertical mathematising can be used to characterise symbolising activity itself. They present an example from a university course in differential equations in which the observed students' uses of symbolising shifted from a means of *recording* and *communicating* their thinking (horizontal symbolising) to that of using symbols for mathematical *reasoning* and *concept development* (vertical symbolising). During horizontal symbolising, particular collections of symbols were used to create a model *of* a particular mathematical situation. Vertical symbolisation then took place, as these symbols became a model *for* thinking about underlying concepts implicit in the original symbolisation. The authors note how a particular student's vertical symbolising facilitated the development (for him) of a new mathematical reality. Gravemeijer, Cobb, Bowers, & Whitenack (2000) discuss the organising notion of "models-of" and "models-for".

The example given in Rasumssen, Zandieh, King, and Teppo (in press), discussed above, is fairly complex and, for brevity, a simpler example of horizontal and vertical symbolising is given here. Consider the geometrical representation for square numbers shown in Figure 9.9. If the *first* square is represented by a 2×2 array, interesting patterns can be examined—notice that the number of dots in each successive figure can be found by adding twice the number of dots in a given row plus one more dot.

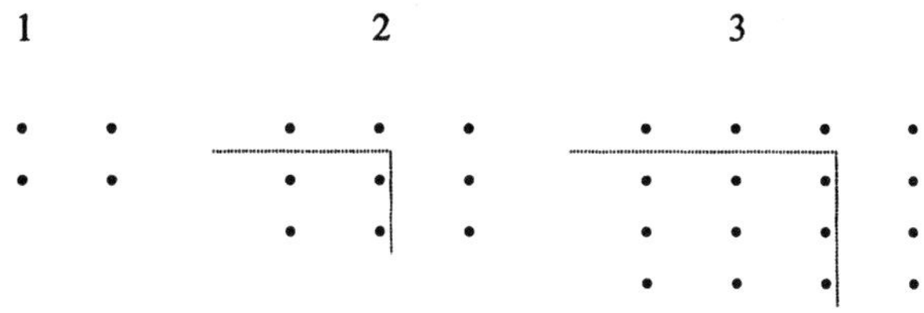

1	$1 + 3$	→	$1 + 2\times1 + 1 = 4$
2	$4 + 5$	→	$4 + 2\times2 + 1 = 9$
3	$9 + 7$	→	$9 + 2\times3 + 1 = 16$
..	. . . .		. . . .
n	$n^2 + (2n + 1)$	→	$n^2 + 2n + 1 = (n + 1)^2$

Figure 9.9. Generating squares by 'adding' rows, columns, and an extra dot.

Horizontal symbolising occurs when the geometric pattern for enlarging one square into the next is expressed numerically—for example, when recording the fact that seven more dots are added to the existing nine to make the 4×4 square array. In addition, the numerical equations following each arrow record the physical actions of extending each *square* by one row and column and adding a final dot to fill in the *corner*. It can be claimed that even though the equality $n^2 + 2n + 1 = (n + 1)^2$ is a generalisation of the numerical examples, the purpose of this move to algebraic symbols is still to record and communicate actions and perceived patterns.

Vertical symbolising occurs when the focus of attention is shifted to the processes used to create a new array by extending on to the sides of an existing square. Consider the following arrays shown in Figure 9.10. The pattern of expressing squares by *adding on to the existing rows and columns* is no longer a recording device, but a way of structuring new, but related, situations. Symbolising is now used to reflect on a particular procedure for squaring. The process of finding the number of dots in each array is symbolised numerically in each case, and as a generalisation as:

$$3^2 + 2\times3 + 1 = (3 + 1)^2$$
$$4^2 + 2(2\times4) + 2^2 = (4 + 2)^2$$
$$a^2 + 2ab + b^2 = (a + b)^2$$

Notice in this activity, symbolisation is used as a model *for* thinking about the process of determining the number of dots in a square array whose sides are expressed as the sum of two numbers. Symbols were used in the first activity to record a particular way of structuring the given situation. However, in the second activity, symbolising made it possible to think about the structuring activity itself. Further vertical symbolising takes place if attention is drawn to the way that each generalised equation in the two activities is used to state a pair of equivalent expressions (or sequence of operations).

a b

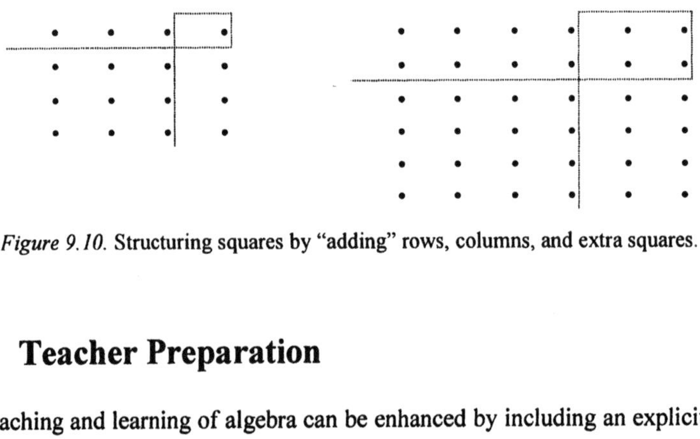

Figure 9.10. Structuring squares by "adding" rows, columns, and extra squares.

9.6 Teacher Preparation

The teaching and learning of algebra can be enhanced by including an explicit focus on the language aspects of the subject. As a front line intervention, teacher preparation represents an important way to facilitate such enhancement. Indeed, if teachers themselves are unaware of how syntax organises algebraic meaning, there is little they can do to encourage linguistic awareness in their students. Menzel (2001, p. 446) points out that "with a language approach, teachers would develop the deep structural knowledge of syntax and the semantic relationships between algebraic expressions in different contexts."

With colleague Esty, Teppo (an author of this chapter) examined a one-semester college course on *The Language of Mathematics* that is designed to deepen students' understanding of mathematics by simultaneously emphasising mathematics (the subject) and mathematics (the language). Versions of this course are required of all pre-service elementary teachers taking a mathematics option within their elementary education degree at Montana State University and all those mathematics majors who will become high school teachers. One of the organising principles of the course is that by naming objects, these conceptual entities can be made the explicit focus of instruction. Variables, expressions, and equations are studied as classes of objects (nouns) with specific mathematical roles, and algebra is presented as being *about operations,* with notice also given to *word* order to focus attention on the role of symbols in abstract methods such as formulas, identities, and theorems. This "naming" utilises the syntactic aspect of a language system "associated with identification and discrimination of notational objects" (Harel & Kaput 1991, p. 89). The following homework questions from the course give a flavour of this linguistic approach (Esty, 1999).

- Write out, in English, the proper pronunciation of
  "$2(x + 3)$", "$2x + 3$", and "$2x = 3$".
- Decide if the given equations fit the problem-pattern $a.b = 0$.
  a) $(x-3)(x+5) = 1$
  b) $(x+8)(x-4) - 2 = 0$
  c) $x(x - 9) = 0$
- Which are dummy variables and which are unknowns?
  $2x + 3 = 12$, $x + x = 2x$
- Which of the following processes applied to both sides of an equation always produce an equivalent equation?
  (a) add 5, (b) subtract 7, (c) multiply by 5, (d) divide by 4, (e) add x, (f) multiply by x, (g) cancel a common factor of "x", (h) square
- Disprove: (a) $|x| < |x + 1|$, (b) $x < 5 \Rightarrow x^2 < 25$,
  (c) $bc > 25 \Rightarrow b > 5$ or $c > 5$

An approach like that taken in the *Language of Mathematics* course can be used to shift students' attention away from situations in which symbols and the rules of syntax are the medium for carrying out an activity to those in which the symbols themselves are the focus of the activity.

A focus on the language aspects of algebra can be incorporated into mathematics lessons at any level. For example, the sample activity described in Section 9.3.2.2 that compared uses of variables (Trigueros & Ursini, 2001), can be easily extended linguistically. Not only do the three symbolic writings $2x + 9 = 0$, $2x + 9$, and $y = 2x + 9$ utilise x in a different manner, they themselves represent different mathematical entities. Students' attention can be drawn to the way in which equations, expressions, and functions express different mathematical ideas and the roles of the variable and the symbol for equality (i.e., " = ") in each type of entity. Such discussions help students begin to construct the rich relational web that connects mathematical objects and algebraic symbols. In a serendipitous way, students' mathematical concept images deepen as the meaning of the mathematical contexts in which they are embedded becomes enriched (Esty & Teppo, 1994).

9.7 Where do we go from here?

There is a need to recognise the particular perspective on symbols and language that we have developed in this chapter as an integral, but at the same time, separable aspect of algebra. We do not advocate that this aspect should be isolated from the complexities of teaching and learning, but rather included as a useful framework for investigating certain parts of the whole. As we conclude the chapter, we speculate about the future and present a brief discussion about possible directions for future research.

Research areas outside of mathematics education offer promising frameworks for new ways to think about and investigate relationships among linguistic aspects, symbol use, and the teaching and learning of algebra. It is useful to explore ways to build upon these methods to meet the unique conditions of mathematics teaching and learning. MacGregor and Price's (1999) research, which explored aspects of language proficiency and algebra learning, is an example of how studies of literacy and learning can be used to investigate parallel notions in algebra literacy. In contrast, Davis and McGowen (2001) illustrate how theoretical frameworks, such as Peirce's semiology, can be interpreted and extended to illuminate levels of symbolic interpretation and mathematical understanding. Hewitt (2001) suggests investigating parallels between the development of language in young children and that of algebraic notation related to algebraic activity. He argues that "algebraic activity is common amongst all children well before they enter formal schooling" and that this activity be "... utilised and formalised in mathematics classrooms" (p. 308). This theme is the topic of Chapter 4.

This chapter and the work of those cited herein illustrate the ways in which research on language aspects can address aspects of the rich complexity of mathematics education (see also Drouhard (2001) for a partial bibliography). In particular, a perspective focused on algebraic symbols and language can be used to explore the following areas of interest (some of which are derived from e-mail messages in the electronic SymCog site, developed to facilitate on-going conversation among members of the Discussion Group on Symbolic Cognition in Advanced Mathematics begun at the PME-25 Conference in Utrecht in 2001).

- Investigate how various semiotic and linguistic frameworks can provide insight and direction for research.
- Observe individuals who exhibit *symbolic activity* in order to characterise this type of activity.
- Document the movement from an iconic to an indexical to a symbolic use of algebraic symbols.
- Examine how an explicit instructional focus on language aspects facilitates student understanding.
- Look for links with brain research on how individuals acquire language and interpret signs.

We stated earlier that a given language enables one to structure thinking in a particular way. The symbolic language of algebra with its unique vocabulary and syntax is the visible sign of *doing algebra.* It may be that, in the future, different ways of thinking may become more useful. The existing system of algebraic language might not be the most effective medium for future mathematical ways of thinking. One impetus for change comes from the emergence of powerful and easily accessible computational power.

> Computer scientists wanting to make more sense of the activity of computing than the mere following of rules have been inventing and re-inventing mathematics. The computational ways of thinking that dominated mathematics in centuries prior to the 20th [century] are making a comeback. Mathematics is being seen, again, as a way of talking about computation. New insights from computing are requiring the creation of new mathematics. It is a process that promises to invigorate the learning of algebra. (Fearnley-Sander 2001, p. 251)

Algebraic language can be viewed as a creation of the human mind, developed to facilitate a particular type of abstract thinking. The language will continue to adapt within the hands of creative mathematicians. Hopefully, classroom instruction will follow, facilitating meaningful engagement with this powerful tool of mathematical communication.

9.8 References

Artigue, M., Abboud, M., Drouhard, J-Ph., & Lagrange, J-B. (1994). Integrating *DERIVE®* in secondary level mathematical teaching: Theoretical potentialities and the real life of teachers and students. In J. P. da Ponte & J. Matos (Eds.), *Proceedings of the 18th annual conference of the International Group for the Psychology of Mathematics Education* (Vol. 1, p. 30). Lisbon, Portugal: Program Committee.

Arzarello, F., Bazzini, L., & Chiappini, G. (1994). Intentional semantics as a tool to analyze algebraic thinking. *Rendiconti del Seminario Matematico dell'Università e del Politecnico di Torino 52*(2), 105-125.

Baker, B., Hemenway, C., & Trigueros, M. (2001). On transformations of basic functions. In H. Chick, K. Stacey, J. Vincent, & J. Vincent (Eds.), *The future of the teaching and learning of algebra* (Proceedings of the 12th ICMI Study Conference, pp. 41-47). Melbourne, Australia: The University of Melbourne.

Bednarz, N. (2001). A problem-solving approach to algebra: Accounting for the reasonings and notations developed by students. In H. Chick, K. Stacey, J. Vincent, & J. Vincent (Eds.), *The future of the teaching and learning of algebra* (Proceedings of the 12th ICMI Study Conference, pp. 69-78). Melbourne, Australia: The University of Melbourne.

Boero, P. (1993). About the transformation function of the algebraic code. In R. Sutherland (Ed.), *Algebraic processes and the role of symbolism (working conference of the ESRC seminar group*, pp. 48-55). London: University of London, Institute of Education.

Brown, T. (1997). *Mathematics education and language: Interpreting hermeneutics and post-structuralism.* Dordrecht, The Netherlands: Kluwer Academic.

Buchler, J. (Ed.) (1955). *The philosophical writings of Peirce.* New York: Dover Books.

Chomsky, N. (1965). *Aspects of the theory of syntax*. Cambridge, MA: MIT Press.

Cobb, P. (2000). From representations to symbolizing: Introductory comments on semiotics and mathematical leraning. In P. Cobb, E. Yackel, & K. McClain (Eds.), *Symbolizing and communicating in mathematics classrooms: Perspectives on discourse, tools, and instructional design* (pp. 17-36). Mahwah, NJ: Lawrence Erlbaum.

Davis, G. E., & McGowen, M. A. (2001). *Embodied objects and the signs of mathematics*. A discussion paper prepared for the PME discussion group "Symbolic Cognition in Advanced Mathematics" at the 25th annual conference of the International Group for the Psychology of Mathematics Education Freudenthal Institute, University of Utrecht, The Netherlands (pp. 1-25). Available at http://www.harpercollege.edu/~mmcgowan/publications

Deacon, T. W. (1997). *The symbolic species: The co-evolution of language and the brain*. New York: W. W. Norton & Company.

Devlin, K. (2000). *The math gene: How mathematical thinking evolved and why numbers are like gossip*. New York: Basic Books.

Donald, M. (1991). *Origins of the modern mind: Three stages in the evolution of culture and cognition*. Cambridge, MA: Harvard University Press.

Dörfler, W. (2000). Means for meaning. In P. Cobb, E. Yackel, & K. McClain (Eds.), *Symbolizing and communicating in mathematics classrooms: Perspectives on discourse, tools, and instructional design* (pp. 99-132). Mahwah, NJ: Lawrence Erlbaum.

Drouhard, J-Ph. (1989). Une étude pro-didactique: l'analyse syntaxique des expressions de l'algèbre élémentaire [Synatactic analysis of elementary algebra writings]. In A. Bessot (Ed.), *Actes du Séminaire de Didactique des Mathématiques et de l'Informatique* (pp. 5-13). Grenoble: Université Joseph Fourier.

Drouhard, J-Ph. (1992). *Les écritures symboliques de l'algèbre élémentaire* [Elementary algebra symbolic writings]. Unpublished Doctoral Dissertation. Université Paris 7.

Drouhard. J-Ph., Léonard F., Maurel M., Pécal M., & Sackur C. (1994). 'Blind calculators', 'denotation' of algebra symbolic expressions, and 'write false' interviews. In D. Kirshner (Ed.), *Proceedings of the Sixteenth Annual Meeting, North American Chapter of the International Group for the Psychology of Mathematics Education* (Vol. 1, pp. 152-158). Baton Rouge, LA: Louisiana State University.

Drouhard, J-Ph. (1995). Algèbre, calcul symbolique et didactique [Algebra, symbolic calculation and didactics]. In R. Noirfalise & M.-J. Perrin-Glorian (Eds.), *Actes de la VIIIème École d'Été de Didactique des Mathématiques* (pp. 325-344). Clermont-Ferrand, France: IREM de Clermont-Ferrand.

Drouhard, J-Ph. (2001). Research in language aspects of algebra: a turning point? In H. Chick, K. Stacey, J. Vincent, & J. Vincent (Eds.), *The future of the teaching and learning of algebra* (Proceedings of the 12th ICMI Study Conference, pp. 238-242). Melbourne, Australia: The University of Melbourne.

Ducrot, O., & Todorov T. (1972). *Dictionnaire encyclopédique des sciences du langage* [Encyclopaedic dictionary of language sciences]. Paris: Le Seuil.

Durkin, K. (1991). Language in mathematical education: An introduction. In K. Durkin & B. Shire (Eds.), *Language in mathematical education: Research and practice* (pp. 3-16). Milton Keynes, UK: Open University Press.

Duval, R. (1999). Representation, vision and visualization: Cognitive functions in mathematical thinking. Basic issues for learning. In F. Hitt & M. Santos (Eds.), *Proceedings of the Twenty First Annual Meeting of the North American Chapter of the International Group for the Psychology of Mathematics Education* (pp. 3-26). Columbus, OH: ERIC Clearninghouse for Science, Mathematics, and Environmental Education.

Duval, R. (2000). Basic issues for research in mathematics education. In T. Nakahara & M. Koyama (Eds.), *Proceedings of 24th annual conference of the International Group for the Psychology of Mathematics Education* (Vol. 1, pp. 55-69). Hiroshima, Japan: Program Committee.

Esty, W. W. (1999). *The Language of Mathematics, 16th Edition*. Bozeman, MT: Montana State University Technical Report.

Esty, W. W., & Teppo, A. R. (1994). A general-education course emphasizing mathematical language and reasoning. *Focus on Learning Problems in Mathematics, 16*(1), 13-35.

Fearnley-Sander, D. (2001). Algebra worlds. In H. Chick, K. Stacey, J. Vincent, & J. Vincent (Eds.), *The future of the teaching and learning of algebra* (Proceedings of the 12th ICMI Study Conference, pp. 243-251). Melbourne, Australia: The University of Melbourne.

Freudenthal, H. (1991). *Revisiting Mathematics Education: China Lectures*. Dordrecht, The Netherlands: Kluwer Academic.

Freudenthal, H. (1980). *Weeding and sowing*. Dordrecht, The Netherlands: D. Reidel Publishing Company.

Filloy, E., & Rojano, T. (1984). From an arithmetical to an algebraic thought. In J. M. Moser (Ed.), *Proceedings of the Sixth Annual Meeting of PME-NA* (pp. 51-56). Madison: University of Wisconsin.

Goldin, G. A., & Kaput, J. J. (1996). A joint perspective on the idea of representation in learning and doing mathematics. In L. P. Steffe, P. Nesher, P. Cobb, G. Goldin, & B. Greer (Eds.), *Theories of Mathematical Learning* (pp. 397-430). Mahwah, NJ: Lawrence Erlbaum.

Goody, J. (1987). *The interface between the written and the oral. (Studies in literacy, family, culture and the state)*. Cambridge, MA: Cambridge University Press.

Goody, J. (1994). *Entre l'Oralité et l'écriture* [The interface between the written and the oral]. Paris: PUF.

Gray, E., & Tall, D. (1994). Duality, ambiguity, and flexibility: A "proceptual" view of simple arithmetic. *Journal for Research in Mathematics Education, 25*(2), 116-140.

Gravemeijer, K., Cobb, P., Bowers, J., & Whitenack, J. (2001). Symbolizing, modeling, and instructional design. In P. Cobb, E. Yackel, & K. McClain (Eds.), *Symbolizing and communicating in mathematics classrooms: Perspectives on discourse, tools, and instructional design* (pp. 225-274). Mahwah, NJ: Lawrence Erlbaum.

Guillerault, M., & Laborde C. (1982). *Ambiguities in the description of a geometrical figure*. New York: Plenum Press.

Harel, G. & Kaput, J. (1991). The role of conceptual entities and their symbols in building mathematical concepts. In D. Tall (Ed.), *Advanced mathematical thinking* (pp. 82-94). Dordrecht, The Netherlands: Kluwer Academic.

Havelock, E. A. (1986). *The Muse learns to write. Reflections on orality and literacy from antiquity to the present.* New Haven & London: Yale University Press.

Hewitt, D. (2001). On learning to adopt formal algebraic notation. In H. Chick, K. Stacey, J. Vincent, & J. Vincent (Eds.), *The future of the teaching and learning of algebra* (Proceedings of the 12th ICMI Study Conference, pp. 305-312). Melbourne, Australia: The University of Melbourne.

Kieran, C., & Sfard, A. (1991). Seeing through symbols: The case of equivalent expressions. *Focus on Learning Problems in Mathematics, 21*(1), 1-17.

Kinzel, M. T. (2001). Analyzing college calculus students' interpretation and use of algebraic notation. In R. Speiser, C. A. Maher, & C. N. Walter, (Eds.), *Proceedings of the 21st Annual Meeting of the North American Chapter of the International Group for the*

Psychology of Mathematics Education (Vol. 1, pp. 109-116). Columbus, OH: ERIC Clearinghouse for Science, Mathematics, and Environmental Education.
Kirshner, D. (1987a). *Linguistic analysis of symbolic elementary algebra*, Unpublished Doctoral Dissertation, University of British Columbia.
Kirshner, D. (1987b). The myth about binary representation in algebra. In J. Bergeron, N. Herscovics, & C. Kieran (Eds.), *Proceedings of the 11th annual conference of of the the International Group for the Psychology of Mathematics Education* (Vol. 1, pp. 308-315). Montréal, Canada: Program Committee.
Kirshner, D. (1989). The visual syntax of algebra. *Journal for Research in Mathematics Education, 20* (3), 276-287.
Laborde, C. (1982). *Langue naturelle et écriture symbolique deux codes en interaction dans l'enseignement mathématique* [Natural language and symbolic writings : Two interacting codes in the teaching of mathematics]. Thèse d'Etat, Université Joseph Fourier, Grenoble.
Laborde, C. (1990). Language and Mathematics. In J. Kilpatrick & P. Nesher (Eds.), *Mathematics and cognition* (pp. 4-19). Cambridge: Cambridge University Press.
Lagrange, J-B. (2001). Tasks for students using computer algebra: a study of the design of educational computer applications. In H. Chick, K. Stacey, J. Vincent, & J. Vincent (Eds.), *The future of the teaching and learning of algebra* (Proceedings of the 12th ICMI Study Conference, pp. 376-383.). Melbourne, Australia: The University of Melbourne.
Linchevski, L., & Livneh, D. (1999). Structure sense: The relationship between algebraic and numerical contexts. *Educational Studies in Mathematics, 40*, 173-196.
Malara, N. A., & Navarra, G. (2001). "Brioshi" and other mediation tools employed in a teaching of arithmetic from a relational point of view with the aim of approaching algebra as a language. In H. Chick, K. Stacey, J. Vincent, & J. Vincent (Eds.), *The future of the teaching and learning of algebra* (Proceedings of the 12th ICMI Study Conference, pp. 412-419). Melbourne, Australia: The University of Melbourne.
McArthur, T. (Ed.) (1992). *The Oxford companion to the English language*. Oxford: Oxford University Press.
MacGregor, M., & Price, E. (1999). An exploration of aspects of language proficiency and algebraic learning. *Journal for Research in Mathematics Education, 30*(4), 449-467.
McGuinness, B. (Ed.) (1984). *Collected papers on mathematics, logic, and philosophy*. Oxford: Blackwell.
Menzel, B. (2001). Language conceptions of algebra are idiosyncratic. In H. Chick, K. Stacey, J. Vincent, & J. Vincent (Eds.), *The future of the teaching and learning of algebra* (Proceedings of the 12th ICMI Study Conference, pp. 446-453). Melbourne, Australia: The University of Melbourne.
Miles, D. D. (1999). Algebra vocabulary as an improvement factor in algebra achievement. *Focus on Learning Problems in Mathematics, 21*(2), 37-49.
Miller, D. L. (1993). Making the connection with language. *Arithmetic Teacher*, (February, 311-316.
Nicaud, J.-F., Bouhineau, D., & Gélis, J.-M. (2001). Syntax and semantics in algebra. In H. Chick, K. Stacey, J. Vincent, & J. Vincent (Eds.), *The future of the teaching and learning of algebra* (Proceedings of the 12th ICMI Study Conference, pp. 475-486). Melbourne, Australia: The University of Melbourne.
Novotná, J., & Kubínová, M. (2001). The influence of symbolic algebraic descriptions in word problem assignments on grasping process and on solving strategies. In H. Chick, K. Stacey, J. Vincent, & J. Vincent (Eds.), *The future of the teaching and learning of algebra* (Proceedings of the 12th ICMI Study Conference, pp. 495-500). Melbourne, Australia: The University of Melbourne.

Panizza, M (2002). Generalizaci6n y control en álgebra [Generalisation and control in algebra]. In C. Crespo (Ed.), *Proceedings of the 15th Conference RELME* (pp. 213-218). México: Grupo Editorial Iberoamericana.

Pirie, S. E. B., & Martin, L. (1997). The equation, the whole equation, and nothing but the equation: One approach to the teaching of linear equations. *Educational Studies in Mathematics, 34*, 159-181.

Quinlan, C. (2001). Importance of view regarding algebraic symbols. In H. Chick, K. Stacey, J. Vincent, & J. Vincent (Eds.), *The future of the teaching and learning of algebra* (Proceedings of the 12th ICMI Study Conference, pp. 507-514). Melbourne, Australia: The University of Melbourne.

Radford, L. (2002). On heroes and the collapse of narratives: A contribution to the study of symbolic thinking. In A. D. Cockburn & E. Nardi (Eds.), *Proceedings of 26th annual conference of the International Group for the Psychology of Mathematics Education* (Vol. 4, pp. 81-88). Norwich, UK: Program Committee.

Rasmussen, C. L., Zandieh, M., King, K., & Teppo, A. (In press). Advancing mathematical activity: A practice-oriented view of advanced mathematical thinking. *Mathematical Thinking and Learning.*

Rojano, T., & Sutherland, R. (2001). Arithmetic world - Algebra world. In H. Chick, K. Stacey, J. Vincent, & J. Vincent (Eds.), *The future of the teaching and learning of algebra* (Proceedings of the 12th ICMI Study Conference, pp. 515-522). Melbourne, Australia: The University of Melbourne.

Sackur, C. (1995). Blind calculators in algebra: Write false interviews. In E. Cohors-Fresenborg (Ed.), *Proceedings of the first European Research Conference on Mathematics Education (ERCME '95)* (pp. 82-85). Osnabrück (Germany): University of Osnabrück.

Sfard, A. (2000). Symbolizing mathematical reality into being - or how mathematical discourse and mathematical objects create each other. In P. Cobb, E. Yackel, & K. McClain (Eds.), *Symbolizing and communicating in mathematics classrooms: Perspectives on discourse, tools, and instructional design* (pp. 37-98). Mahwah, NJ: Lawrence Erlbaum.

Sfard, A., & Linchevski, L. (1994). The gains and pitfalls of reification - the case of algebra. *Educational Studies in Mathematics, 26*, 191-228.

Sfard, A., & Thompson, P. (1994). Problems of reification: Representations and mathematical objects. In D. Kirshner (Ed.), *Proceedings of the Sixteenth Annual Meeting, North American Chapter of the International Group for the Psychology of Mathematics Education* (Vol. 1, pp. 3-34). Baton Rouge, LA: Louisiana State University.

Sutherland, R. (2001). Algebra as an emergent language of expression. In H. Chick, K. Stacey, J. Vincent, & J. Vincent (Eds.), *The future of the teaching and learning of algebra* (Proceedings of the 12th ICMI Study Conference, pp. 570-576). Melbourne, Australia: The University of Melbourne.

Teppo, A. R., & Esty, W. W. (2001). Mathematical contexts and the perception of meaning in algebraic symbols. In H. Chick, K. Stacey, J. Vincent, & J. Vincent (Eds.), *The future of the teaching and learning of algebra* (Proceedings of the 12th ICMI Study Conference, pp. 577-581). Melbourne, Australia: The University of Melbourne.

Trigueros, M., & Ursini, S. (2001). Approaching the study of algebra through the concept of variable. In H. Chick, K. Stacey, J. Vincent, & J. Vincent (Eds.), *The future of the teaching and learning of algebra* (Proceedings of the 12th ICMI Study Conference, pp. 598-605). Melbourne, Australia: The University of Melbourne.

Vergnaud, G. (1996). The theory of conceptual fields. In L. P. Steffe, P. Nesher, P. Cobb, G. Goldin, & B. Greer (Eds.), *Theories of mathematical learning* (pp. 219-239). Mahwah, NJ: Lawrence Erlbaum.

Villarreal, M. (2000). Mathematical thinking and intellectual technologies: The visual and the algebraic. *For the Learning of Mathematics, 20*(2), 2-7.

The Working Group on Teachers' Knowledge and the Teaching of Algebra

Leader: Helen M. Doerr

Working Group Members:
Cecilia Agudelo, Elizabeth Belfort, Mary Enderson, George Gadanidis, Brigitte Grugeon, Sylvia Johnson, Vilma Mesa, and Sheryl Stump.

The Working Group on Teachers' Knowledge and the Teaching of Algebra. Seated (L to R): Sylvia Johnson, Cecilia Agudelo. Standing (L to R) George Gadanidis, Brigitte Grugeon, Mary Enderson, Sheryl Stump, Helen Doerr, Elizabeth Belfort. Absent: Vilma Mesa.

Prior to the conference, members of the Working Group on Teachers' Knowledge and the Teaching of Algebra reviewed aspects of the research literature that related to teachers' knowledge for the teaching of algebra. Helen Doerr compiled these reviews and circulated them electronically to all of the members. In addition, each member of the Working Group prepared a paper for the 12th ICMI Study Conference Proceedings. The authors (sometimes with co-authors) are listed together with the titles of their papers:

- M. Artigue, T. Assude, Brigitte Grugeon, & A. Lenfant: *Teaching and learning algebra: Approaching complexity through complementary perspectives* (pp. 21-32).
- Elizabeth Belfort, L. Guimaraes, & R. Barbastefano: *Tertiary algebra and secondary classroom practices in number and algebra: Closing the gap* (pp.79-86).

- Helen Doerr: *Learning algebra with technology: The affordances and constraints of two environments* (pp. 199-206).
- Sylvia Johnson: *Learning to teach algebra in the UK: Trainee teachers' experiences* (pp. 336-343).
- A. Manouchehri & Mary Enderson: *Learning to teach reformed algebra: The challenge of educating future mathematics teachers* (pp. 420-424).
- Vilma Mesa: *Functions in middle school mathematics textbooks: Implications for a functional approach to algebra* (pp. 454-461).
- Sheryl Stump & J. Bishop: *Framing the future: Inventing an algebra course for pre-service elementary and middle school teachers* (pp. 564-569).

During the conference, the members established a framework for working together in order to achieve consensus on issues. All members participated in whole group discussions that enabled a wide range of different cultural perspectives to be heard and accommodated. During these discussions the research papers on teachers' knowledge and pedagogy were reviewed, carefully analysed, and synthesised. Key issues related to the nature and development of teachers' knowledge and teaching practices were identified together with areas in need of further research. The members also identified difficulties associated with articulating the issues and synthesising the research.

This chapter reflects the Working Group members' discussions that were primarily based on the research literature that had been nominated and brought to the conference and the papers that the members had prepared for the Conference Proceedings, as well as additional comments by members after they reviewed draft copies of the chapter.

All members are thanked for contributing to the success of the Working Group on Teachers' Knowledge and the Teaching of Algebra. Special thanks are extended to Helen Doerr who capably led the group and to Mary Enderson who graciously undertook the role of recorder.

Chapter 10

Teachers' Knowledge and the Teaching of Algebra

Helen M. Doerr
Syracuse University, New York State, USA

Abstract: In this chapter an analysis of the research on teachers' knowledge and practice and its development with respect to the teaching of algebra is presented. The chapter begins with a brief discussion of four dilemmas that were confronted during this analysis. The findings from research on teachers' knowledge are reported in three areas: (a) teachers' subject matter knowledge and pedagogical content knowledge, (b) teachers' conceptualisations of algebra, and (c) teachers learning to become teachers of algebra. The chapter concludes with a discussion of critical issues and suggestions for further research.

Key words: Teachers' knowledge, teachers' learning, teachers' professional development, teachers' conceptions of algebra, pedagogical content knowledge

10.1 Introduction

The last thirty years of research in mathematics education have resulted in a substantial body of research on theoretical conceptualisations of the meaning of school algebra, on children's learning of algebra and, most recently, on how children's learning is influenced by computational technologies. Some of the results of this research can be seen in curricular changes that have happened in varying degrees in different countries and cultures (also see Chapter 13 in this book), including, for example, an increased emphasis on graphing and an earlier introduction of exponential functions into the curriculum. Despite the rather large body of research on children's learning that demonstrates the ineffectiveness of learning algebra as procedures that are disconnected from meaning and purpose, much of school algebra is still taught as such disconnected procedures. This disconnection suggests that the research on children's algebra learning has had limited impact in schools. One of the major impediments to change in how algebra is taught in schools would appear to be the lack of a substantial body of research on teachers' knowledge and practice in the teaching of algebra.

Teachers' knowledge and practices and their development for the teaching of algebra have been largely unexamined in the research literature. In the early 1990s, Kieran's review of the literature on the learning and teaching of school algebra pointed to the grave scarcity of the literature on teaching algebra and posed the need to describe "the ways in which the teaching of algebra ought to be considered in a different light from, say, the teaching of geometry or arithmetic" (Kieran, 1992, p. 394). The scarcity of research on teachers' beliefs and cognitions about algebra is accompanied by a lack of knowledge on how teachers interpret and adapt materials in the textbooks from which they teach and on their understandings of students' cognitions related to algebra. Several researchers (Cooney & Wilson, 1993; Leinhardt, Zaslavsky, & Stein, 1990; Norman, 1993) have pointed to the lack of research on teachers' knowledge and beliefs with respect to functions, graphs, and graphing, which are central topics in the teaching and learning of algebra. Norman pointed out that the emergence of computing technology has profound implications for teachers who are "relatively inexperienced" in the context of new technologies. They are unlikely to have had "formative computing experiences as learners," and may suffer from weak conceptual knowledge of the function concept (1993, p. 169). This suggests that teaching and learning with technology are important areas for research on teacher learning, not just student learning. Progress in this regard, however, seems to be slow. The same lack of research on teachers' knowledge that Leinhardt, Zaslavsky, and Stein (1990) found in their review of the literature is reflected again in Penglase and Arnold's (1996) review of the literature on the role of the graphics calculator in teaching and learning mathematics.

While lamenting the lack of research on teachers' knowledge and practice, we also found that all too often the practical wisdom of teacher educators is fragmented and not systematically codified in ways that are usable by other teacher educators. In others words, as teacher educators have engaged in their work of preparing mathematics teachers and supporting the continued professional development of practising teachers, their work and wisdom is largely shared in ways that are anecdotal and embedded in the specificity of particular cultural situations. This means that the work of teacher educators has not itself been subject to the scrutiny of research and subsequent revision by the larger community of researchers and practitioners. Taken together with the lack of research on teachers' knowledge, this suggests that there is a serious need for theory building to describe and explain what it is that teachers need to know to teach algebra, and how it is that this knowledge is developed by novice teachers and by experienced teachers. We further argue that teacher educators need both principles and cases of practice that can systematically illuminate the work of teacher professional development.

In this chapter, therefore, we will attempt to accomplish three goals related to the teaching of algebra:

1. Identify what is known from research about the nature of teachers' knowledge and its development.
2. Identify critical issues related to the nature and development of teachers' knowledge and practice.
3. Discuss and analyse these issues in light of the research and the practical wisdom of teacher educators in order to suggest areas in need of future research.

Before proceeding with the above, however, we wish to discuss four background dilemmas that emerged from the analysis of teachers' knowledge and practice that took place in the meeting of the Working Group at the conference and that underlie the ideas and perspectives presented later in this chapter.

The first dilemma is what we call the "dilemma of experience". By this, we mean that teacher educators are faced with the difficulty of how to simultaneously build on pre-service teachers' experience as pupils in schools and to break the mould of that experience (Ball, 1988). Pre-service teachers come to their preparation programs with years of experience observing what it is that mathematics teachers *do* in classrooms. However, seeing what experienced teachers *did* in the classroom does not necessarily yield any insight into *why* they acted as they did or what alternative courses of action they considered in particular situations (Doerr & Lesh, 2003). Pre-service teachers often believe, at least initially, that the methods that they were taught by were effective because they were able to learn mathematics. Pre-service teachers often fail to recognise that many of their peers did not succeed in learning mathematics as well as they did. Teacher educators are faced with the challenge of helping pre-service teachers to understand their own experiences and observations of teaching in ways that move beyond the surface features of what happened in the classroom and to understand alternative courses of action that might be more effective in developing conceptual understandings of mathematics for a greater number of students. The collective experiences of teacher educators and professional developers in facing the dilemma of experience have not been systematically investigated and reported.

The second dilemma is that of "what algebra?" In other words, what *is* the algebra that should be taught in school? In our deliberations about teachers and teaching, we took school algebra itself as non-problematic. In part, this was due to the fact that other authors are addressing this important topic in other chapters in this book (e.g., Chapter 12 with a focus on curriculum goals for compulsory schooling) and, in part, this was a reflection of our own understanding that the research base on teachers' knowledge and its development is still in its infancy. Understanding teachers' conceptualisations of algebra and their interpretations of the concomitant instantiation of algebra in curricular materials are areas in need of further research. Nonetheless, we want to be clear that in this chapter there are some implicit assumptions about what algebra is that we will not attempt to make explicit.

The third dilemma is the difficulty in speaking and writing about teachers' knowledge as a noun, as if it were a static set of things that teachers possess. Rather, we would prefer to speak and write about teachers' knowledge using the verb form: teachers' *knowing*. This suggests that the nature of what it is that teachers need to know is dynamic, fluid, situational, reflected in action, and situated in specific cultures and in specific social settings within those cultures. These characteristics of the nature of teachers' knowing also suggest a focus on teachers' learning and on teachers' reasoning in context. In addition to describing what it is that teachers might *see* in a particular instance of teaching, we are interested in understanding how teachers reason about that particular instance and how teachers learn from such instances. In other words, we want to characterise the change and development of teachers' thinking over time and across settings. So, while the limitations of language will lead us to write about "teachers' knowledge," we urge the reader to keep in mind an active, participatory sense of what it means to know in the work that follows.

The final dilemma in articulating the nature and development of teachers' knowing with respect to the teaching of algebra is the difficulty of situating any claims to be made within the larger body of research on teacher development. It is beyond the scope of this chapter to review the competing and emerging theories of teachers' knowledge and its development (e.g., Ball, Lubienski, & Mewborn, 2002; Cooney, 1999; Doerr & Lesh, 2003; Munby, Russell, & Martin, 2002), yet at the same time, we find it important that we draw on that work in understanding the teaching of algebra. Hence, in this chapter, we will occasionally refer to that larger body of work, but our intention here is to focus more clearly on the nature and development of teachers' knowing with respect to the teaching and learning of algebra. These four dilemmas reflect the deliberations of the Working Group and are intended to provide the reader of this chapter with some sense of the limitations and difficulties inherent in the present work.

10.1.1 Findings from research

The research on algebra learning has tended to focus on the algebraic nature of the mathematical tasks, the development of ideas by the learners, and, in some cases, on the influence of the technology, but rarely are the teachers, and the nature and development of their knowledge and teaching practices, the focus of the study. We have organised our discussion of the research that does focus on the teacher into three broad areas: (a) teachers' subject matter knowledge and pedagogical content knowledge, (b) teachers' conceptualisations of algebra, and (c) teachers learning to become teachers of algebra. We will use this framing to discuss what is known from research, the limitations of that research, and future directions for research. We recognise that there continues to be a need for research on teachers' understandings of technology and on the implications of technology for changes in the role of the teacher. The research in this area (e.g., Doerr & Zangor, 1999, 2000; Heid, 1995; Slavit, 1996; Tharp, Fitzsimmons, & Ayers, 1997) is discussed in the chapters on technology in this volume (Chapter 6 focuses on a range of technological environments and Chapter 7 focuses particularly on CAS), and so will not be considered here.

10.1.2 Teachers' content knowledge

Much of the research on teachers' subject matter knowledge and pedagogical content knowledge has focused on the conceptions and misconceptions that secondary teachers have about the concept of function, with more recent work addressing teachers' knowledge about the concepts of slope, variables, and expressions. Researchers appear to agree uniformly that the subject matter knowledge needed for teaching includes much more than the knowledge and understanding of the mathematical concepts. In her 1990 study of the subject matter knowledge of functions for teaching, Even put forward a framework that includes the following essential features: different representations, alternative ways of approaching the concept, strength of the conceptual knowledge, a basic repertoire of examples, knowledge and understanding of the concept, and knowledge of mathematics. This framing of subject matter knowledge includes aspects that might be described as pedagogical content knowledge: such as, alternative ways of approaching the concept and a basic repertoire of examples. Assuming that subject matter knowledge and pedagogical content knowledge are inter-related, there is little research evidence that illustrates the relationships between the two, and that describes and explains how content knowledge is transformed into the more powerful ways of knowing the subject matter so that others can be helped to learn.

Norman (1992) found that the secondary teachers in his study (all of whom were working towards a Masters' degree in mathematics education) tended to have inflexible images of the concept of function that restricted their abilities to identify functions in unusual contexts and to shift among representations of functions. These teachers expressed preferences for the graphical representations of functions. They were able to give formal definitions of a function, were able to distinguish functions from relations, and were able to correctly identify whether or not a given situation was functional. However, the teachers did not show strong connections between their informal notions of function and formal definitions and were not comfortable with generating contexts for functions. Difficulty in constructing functions was also observed by Hitt (1994) who found that teachers had difficulty in constructing functions that were not continuous or were defined by different algebraic rules on different parts of the domain. Consistent with Norman's findings, Chinnappan and Thomas (2001) found that all four pre-service secondary teachers in their study had a preference for thinking about functions graphically, weak understandings of representational connections, and limited ability to describe applications of functions.

Unlike the teachers in Norman's (1992) study, Even (1993) found that many prospective secondary teachers did not hold a modern conception of a function as a univalent correspondence between two sets. These teachers tended to believe that functions are always represented by equations and that their graphs are well behaved. None of the teachers had a reasonable explanation of the need for functions to be univalent and over-emphasised the procedure of the "vertical line test" without concern for understanding. Given the often weak and fragile understanding of secondary mathematics teachers about the concept of function, it is not surprising to find that the knowledge of an experienced 5th grade teacher was missing several key ideas (such as univalence and unclear notions of dependency) and lacked a notion of the connectivity among representations (Stein, Baxter, & Leinhardt, 1990). As these researchers noted, "limited, poorly organised teacher knowledge often leads to instruction characterised by few, if any, conceptual connections, less powerful representations, and over-routinised student responses" (p. 659). Such instruction is not likely to lead to the kind of student learning envisioned by those researchers who are investigating early algebra instruction. Collectively, this work would suggest that teachers' knowledge about functions tends to be instrumental (or procedural), rather than relational (or conceptual), and to lack the kinds of connectedness and flexibility that would lead to teaching strategies that would in turn promote conceptual understanding by students.

More recent research has examined teachers' knowledge about other important concepts in school algebra, such as expressions, equations, and slope. In a study that compared the connectedness of lessons for two novice secondary teachers and an expert teacher when teaching equivalent algebraic expressions, Even, Tirosh, and Robinson (1993) found that only the expert teacher used connections between lessons and connections in mathematical content to guide her lesson. This finding suggests three dimensions of expertise in teaching algebra: (a) planning for connections across lessons, (b) teaching in ways that make and exploit mathematical connections, and (c) seeing connectedness as a major goal of instruction.

Within the context of school algebra, the concept of slope is usually an early topic. This concept builds on students' earlier experiences with steepness and angle (encountered in geometry), and with ratios and rates (encountered throughout the later years of primary schooling). At the same time, the notion of slope foreshadows central ideas in calculus (namely the derivative) and connects with the students' emerging ideas about functions, in particular linear functions. This suggests that the knowledge of slope from a teachers' perspective needs to be one that has significant connections across the curriculum. In her investigation of 18 pre-service and 21 in-service teachers' conceptions of slope, Stump (1999) found that most of the teachers thought of slope as a geometric ratio, with most of the in-service teachers also describing slope as a physical property, such as the slope of a ramp. Less than one-fifth of each group, however, thought of slope as the rate of change involving two variables, namely as a functional concept. While both groups of teachers expressed concern for students' conceptual understanding, their identification of students' difficulties with the slope concept reflected an emphasis on procedural aspects (such as correctly using the slope formula), and on the algebraic and geometric representations of slope. In a later study, which also examined the lessons taught by three pre-service teachers to college algebra students, Stump (2001) found that while "their knowledge of representations for teaching slope was dominated by graphs and physical situations" (p. 224), the teachers' actual lessons were focused more on graphs and equations. Only one of the pre-service teachers asked students to interpret the meaning of slope in physical situations. In a larger scale study with 162 pre-service teachers, Even and Tirosh (1995) found that the teachers recognised a common student mistake of assuming a proportional relationship between the slope of a linear function and the angle that the line makes with the x-axis. While many of the pre-service teachers saw the student error as a difficulty in estimation, about half of the teachers were able to give an accurate description of the source of the students' thinking. These researchers concluded that many teachers "made no attempt at understanding the sources of students' responses" and that one cannot assume that teachers' subject matter knowledge is sufficient for understanding and explaining students' reasoning.

Another aspect of teachers' knowledge of algebra is the understanding of students' conceptions and misconceptions. One well-established finding from the research on student learning is the tendency for students to conjoin or finish expressions such as $3x + 5$ to get $8x$. In examining the knowledge of four teachers on this aspect of simplifying algebraic expressions, Tirosh, Even, and Robinson (1998) found that the two novice teachers were unaware of this tendency and that the two experienced teachers anticipated this student difficulty (although they did not give any of the explanations put forward in the research literature) and planned their lessons accordingly. The lessons taught by the novice teachers used a version of the "collecting like terms" approach and when encountering student difficulties appealed to the application of rules and "fruit salad" analogies. One of the experienced teachers, in anticipation of the student difficulties, spent the first part of the lesson on identifying like terms, before proceeding to collecting like terms. The other experienced teacher used multiple strategies (substitution, order of operations, and going backwards) to create conflicts in her students' thinking that they would need to resolve. This study points to the need for teachers to understand students' difficulties, and also to understand alternative approaches to concepts and the pros and cons of such approaches in different contexts with different students.

The research on teachers' subject matter and pedagogical content knowledge has tended to be dominated by investigations that are framed in a deficit or deficiency view of teachers and teaching. That is, such studies have tended to focus on the inadequacies of teachers' knowledge for the activities of teaching. We would like to argue that this focus of research needs to shift to an emphasis on what it is teachers do know and how they interpret the complex situations of teaching (Doerr & Lesh,

2003). Within such a framing, it is clear that there are many areas of teachers' content knowledge that are yet to be fully investigated (such as the concepts of variable and equivalence; algebra as a language for modelling phenomena; and aspects of the function concept including the representations and contexts of exponential, trigonometric, and probability distribution functions). We would argue that such investigations need to examine how teachers' content knowledge is transformed into useful knowledge for teaching in a range of social and cultural settings.

10.1.3 Teachers' conceptualisations of algebra

The research on the learning of algebra has suggested that there are four conceptualisations of algebra (Bednarz, Kieran, & Lee, 1996; Usiskin 1988). See also Kaput and Blanton (2001 and Chapters 5 and 7 in this volume. Algebra may be conceptualised as: (a) generalised arithmetic, (b) a means to solve certain problems, (c) a study of relationships, and (d) structure. The conceptualisations of algebra put forward by researchers, curriculum developers, and mathematics educators do not necessarily reflect the conceptualisations of algebra that are held by teachers. In her study of pre-service teachers, Johnson (2001) found that many of them were unable to articulate their own understandings of algebra or discuss the nature of the subject matter, and Menzel (2001) found similar difficulties among experienced teachers of algebra. In his study of teachers' conceptions of algebra, Gadanidis (2001) found that the teachers saw the formula for area as the key piece of algebra when using a technology-based lesson on maximising area through an interactive exploration of graphs. These teachers did not see the graphs or the context as part of the algebraic content in the activity.

Other researchers (Bishop & Stump, 2000; Haimes, 1996; Menzel & Clarke, 1998) have found evidence that teachers tended to emphasise procedural knowledge for solving equations as the core of algebra. Many of the elementary and middle school teachers in Bishop and Stump's study saw a pattern generalisation task as primarily a problem-solving activity rather than as an opportunity to generalise the familiar patterns of arithmetic. The in-service teachers in Menzel and Clarke's study emphasised procedural knowledge for the study of algebraic equations and considered graphing techniques and skills to be the most important aspect of functions for students to know about. The in-service teacher in Haimes' study saw that the difference between a generalised arithmetic approach and a functions-based approach to algebra was the omission of the repetitive work in the latter. These findings suggest that the teachers placed a diminished value on the conceptual use of graphical representations and on the use of meaningful contexts, as well as a disconnection between those types of activities and the development of students' algebraic thinking. One consequence of the disconnection from graphs and contexts, as shown when teachers teach the rules for manipulating algebraic expressions, is that students do not seem to be able to use symbolic expressions as tools for meaningful mathematical communication (Kieran & Sfard, 1999).

In contrast to these findings, other researchers (Chazan, Larriva, & Sandow, 1999) described the knowledge of one teacher who conceptualised a functions-based approach to algebra and of another teacher whose concepts were ambiguous. The pre-service teacher whose preparation program provided a strong foundation in a functions-based approach to algebra was able to articulate clearly the dilemmas and tensions that she experienced in simultaneously understanding a variable both as an unknown quantity and as a quantity that changes (Leikin, Chazan, & Yerushalmy, 2001). This pre-service teacher recognised the difficulties that she encountered instructionally when considering the solution of an equation such as $3x + 7 = 2(x + 5) + x - 1$ as the solution of a system of equations $y = 3x + 7$ and $y = 2(x + 5) + x - 1$. These findings suggest the need for teachers to be involved in programs of professional development that support these deeper conceptualisations of algebra.

One aspect of teachers' conceptualisation of algebra that has received careful attention in several recent studies (among the few that involve a significant number of teachers) is the difference between arithmetic and algebraic problem-solving strategies by teachers in their own problem-solving behaviour and in their evaluations of students' work. This issue was first investigated by Schmidt and Bednarz (1997). They studied three groups of pre-service teachers (N=66 elementary, N=33 secondary, and N=65 learning-difficulties teachers—elementary and secondary) at the beginning of their preparation programs. The pre-service teachers were given a written test with 4 arithmetic problems and 4 algebra problems. For the arithmetic problems, arithmetic strategies were chosen by more than half (55%) of pre-service elementary group, 74% of the learning-difficulties group, and by 35% of the secondary group. For the algebra problems, algebraic strategies were used by 62% of the elementary group, by 27% of the learning-difficulties group, and by 92% of the secondary group. These results are explained in terms of the academic preparation of the participants, with those who have more mathematical preparation being more inclined to use algebra and those with the least preparation (the learning-difficulties group) choosing the arithmetic approaches. In follow up interviews, eight of the subjects (four of whom had an arithmetical profile and four of whom had a structural profile) were given five problems which varied in complexity from arithmetic to algebraic and three solutions to two problems. The researchers found that subjects with an arithmetic profile tended to characterise algebra as a repertoire of arbitrary rules over which students have little or no control and tended to recognise the superiority of algebraic solutions over arithmetic solutions even for arithmetic problems. The subjects with a structure profile tended to view arithmetic as a more effective tool for arithmetic problem solution, with only the more complex problems needing to be solved by algebra.

In a more recent study, van Dooren, Verschaffel, and Onghena (2002) extended the findings of Schmidt and Bednarz by examining the strategies of pre-service teachers at the beginning and at the end of their teacher preparation programs and further examined the pre-service teachers' evaluations of students' work. These researchers found strong similarities between the solution patterns of their pre-service teachers and those in Schmidt and Bednarz's study. However, they also found for both primary and secondary pre-service teachers that there was a "surprisingly small difference in pre-service teachers' problem-solving behaviour at the beginning and at the end of their teacher training: The third-year pre-service teachers largely used the same strategies as their first-year counterparts, but they were more skilled in applying them to complex problems" (p. 332). These findings suggest that the preparation program was successful in improving the skills of the pre-service teachers, but that their preferences for solution strategies were resistant to change.

The second part of van Dooren and colleagues' study (van Dooren, Verschaffel, & Onghena, 2002) examined the evaluations that primary and secondary pre-service teachers made about students' solutions to arithmetic and algebraic problems. They found that the pre-service teachers' evaluations of student strategies tended to reflect the ways in which the teachers had solved the problems themselves. The pre-service secondary teachers gave the highest scores to the algebraic solutions and the pre-service primary teachers generally gave higher evaluations to arithmetical solutions. However, the pre-service primary teachers tended to take notice of the nature of the problem (as arithmetic or algebraic) to justify the appropriateness of the solution method, whereas the secondary pre-service teachers generally "referred to the overall superiority of the algebraic method as such, regardless of the nature of the problem to which it was applied" (van Dooren, Verschaffel, & Onghena, 2002, p. 345). These findings led the researchers to focus on two problematic areas for the teaching of algebra. First, about half of the primary pre-service teachers were unable to solve the more difficult problem on the test because of their lack of mastery of algebra. They experienced difficulty in understanding students' algebraic solutions and negatively evaluated such solutions. This led the researchers to doubt that these pre-service teachers had either the disposition or the mathematical readiness to prepare primary school students for the transition from arithmetic reasoning to algebraic reasoning. Second, for the pre-service secondary teachers, many of them tended to use algebraic methods in a stereotyped way, accompanied by perceptions of arithmetic strategies as inferior, even when more appropriate. The researchers questioned whether these future teachers, too, would have the empathy, insight, and support for students making the transition from arithmetic to algebra.

In a similar study of practising teachers (N=105 elementary, middle, and high school teachers), Nathan and Koedinger (2000a) investigated the match between teachers' beliefs about the development of students' algebraic reasoning and the actual performance of students on a set of beginning algebra problems. This study provides insight into teachers' conceptualisation of algebra as well as into their perceptions of students' performance. Nathan and Koedinger used 6 items that differed along two dimensions. The first dimension addressed the underlying mathematical structure of the problem as arithmetic (result unknown) versus algebraic (start unknown) (cf. Stacey & MacGregor, 2000). The second dimension addressed the format or presentation type, consisting of two verbal formats, story problems and word problems (e.g., "guess my number" problems), and one symbolic format (e.g., "solve for x: $6x + 66 = 81.9$"). In an earlier study, Nathan and Koedinger had analysed actual student performance for this set of problems (Nathan & Koedinger, 2000b) as well as the classification schemes for their difficulty that were developed by secondary mathematics teachers (N=67) and mathematics educators (N=35). The findings across grade levels are particularly illuminating with regard to teachers' knowledge of students' performance. At all levels, teachers recognised that arithmetic problems (result unknown) are easier for students than algebraic problems (start unknown) (Nathan & Koedinger, 2000a). However, the elementary teachers and the high school teachers both regarded the symbolic format equations as easier than the algebra word and story problems, a result that directly contradicted the actual performance of students. The middle school teachers were the most accurate in terms of their predictions of student performance.

In discussing the results of their study, Nathan and Koedinger attributed the success of middle school teachers in assessing students' early algebra problem difficulties to the greater number of "opportunities to observe how students make the transition from arithmetic to algebraic reasoning" (2000a, p. 229). Elementary-level teachers may simply assume that by high school students have mastered the formal procedures as they are presented in textbooks. The high school teachers, who were the least aware of the efficacy of students' invented algebra strategies (notably guess and check, and unwinding or backtracking), may be more distant from the difficulties first encountered by students and may be blinded by their own expertise in symbolic manipulation. This interpretation of the high school teachers is consistent with Menzel and Clarke's (1998) finding that secondary teachers were divided in their views as to whether the verbal description of mathematical relationships that can be generalised should be regarded as algebraic activity. Some teachers regarded verbal generalisations about numeric relationships as algebraic activity; others felt that the algebraic activity was only in the symbolising. Nathan and Koedinger's results further suggest that the textbook exerts a particularly strong influence on teachers' perceptions of the difficulties involved in learning algebra. Typical textbook materials place the learning of symbolic procedures for solving equations before the application of these procedures to word and story problems, suggesting that this is an appropriate learning sequence. However, the evidence from their research on student performance showed that the opposite was true. It would appear that teachers at the middle school level are most perceptive about the different strategies that students might take and how this would affect their performance on different classes of problems.

We found no research evidence in the USA that would suggest that teachers see the concept of function as an integrating theme for algebra instruction across the curriculum, despite this being envisioned in the curriculum standards of the National Council of Teachers of Mathematics (1989, 2000). This is not to say that individual teachers do *not* see functions as an integrating theme for the algebra curriculum and instruction. Furthermore, we could ask the following question: Why should functions be an integrating theme for school algebra? Nor is there any evidence that would suggest that the frameworks that researchers have articulated for understanding the nature of school algebra are useful for teachers. It is also unclear how teachers frame their understanding of algebra. What would be a useful framing of school algebra that would serve teachers in understanding algebra and students' learning of algebra? How would such a framing support a view of algebra as a broad and unifying theme for the mathematics curriculum? How can the existing body of research on students' conceptions and students' learning be articulated within that framework? Currently, our effort to answer these questions is hampered by the small number of studies (with limited numbers of subjects) that have been conducted on teachers' conceptualisations of algebra. A potentially fruitful arena for research would be to investigate the conceptions of algebra that are held by teachers with varying degrees of expertise in a diversity of cultural and social settings.

10.1.4 Becoming teachers of algebra

All of the studies on teachers' content knowledge and on teachers' conceptualisations of algebra point to a larger difficulty. How do novice teachers become expert teachers of algebra? This question has continuing importance for secondary teachers as well as new importance as the research on the early learning of algebra highlights the importance of elementary teachers being involved in algebraic instruction. While there is a substantial body of literature on elementary teachers' development with respect to number and ratio, there is almost no research on elementary teachers' understanding of the shift from arithmetic to algebraic reasoning (see Chapter 4 on early development of algebraic reasoning). But the situation with respect to the learning and development of secondary teachers is not much better. Given the considerable differences in mathematical content knowledge between elementary and secondary teachers, we would expect that the learning and development of these two groups of teachers could be very different along many dimensions.

The few studies that have focused on elementary teachers' algebraic knowledge provide us with some insights into the preparation of elementary teachers and the continuing professional development of practising teachers. In their work with pre-service elementary teachers, McGowen and Davis (2001) focused on enhancing teachers' ability to think flexibly and to see and value mathematical connections. Some of the pre-service teachers were able to establish the connections across a series of tasks beginning with building towers with coloured cubes and culminating in the generalised binomial expansion of $(a + b)^n$. The teachers engaged in examining children's thinking about the tower problem and in explaining their own solutions to each other. Stump and Bishop (2001) suggested that the preparation of elementary and middle school teachers for the teaching of algebra should include learning about generalisation, problem solving, modelling, and functions. Kaput and Blanton (2001) emphasised the need to support elementary teachers in finding and extending the generalising and formalising opportunities that exist within the curriculum for their students.

In a case study on one 3rd grade teacher, Blanton and Kaput (2001) described the development of generalisation that occurred in the classroom across a series of tasks. Extending arithmetic conversations into algebraic ones, adapting other resources, and extending other lessons to build on algebraic themes increasingly characterised the teacher's practice. This teacher belonged to a cohort of teachers involved in a professional development program that was grounded in acting and reflecting on the teachers' existing instructional materials base and existing practice. The teachers engaged in activities to support their own mathematical learning as well as an examination of student work. The researchers intend to move this model of professional development to 350 teachers in an entire school district, where one could anticipate that new difficulties and obstacles will be encountered in terms of school culture and policies (Arens & Meyer, 2000; see also Chapter 4 of this book).

The research on secondary mathematics teachers strongly suggests that changes in pedagogy that reflect an increased emphasis on conceptual learning can only come about when teachers possess strong subject matter knowledge. Several studies have shown how limitations in understanding the concept of function can limit the kinds of tasks that teachers choose for students to engage with, the depth of questions that teachers pose, and the connections that are made within the curriculum (Haimes, 1996; Heid, Blume, Zbiek, & Edwards, 1999; Stein, Baxter, & Leinhardt, 1990; Wilson, 1994). The well-connected knowledge of the teacher studied by Lloyd and Wilson (1998) clearly illustrates how one teacher used his strong content knowledge to shift from a procedural to a conceptual approach to the teaching of algebra. The conceptual approach was functions-based and characterised by an emphasis on a co-variational as well as dependence approach to functions, the use of graphs to understand patterns of co-variation and characteristics of families of functions, the flexible use of multiple representations to provide increased opportunities for student understanding, and the use of meaningful discussions with students. Similarly, Chazan (1999, 2000) found that a functions-based approach to algebra provided "the type of knowledge of the subject matter which supported involvement of students in the exploration of the subject" (1999, p. 122). Subject matter knowledge that is well connected appears to be a necessary (but not sufficient) condition for expertise in teaching algebra.

Several researchers have argued that professional development experiences should address teachers' subject matter knowledge and their pedagogical content knowledge. As Agudelo-Valderrama (2000) found in her study with 7th and 8th grade teachers of algebra, affective factors such as the teachers' conceptions of themselves as mathematics teachers played a role in the development of expertise in teaching the concept of variable in ways that led to improved student performance. Brown and Smith (1997) described a model for professional development that focused on the development of a knowledge base for teaching by addressing the teachers' learning through planning, actual instruction, and reflection on the instruction. This program led to changes in the practice of an experienced algebra teacher that shifted toward student discussion, the use of multiple representations (algebra pieces or tiles, symbols, sketches), and questions that encouraged exploration. Another example of teachers' learning through their instructional practices was found by Miller (1992) who investigated the sustained use of informal writing prompts in algebra classes. Student writing made visible to the teachers that while the students knew the correct computational rules for dividing by zero and into zero, they were unable to give any explanations as to why these rules were true. This led the teachers to make adaptations in their instructional practice that included re-teaching, reviewing, and working with individual students, but not the kinds of profound changes envisioned by reformers. In all of these studies, teachers were part of a larger cohort group that was engaged in a sustained professional development program of a year or more.

As Stump (1999) concluded from her work with pre-service and in-service teachers, "we cannot assume that teachers will make connections among various representations of slope on their own" (p. 141). Stump argued that when curriculum resources (primarily textbooks) present topics (such as slope, linear equations, rate of change, and trigonometric functions) as unrelated concepts, then it is likely that teachers will teach them as unrelated concepts. Similarly, Mesa (2001) found that different textbooks implicitly suggest different teaching practices that reflected different conceptions of function. Mesa argued that because the practices were in general unrelated and disconnected this would explain why teachers and students found it difficult to find the connections or algebraic ideas that transcended different practices and conceptions. Nor can we assume that pre-service teachers will bridge the gap between their tertiary courses in algebra and school algebra (Belfort, Guimaraes, & Barbastefano, 2001). The difficulties of learning mathematical and pedagogical connections are exacerbated by the isolating experiences of many first year teachers that limit their opportunities for debate and discussion with colleagues (Johnson, 2001). Because teachers necessarily acquire much of their knowledge for teaching from their own practice, the research findings just discussed—about the disconnections in curricular materials—raise important questions for research on the teaching of algebra: (a) How can teachers learn a more connected approach to the teaching of algebra? and (b) How do the curricular resources and textbooks that the teachers use influence this learning?

There is clear evidence, which is supported by the data on teacher development from the broader base of research on teaching, that the current mathematical preparation of teachers leaves most teachers inadequately prepared with the kinds of subject matter knowledge that are powerful for teaching school algebra. Furthermore, there is substantial evidence that when working in depth with a relatively small numbers of teachers (often as few as one or two) over an extended amount of time teacher learning, accompanied by significant changes in practice, can occur. However, we found no reported studies of teacher learning and concomitant changes in practice on a medium or large scale.

10.2 Critical Issues

These research findings have led to the identification of the following three sets of critical issues related to the nature and development of teachers' knowledge and practice. Here we want to remind the reader of our earlier dilemma about knowledge and *knowing*. We do not mean to imply by this question that there is a fixed body of knowledge that teachers need to acquire. Rather, we wish to suggest that there is a need to develop a knowledge base for the teaching of algebra that captures the dynamics, fluidity, and situated character of knowing in action.

10.2.1 Teachers' knowledge about algebra, teaching, and learning

Earlier in this chapter (Section 10.1.2), we discussed the findings from research with respect to teachers' content knowledge of algebra. These findings were largely organised by the distinction between subject matter knowledge and pedagogical content knowledge. In this section, we wish to refine and extend that distinction by drawing on the three dimensions of the framework put forward by Artigue and colleagues (Artigue, Assude, Grugeon, & Lenfant, 2001): epistemological, cognitive, and didactic/pedagogical. These interrelated dimensions of knowing help us, as researchers, to organise our descriptions of the knowledge for teaching algebra and to suggest research perspectives for investigating that knowledge.

The epistemological dimension involves knowing: (a) content of algebra, (b) the structure of algebra, (c) the role and place of algebra within mathematics, (d) the nature of valuable algebra tasks for learners, and (e) the connections between algebra and other areas of mathematics and to physical phenomena. The epistemological dimension enables us to be more specific about what we mean by subject matter knowledge for the teaching of algebra. At the same time, we wish to describe the cognitive dimension separately from the epistemological dimension.

This enables us to emphasise the importance of knowing subject matter in ways that can lead to effective teaching of that subject matter.

The cognitive dimension includes knowing: (a) the development of students' algebraic thinking, (b) students' interpretations of algebraic concepts and notation, (c) students' misconceptions and difficulties in algebra (often referred to as epistemological and cognitive obstacles), (d) different approaches taken by learners, (e) ways to motivate learners, and (f) theories of learning.

The didactic/pedagogical dimension includes knowing: (a) the curriculum (including the dynamic interrelations between the mathematical content, the specific teaching goals, the teaching methods or strategies, and the assessment practices), (b) the resources (textbooks, technology, manipulatives, and other curriculum materials), (c) different instructional representations, (d) different practices and approaches taken by other teachers, (e) the connections across the grade levels, and (f) the nature and development of effective classroom discourse. We particularly wish to emphasise here that curricular knowledge includes the knowledge of teaching methods and the relationship between those methods and the mathematical content of algebra. We also wish to point to the importance of teachers knowing the practices and approaches taken by other teachers. Such shared knowledge of teaching becomes the basis for building a professional knowledge that moves beyond the particular knowledge of one teacher.

10.2.2 How teachers learn to teach algebra

We have many pieces of the picture of appropriate practices for the teaching of algebra (e.g., Bednarz, Kieran & Lee, 1996; Blanton & Kaput, 2001; Chazan, 1999; Lloyd & Wilson, 1998; Pirie & Martin, 1997; Raymond & Leinenbach, 2000; Sutherland, Rojano, Bell, & Lins, 2001, and other chapters in this volume). Yet there are two parts of the picture that are incomplete. First, we need to know far more about the variations in practice across different classrooms with different students in different cultures. From the pieces of the picture that we do have, it would appear that teachers must have knowledge of the epistemological, cognitive, and didactic dimensions of knowing algebra for teaching that we outlined above. However, such a list of attributes does not capture the complex interrelationships of these dimensions nor does it give us a sense of how they are related to student learning in various contexts and cultures.

Second, there is a significant shortage of research about how teachers learn to teach algebra, how they understand their own practice, and how they form and are formed by their own practice within their own specific cultural contexts. For example, we need to understand more about how the teacher mediates curricular change and conceptualises the development of students' algebraic reasoning. We wish to argue that there needs to be a research focus on teacher learning. Such a research focus will enable us to better understand the complex interrelationships of a teacher's knowledge about the epistemological elements of algebra, about cognitive aspects, and about the didactic dimension. The focus of such research needs to be not only on the nature of this knowledge, but most importantly how this knowledge is acquired and developed. Furthermore, we wish to argue that if such research is to inform change in practice, then it must be done in collaboration with teachers as professional partners in the acquisition of knowledge about practice. Evidence to date suggests that this activity needs to be longer term (of the scope of years) rather than shorter term (weeks or months). We see as an important research goal the development of theoretical frameworks to interpret teachers' development in the teaching and learning of algebra. We see the need for more research data in order to build such theoretical frameworks.

10.2.3 Investigating the practice of teachers of algebra

There is little critical discussion about methodological approaches for investigating both teacher knowledge and teacher practice. While there are clear advantages to multiple methodologies, it would also appear that in some cases we as a field are hindered by a lack of shared methodologies. Many current studies are now a combination of surveys or questionnaires, interviews, and observations. However, the protocols for interviews and observations, and the methods for analysing the resulting data corpus appear to be almost as varied as the number of researchers in the field. Several recent studies have examined the use of concept maps as a methodological approach (Doerr & Bowers, 1999; Leikin, Chazan, & Yerushalmy, 2001), while other researchers have used varying forms of action research and practitioner research (Bishop & Stump, 2000; Chazan, 2000; Raymond & Leinenbach, 2000). Much of the work of teacher educators, in preparing teachers for practice, is not part of this larger body of research but rather resides as the uncodified collective wisdom of practitioners and is not subject to the scrutiny of research. Greater coherence across research methodologies and greater involvement of practitioners in that research could result in the establishment of a professional knowledge base for investigating the teaching of algebra.

10.3 Directions for Future Research

We highlight below some of the areas where research is needed if we are to know more about how teachers learn to teach algebra. These are selected as representative of issues that research suggests are significant in the development of teachers and their practice. Priorities are likely to vary from country to country and culture to culture. These areas aim to address aspects of how teachers acquire epistemological, cognitive, and didactic knowledge. Here we have organised the research questions along two dimensions: theory and practice. However, as we have indicated above, actual research investigations must be embedded in practice. The separation here is only for the convenience of articulating both the theoretical and practical aspects of the knowledge base for teaching.

10.3.1 Investigations for theory

- While some components for teacher knowledge have been identified, what could serve as a framework to provide a more complete picture of that knowledge?
- What are the principles that guide the selection of issues/elements to investigate?
- With such a framework, what kinds of phenomena can we observe, explain and make sense of?
- What are the theoretical models that help us explain how teachers become teachers of algebra?
- What are the links between students' and teachers' misconceptions of algebra?
- How can we investigate teacher learning and reasoning in ways that are based on teachers' experiences and training, and in ways that continue to help them make appropriate teaching decisions?
- What kinds of new methodologies are needed so that can we carry out larger scale research on practice?

10.3.2 Investigations of practice

- How do teachers gain a rich view of algebra and algebraic activity?
- How do teachers learn to make connections between their own knowledge of algebra and the algebraic activity they do with pupils?
- What motivates teachers to change their teaching of algebra?
- What do teachers actually do to promote successful learning of algebra?
- How do teachers make decisions in the algebra classroom?
- What aspects of teaching and learning do teachers attend to in algebra classrooms?
- What questions from learners challenge teachers in algebra classrooms?
- What kinds of incidents are critical for the learning of pre-service teachers and experienced teachers? Why?
- What prevents teachers from learning from their experiences of teaching algebra? When is experiential learning not effective learning?
- What are the ways in which reflection specifically supports the development of improved practice in algebra teaching?
- What is the nature and role of classroom discourse about algebraic activity?
- What is the role of the professional community in teacher development?
- What is the relationship of teachers to classroom resources, including technology, and how these are used effectively?

- How do teachers conceptualise algebra topics across the curriculum with respect to students' understanding of algebra?
- How do teachers teach particular topics with or without technology?

10.4 Conclusions

Overall, there has been little research about teachers' knowledge and practice and its development with respect to the teaching and learning of algebra. Much of what is known about teachers' knowledge points to some of the deficiencies in teachers' mathematical knowledge and to complexities of translating subject matter knowledge into a useful form for teaching. The development of a theoretical base for understanding the epistemological, cognitive, and didactic/pedagogical dimensions of teachers' knowledge is largely in its infancy. While we have made some progress in understanding some aspects of this knowledge, there are many areas of teachers' thinking that have not, in the main, been investigated. In particular, while we are aware that many teachers appear to have conceptualisations of algebra that are dominated by procedures, we have only a few accounts of the ways of thinking about algebra that characterise the understandings of more expert teachers. We certainly do not yet have a rich set of portrayals of expertise in the algebraic thinking and practices of teachers across a range of social and cultural settings.

There is a critical need to build theory about the nature of teachers' knowledge and practice and their development. While we know that traditional mathematics courses alone will not be sufficient in preparing teachers' understanding of school algebra, the practical knowledge of teacher educators in preparing teachers has only sporadically contributed to the building of a professional knowledge base for teaching in ways that are able to be shared, generalised, and reused by others in the field. As we have indicated above, there are many unanswered practical and theoretical questions about how teachers become effective teachers of algebra. Furthermore, continued research on student learning and technology needs to keep the teachers and their perspectives in focus.

As we continue to find and develop new methodological approaches to effectively investigate the practices of teachers of algebra, we need to do this in a way that will contribute to a growing and sustainable knowledge base for teaching. This implies that research investigations will need to become more deeply embedded in the practical knowledge of teachers. This is, of course, fraught with many difficulties, including the limitations of experience we referred to in the opening pages of this chapter. We need to investigate and understand the current experiences of teachers, including the many practical factors that constrain teachers' continued learning, while at the same time moving those experiences forward in a way that will create a self-sustaining and growing professional knowledge base for teaching.

10.5 References

Agudelo-Valderrama, C. (2000). *Una innovación curricular que enfoca el proceso de transición entre el trabajo aritmético y el algebraico.* [A curricular innovation that focuses on the transition between arithmetical and algebraic work.] Tunja, Colombia: Universidad Pedagógica y Tecnológica de Colombia.

Arens, S., & Meyer, R. (2000). *Algebraic thinking: Implications for rethinking pedagogy and professional development.* Aurora, CO: Mid-Continent Research for Education and Learning. (ERIC Document Reproduction Service No. ED 449 983).

Artigue, M., Assude, T., Grugeon, B., & Lenfant, A. (2001). Teaching and learning algebra: Approaching complexity through complementary perspectives. In H. Chick, K. Stacey, J. Vincent, & J. Vincent (Eds.), *The future of the teaching and learning of algebra* (Proceedings of the 12th ICMI Study Conference, pp. 21-32). Melbourne, Australia: The University of Melbourne.

Ball, D. L. (1988). Unlearning to teach mathematics. *For the Learning of Mathematics, 8*(1), 40-48.

Ball, D., Lubienski, S., & Mewborn, D. (2002). Research on teaching mathematics: The unsolved problem of teachers' mathematical knowledge. In V. Richardson (Ed.), *Handbook of research on teaching* (4th ed., pp. 433-456). Washington, DC: American Educational Research Association.

Bednarz, N., Kieran, C., & Lee L. (Eds.) (1996). *Approaches to algebra: Perspectives for research and teaching.* Dordrecht, The Netherlands: Kluwer Academic.

Belfort, E., Guimaraes, L., & Barbastefano, R. (2001). Tertiary algebra and secondary classroom practices in number and algebra: Closing the gap. In H. Chick, K. Stacey, J. Vincent, & J. Vincent (Eds.), *The future of the teaching and learning of algebra* (Proceedings of the 12th ICMI Study Conference, pp. 79-86). Melbourne, Australia: The University of Melbourne.

Bishop, J. W., & Stump, S. L. (2000). Preparing to teach in the new millennium: Algebra through the eyes of pre-service elementary and middle school teachers. In M. Fernandez (Ed.), *22nd annual conference of the North American Chapter of the International Group for the Psychology of Mathematics Education* (pp. 107-113). Columbus, OH: ERIC Clearinghouse for Science, Mathematics, and Environmental Education.

Blanton, M., & Kaput, J. (2001). Algebrafying the elementary mathematics experience Part II: Transforming practice on a district-wide scale. In H. Chick, K. Stacey, J. Vincent, & J. Vincent (Eds.), *The future of the teaching and learning of algebra* (Proceedings of the 12th ICMI Study Conference, pp. 87-95). Melbourne, Australia: The University of Melbourne.

Brown, C., & Smith, M. (1997). Supporting the development of mathematical pedagogy. *The Mathematics Teacher, 90*(2), 138-143.

Chazan, D. (1999). On teachers' mathematical knowledge and student exploration: A personal story about teaching a technologically supported approach to school algebra. *International Journal of Computers for Mathematical Learning, 4*(2-3), 121-149.

Chazan, D. (2000). *Beyond formulas in mathematics and teaching: Dynamics of the high school algebra classroom.* New York: Teachers College Press.

Chazan, D., Larriva, C., & Sandow, D. (1999). What kind of mathematical knowledge supports teaching for "conceptual understanding"? Preservice teachers and the solving of equations. In O. Zaslavsky (Ed.), *Proceedings of the 23rd annual conference of the International Group for the Psychology of Mathematics Education* (Vol. 2, pp. 197-205). Haifa, Israel: Program Committee.

Chinnappan, M., & Thomas, M. (2001). Prospective teachers' perspectives on function representation. In J. Bobis, B. Perry, & M. Mitchelmore (Eds.), *Numeracy and beyond* (Proceedings of the 24th annual conference of the Mathematics Education Research Group of Australasia, pp. 155-162). Sydney: MERGA.

Cooney, T. (1999). Conceptualizing teachers' ways of knowing. *Educational Studies in Mathematics, 38*, 163-187.

Cooney, T. J., & Wilson, M. R. (1993). Teachers' thinking about functions: Historical and research perspectives. In T. A. Romberg, E. Fennema, & T. P. Carpenter (Eds.), *Integrating research on the graphical representation of functions* (pp. 131-158). Hillsdale, NJ: Lawrence Erlbaum Associates.

Doerr, H. M., & Bowers, J. S. (1999). Revealing pre-service teachers' thinking about functions through concept mapping. In F. Hitt & M. Santos (Eds.), *Proceedings of the 21st annual meeting of the North American Chapter of the International Group for the Psychology of Mathematics Education* (Vol. 1, pp. 364-369). Columbus, OH: ERIC Clearinghouse for Science, Mathematics, and Environmental Education.

Doerr, H. M., & Lesh, R. (2003). A modeling perspective on teacher development. In R. Lesh & H. M. Doerr (Eds.), *Beyond constructivism: A models and modeling perspective on mathematics teaching, learning and problem solving* (pp. 125-139). Mahwah, NJ: Lawrence Erlbaum Associates.

Doerr, H. M., & Zangor, R. (1999). The teacher, the tasks and the tool: The emergence of classroom norms. *International Journal of Computer Algebra in Mathematics Education, 6*(4), 267-279.

Doerr, H. M., & Zangor, R. (2000). Creating meaning for and with the graphing calculator. *Educational Studies in Mathematics, 41*(2), 143-163.

Even, R. (1990). Subject matter knowledge for teaching and the case of functions. *Educational Studies in Mathematics, 21*(6), 521-544.

Even, R. (1993). Subject-matter knowledge and pedagogical content knowledge: Prospective secondary teachers and the function concept. *Journal for Research in Mathematics Education, 24*(2), 94-116.

Even, R., & Tirosh, D. (1995). Subject-matter knowledge and knowledge about students as sources of teacher presentations of the subject-matter. *Educational Studies in Mathematics, 29*(1), 1-20.

Even, R., Tirosh, D., & Robinson, N. (1993). Connectedness in teaching equivalent algebraic expressions: Novice versus expert teachers. *Mathematics Education Research Journal, 5*(1), 50-59.

Gadanidis, G. (2001). Web-based multimedia activities as pedagogical models. In W. Yang, S. Chu, Z. Karian, & G. Fitz-Gerald (Eds.), *Proceedings of the Sixth Asian Technology Conference in Mathematics* (pp. 223-232). Melbourne, Australia: RMIT University.

Haimes, D. H. (1996). The implementation of a "function" approach to introductory algebra: A case study of teacher cognitions, teacher actions, and the intended curriculum. *Journal for Research in Mathematics Education, 27*(5), 582-602.

Heid, M. K. (1995). Impact of technology, mathematical modeling, and meaning on the content, learning, and teaching of secondary-school algebra. *Journal of Mathematical Behavior, 14*(1), 121-137.

Heid, M. K., Blume, G. W., Zbiek, R. M., & Edwards, B. S. (1999). Factors that influence teachers learning to do interviews to understand students' mathematical understandings. *Educational Studies in Mathematics, 37*(3), 223-249.

Hitt, F. (1994). Teachers' difficulties with the construction of continuous and discontinuous functions. *Focus on Learning Problems in Mathematics, 16*(4), 10-20.

Johnson, S. (2001). Learning to teach algebra in the UK: Trainee teachers' experiences. In H. Chick, K. Stacey, J. Vincent, & J. Vincent (Eds.), *The future of the teaching and learning of algebra* (Proceedings of the 12th ICMI Study Conference, pp. 336-343). Melbourne, Australia: The University of Melbourne.

Kaput, J., & Blanton, M. (2001). Algebrafying the elementary mathematics experience Part I: Transforming task structures. In H. Chick, K. Stacey, J. Vincent, & J. Vincent (Eds.), *The future of the teaching and learning of algebra* (Proceedings of the 12th ICMI Study Conference, pp. 344-351). Melbourne, Australia: The University of Melbourne.

Kieran, C. (1992). The learning and teaching of school algebra. In D. Grouws (Ed.), *Handbook of research on mathematics teaching and learning* (pp. 390-419). New York: Macmillan Publishing Company.

Kieran, C., & Sfard, A. (1999). Seeing through symbols: The case of equivalent expressions. *Focus on Learning Problems in Mathematics, 21*(1), 1-17.

Leikin, R., Chazan, D., & Yerushalmy, M. (2001). Understanding teachers' changing approaches to school algebra: Contributions of concept maps as part of clinical interviews. In M. van den Heuvel-Panhuizen (Ed.), *Proceedings of the 25th annual conference of the International Group for the Psychology of Mathematics Education* (Vol. 3, pp. 289-296). Utrecht, Netherlands: Program Committee.

Leinhardt, G., Zaslavsky, O., & Stein, M. K. (1990). Functions, graphs, and graphing: Tasks, learning, and teaching. *Review of Educational Research, 60*(1), 1-64.

Lloyd, G. M., & Wilson, M. (1998). Supporting innovation: The impact of a teacher's conceptions of functions on his implementation of a reform curriculum. *Journal for Research in Mathematics Education, 29*(3), 248-274.

McGowen, M., & Davis, G. (2001). Changing pre-service teachers' attitudes to algebra. In H. Chick, K. Stacey, J. Vincent, & J. Vincent (Eds.), *The future of the teaching and learning of algebra* (Proceedings of the 12th ICMI Study Conference, pp. 438-445). Melbourne, Australia: The University of Melbourne.

Menzel, B. (2001). Language conceptions of algebra are idiosyncratic. In H. Chick, K. Stacey, J. Vincent, & J. Vincent (Eds.), *The future of the teaching and learning of algebra* (Proceedings of the 12th ICMI Study Conference, pp. 446-453). Melbourne, Australia: The University of Melbourne.

Menzel, B., & Clarke, D. (1998). Teachers interpreting algebra: Teachers' views about the nature of algebra. In C. Kanes, M. Goos, & E. Warren (Eds.), *Teaching mathematics in new times* (Proceedings of the 21st annual conference of the Mathematics Education Research Group of Australasia, pp. 365-372). Gold Coast, Australia: MERGA.

Mesa, V. (2001, April). *Conceptions of function present in seventh- and eighth-grade textbooks from fifteen countries.* Paper presented at the annual meeting of the American Educational Research Association. Seattle, WA.

Miller, L. D. (1992). Teacher benefits from using impromptu writing prompts in algebra classes. *Journal for Research in Mathematics Education, 23*(4), 329-340.

Munby, H., Russell, T., & Martin, A. (2002). Teachers' knowledge and how it develops. In V. Richardson (Ed.), *Handbook of research on teaching* (4th ed., pp. 877-904). Washington, DC: American Educational Research Association.

Nathan, M. J., & Koedinger, K. R. (2000a). An investigation of teachers' beliefs of students' algebra development. *Cognition and Instruction, 18*(2), 209-237.

Nathan, M. J., & Koedinger, K. R. (2000b). Teachers' and researchers' beliefs about the development of algebraic reasoning. *Journal for Research in Mathematics Education, 31*(2), 168-190.

National Council of Teachers of Mathematics. (1989). *Curriculum and evaluation standards for school mathematics*. Reston, VA: Author.

National Council of Teachers of Mathematics. (2000). *Principles and standards for school mathematics*. Reston, VA: Author.

Norman, A. (1992). Teachers' mathematical knowledge of the concept of function. In G. Harel & E. Dubinsky (Eds.), *The concept of function: Aspects of epistemology and pedagogy* (Vol. 25, MAA Notes, pp. 215-232). Washington, DC: Mathematical Association of America.

Norman, A. (1993). Integrating research on teachers' knowledge of functions and their graphs. In T. A. Romberg, E. Fennema, & T. P. Carpenter (Eds.), *Integrating research on the graphical representation of functions* (pp. 159-187). Hillsdale, NJ: Lawrence Erlbaum.

Penglase, M., & Arnold, S. (1996). The graphics calculator in mathematics education: A critical review of recent research. *Mathematics Education Research Journal, 8*(1), 58-90.

Pirie, S. E. B., & Martin, L. (1997). The equation, the whole equation, and nothing but the equation! One approach to the teaching of linear equations. *Educational Studies in Mathematics, 34* (2), 159-181.

Raymond, A. M., & Leinenbach, M. (2000). Collaborative action research on the learning and teaching of algebra: A story of one mathematics teacher's development. *Educational Studies in Mathematics, 41*(3), 283-307.

Schmidt, S., & Bednarz, N. (1997). Raisonnements arithmetiques et algebriques dans un contexte de resolution des problemes: difficultes rencontrees par les futurs enseignants [Arithmetical and algebraic reasoning within a problem-solving context: Difficulties encountered by future teachers]. *Educational Studies in Mathematics, 32*(2), 127-155.

Slavit, D. (1996). Graphing calculators in a "hybrid" Algebra II classroom. *For the Learning of Mathematics, 16*(1), 9-14.

Stacey, K., & MacGregor, M. (2000). Learning the algebraic method of solving problems. *Journal of Mathematical Behavior, 18*(2), 149-167.

Stein, M. K., Baxter, J., & Leinhardt, G. (1990). Subject-matter knowledge and elementary instruction: A case from functions and graphing. *American Educational Research Journal, 27*(4), 639-663.

Stump, S. (1999). Secondary mathematics teachers' knowledge of slope. *Mathematics Education Research Journal, 11*(2), 124-144.

Stump, S. L. (2001). Developing preservice teachers' pedagogical content knowledge of slope. *Journal of Mathematical Behavior, 20*, 207-227.

Stump, S., & Bishop, J. (2001). Framing the future: Inventing an algebra course for pre-service elementary and middle school teachers. In H. Chick, K. Stacey, J. Vincent, & J. Vincent (Eds.), *The future of the teaching and learning of algebra* (Proceedings of the 12th ICMI Study Conference, pp. 564-570). Melbourne, Australia: University of Melbourne.

Sutherland, R., Rojano, T., Bell, A., & Lins, R. (Eds.) (2001). *Perspectives on school algebra*. Dordrecht, The Netherlands: Kluwer Academic.

Usiskin, Z. (1988). Conceptions of school algebra and uses of variables. In A. F. Coxford (Ed.), *The ideas of algebra, K-12* (pp. 8-19). Reston, VA: National Council of Teachers of Mathematics.

Tharp, M. L., Fitzsimmons, J. A., & Ayers, R. L. B. (1997). Negotiating a technological shift: Teacher perception of the implementation of graphing calculators. *Journal of Computers in Mathematics and Science Teaching, 16*(4), 551-575.

Tirosh, D., Even, R., & Robinson, N. (1998). Simplifying algebraic expressions: Teacher awareness and teaching approaches. *Educational Studies in Mathematics, 35*(1), 51-64.

van Dooren, W., Verschaffel, L., & Onghena, P. (2002). The impact of preservice teachers' content knowledge on their evaluation of students' strategies for solving arithmetic and algebra word problems. *Journal for Research in Mathematics Education, 33*(5), 319-351.

Wilson, M. R. (1994). One pre-service secondary teacher's understanding of function: The impact of a course integrating mathematical content and pedagogy. *Journal for Research in Mathematics Education, 25*(4), 346-370.

The Working Group on Teaching and Learning Tertiary Algebra

Leaders: Dave Carlson and Ed Dubinsky

Working Group Members:
Ivan Cnop, Caroline Lajoie, Nguyen Xuan Tuyen, Asuman Oktaç, Man-Keung Siu, Sri Wahyuni, and Leigh Wood.

The Working Group on Teaching and Learning Tertiary Algebra. Seated (L to R): Nguyen Xuan Tuyen, Ed Dubinsky, Sri Wahyuni, David Carlson. Standing (L to R): Asuman Oktaç, Leigh Wood, Ivan Cnop, Man-Keung Siu, Caroline Lajoie.

The Working Group on Teaching and Learning Tertiary Algebra was a diverse group with participants from Australia (Leigh Wood), Belgium (Ivan Cnop), Canada (Caroline Lajoie), China (Man-Keung Siu), Indonesia (Sri Wahyuni), Mexico (Asuman Oktaç, originally from Turkey, who is also affiliated to a Canadian university), USA (David Carlson and Ed Dubinsky), and Vietnam (Nguyen Xuan Tuyen). Between them the members of the group spoke a wide range of languages: Chinese, Dutch, English, Flemish, French, Indonesian, Spanish, and Turkish, with Tuyen speaking only Vietnamese and Russian, languages not spoken by any other person. Group discussions took place in English.

The group was diverse in other ways. Several members came from university mathematics departments but were not involved in mathematics education. In contrast, Ed Dubinsky came with considerable experience in mathematics education and a very broad knowledge of the associated literature. On occasions, lively debates developed when specific mathematics issues were considered in opposition to the broader overview.

The Working Group held comprehensive discussions that were informed by members' prior experiences in teaching and researching tertiary algebra, as evidenced by their contributions to the 12th ICMI Study Conference Proceedings.

- Ivan Cnop: *New insight in mathematics by live CAS documents* (pp. 187-191).
- Ed Dubinsky: *Teaching and learning abstract algebra and linear algebra: A unified approach* (pp. 705-712).
- Caroline Lajoie: *Students difficulties with the concepts of groups, subgroup and group isomorphism* (pp. 384-391).
- Asuman Oktaç: *The teaching and learning of linear algebra: Is it the same at a distance?* (pp. 501-506).
- Man-Keung Siu: *Why is it difficult to teach abstract algebra?* (pp. 541-547).
- Sri Wahyuni: *The relevance of algebraic structure courses in tertiary mathematics education in Indonesia* (pp. 628-633).
- Leigh Wood & G. Smith: *Assessment in linear algebra* (pp. 663-667).

Early conference sessions were used for brainstorming ideas that David Carlson subsequently compiled and categorised. The members then split into two groups and tackled different categories of questions. After reporting back to the Working Group, members elected to work on specific issues. In particular, Caroline Lajoie, Asuman Oktaç, Sri Wahyuni, and Leigh Wood focused on two issues—the relevance of tertiary algebra and students' motivation—while Ivan Cnop, Mann-Keung Siu, and Ed Dubinsky evaluated a wide variety of pedagogical approaches for teaching tertiary algebra. Finally, each member helped develop at least one written brief. These briefs greatly assisted David Carlson to construct the chapter on tertiary algebra. The authors (who may be contacted from their e-mail addresses listed at the back of the book) and titles of the briefs are:

- Ivan Cnop: *Computer Algebra Systems.*
- Ed Dubinsky & Man-Keung Siu: *Pedagogy.*
- Caroline Lajoie, Asuman Oktaç, Sri Wahyuni, & Leigh Wood: *Conceptual difficulties of students.*
- Siu Man-Keung: *The treatment of student difficulties.*

Thanks are extended to all of the members of the Working Group on Teaching and Learning Tertiary Algebra for their invaluable contributions to the discussions and their conscientious work in preparing the briefs. In addition, Ed Dubinsky, Caroline Lajoie, and Douglas McLeod are thanked for reviewing earlier versions of the chapter. Finally, David Carlson and Ed Dubinsky are congratulated for their leadership of this very diverse, international Working Group.

Chapter 11

The Teaching and Learning of Tertiary Algebra

David Carlson
San Diego State University, San Diego, USA

Abstract: This chapter reports on some current educational issues related to the teaching and learning of tertiary algebra—in particular, abstract algebra, discrete mathematics, linear algebra, and number theory. The causes of conceptual difficulties experienced by many students are identified and possible ways of overcoming them, sometimes using a specific pedagogical framework, are discussed. Issues related to students' motivation are explored and pedagogical possibilities for overcoming some of the problems in both these areas are also explored. This report also addresses issues associated with the dissemination of educational work to tertiary instructors who are typically mathematicians rather than mathematics educators. Furthermore, the role of computers in tertiary algebra courses is considered, focusing on the use of Computer Algebra Systems (at the tertiary level) and the use of the programming language ISETL that helps students construct and work with algebraic objects. This chapter makes recommendations for improving practices for teaching tertiary algebra and proposes areas for further research.

Key words: Tertiary algebra teaching, abstract algebra, linear algebra, programming, CAS, motivation, proof, abstractions, definitions, curriculum reform, dissemination

11.1 Introduction

Whilst most mathematics educators work on issues related to primary and secondary education, there has been an increase—particularly in the last decade—in research activity at the tertiary level conducted by both mathematics educators and mathematicians. This is expected to continue and increase in the future. Evidence of this growing interest can be seen in the substantial book recently edited by Holton (2001), which reported on the ICMI study on the *Teaching and Learning of Mathematics at the University Level*. That report focused on a wide range of issues for tertiary mathematics educators, teachers and researchers, whereas the present chapter specifically focuses on the teaching and learning of tertiary algebra.

The issues discussed in this chapter are considered in relation to the teaching and learning of linear algebra, abstract algebra, number theory, and discrete mathematics at the tertiary level. Courses that are an extension of secondary algebra courses, such as "college algebra" in the USA (a pre-calculus tertiary algebra course), are not considered in what follows. We indicate here, however, that the literature provides evidence that there is faculty and student dissatisfaction with such courses in several countries and that some efforts towards reform are being made.

The members of the Working Group on Tertiary Algebra, in exploring the literature in this area, noted that some of the recent tertiary-level education-related research work involved computers. There have been many workshops held for tertiary algebra instructors on using computers in teaching, and materials produced on applications involving computing, both in conjunction with specific texts and as stand-alone materials. One example is the book of computer exercises for linear algebra by Leon, Herman, and Faulkenberry (1997). Since several specific areas in the use of computing in teaching are dealt with in other chapters of this volume, in this chapter we will limit our discussion of computing to the use of Computer Algebra Systems (CAS) specifically at the tertiary level and to the use of the *Interactive Set Language* for programming (ISETL) for helping students construct and work with the algebraic objects studied in tertiary-level algebra courses.

After consideration of the current research into tertiary algebra, the Working Group focused its attention on identifying conceptual difficulties of students and on the treatment of these student difficulties, sometimes using a specific pedagogical framework. Our two principal sections deal with these areas. In addition, this report includes sections on the motivational difficulties of students, and on the dissemination of education-related work to tertiary algebra instructors, whose main professional interest may not be in education.

11.2 Conceptual Difficulties of Students

As is described in the report of the Working Group on Tertiary Algebra (see immediately prior to this chapter) the discussions of the members led to a brief prepared by Lajoie, Oktaç, Wahyuni, and Wood that outlined the conceptual difficulties encountered by tertiary students. These conceptual difficulties are classified into several broad categories that are described below.

11.2.1 Conceptual difficulties related to symbolic logic

For tertiary algebra students, the lack of an effective working knowledge of symbolic logic causes serious difficulties. Sierpinska, Nnadozie, and Oktaç (2002), for example, note "if one confuses quantifiers and negates them incorrectly, then one is bound to have difficulties in, for example, the use of the definitions of linear dependence and independence in formal proofs" (p. 154). Texts in tertiary algebra courses are often written, at least in part, in the formal language of modern mathematics, using the precision of predicate calculus and quantifiers. Students in these courses may have previously studied only more classical areas of mathematics, such as calculus, where texts are usually written in a more informal style, downplaying the role of quantified statements and their syntax. Even students who have studied symbolic logic as a topic in a discrete mathematics course find this new way of saying things difficult, and often do not have an effective understanding of the techniques and the importance of symbolic logic. Dubinsky, Elterman, and Gong (1988), Dubinsky (1997), and Dubinsky and Yiparaki (2000) discuss student difficulties in symbolic logic and investigated the ways in which computer-based learning experiences could assist understanding.

11.2.2 Conceptual difficulties related to definitions

Among the fundamental components in understanding modern mathematics are the recognition of the importance of definitions and the ability to work with them. Tertiary algebra students' understandings of definitions and their roles are generally imperfect. Students often do not form precise notions of definitions and cannot reason from definitions to their consequences. A seminal paper in this regard is by Tall and Vinner (1981). In this work, "concept definition" is contrasted with "concept image", the totality of understanding of an individual about a concept. For students to truly understand a concept, their concept image must be consistent with and be built upon their concept definition (see also Vinner, 1991). Lajoie (2001) notes that abstract algebra students sometimes have weak personal definitions that are very different from the concept definitions they were taught, and have concept images that are not internally consistent. This may explain some of the difficulties that the students experience. Sierpinska, Nnadozie, and Oktaç (2002), in their research on theoretical thinking in linear algebra students, found that students had difficulty in distinguishing between statements that are definitions and statements that are implications. They noticed, for example, that students very often treated definitions only as descriptions of some properties of the concept. The researchers do not recommend, however, that students should have formal training in critical thinking, formal logic, or proof methods prior to taking a course in linear algebra, because these are only a few of the necessary components. Wood (1999) analysed student difficulties with definitions and discussed ways of dealing with these difficulties. She contends that tertiary teachers could spend more time helping students understand the need for precise definitions in mathematics. She discusses how definitions may change for different audiences and gives some examples of activities for students. These activities highlight the similarities and differences between examples, illustrations, and definitions and draw on sources from within and outside mathematics. She also notes the importance of instructors clearly signalling to students the status of mathematical statements. For example, the first of the following statements is a definition and the second is a theorem:

A series $\sum u_n$ is absolutely convergent if $\sum |u_n|$ is convergent.

A series $\sum u_n$ is convergent if it is absolutely convergent.

Attention to more explicit language clues could help students see the differences.

The Working Group recommends that further research be conducted into the following areas:

1. Definition-building by students.
2. Intuitive and formal definitions: Which should come first?
3. Historical development of definitions.
4. Assessing student understanding of definitions.
5. Ways to help students deal with non-constructive definitions in mathematics.

11.2.3 Conceptual difficulties related to proofs

Although tertiary mathematics students are expected to produce coherent arguments while validating mathematical statements, proof making has long been a problem area for many students. Selden and Selden (1987) studied proof-making errors in abstract algebra courses. Hart (1994) studied the differences between expert and novice students' proof techniques in elementary group theory. Harel and Sowder (1998) and Harel (1999) have also done work on students' understanding of proofs, and identified that students do not always appreciate proof for its power to ascertain or persuade the validity of a result, relying instead on empirical evidence or teacher authority. Sowder and Harel (2003) suggest that the abilities of students in areas such as proof, understanding, production, and appreciation of skills do not always develop in the course of their mathematical studies, and suggest that there are teaching strategies that can help them develop.

The Working Group recommends that research be conducted into the following areas:

1. The relationship between the understanding of definitions and proof making.
2. The effects of different pedagogical approaches on students' understanding, using, and valuing of proofs.

11.2.4 Conceptual difficulties related to abstraction

Students in tertiary algebra courses need to learn to work at an abstract level. There is evidence that students can achieve this; see in particular the work in this area by members of RUMEC (Research in Undergraduate Mathematics Education Community). Information about the RUMEC Linear Algebra project is available from http://www.ilstu.edu/~jfcottr/linear-alg/. Hazzan (1999) presents evidence that shows that students try to reduce the level of abstraction in tertiary algebra courses, often by working with prototypical examples rather than definitions. Dubinsky (1991) suggests that a source of students' difficulties relates to them working with hypothetical situations. "How can anyone really define something which may not exist?" During the Working Group discussion, Dubinsky mentioned his current work to overcome this problem: the construction of computer mini-worlds, to assist students to think in mathematical ways. Another source of students' difficulties relates to theoretical approaches to teaching abstraction. To overcome these problems, different pedagogical approaches need to be employed.

Research in this area is complicated by the fact that different researchers have different conceptions of the meaning of abstraction. Whereas Dubinsky (1991) bases his notion of reflective abstraction in the coordination of sensory structures (initially suggested by Piaget), Dreyfus (1991) believes that representing, generalising, and synthesising are essential prerequisite skills for developing the ability to abstract. In order to gauge the breadth of the abstraction problem, Artigue, in discussion with the Working Group, recommended that research be conducted to find answers to the following questions:

1. What level of abstraction is required in tertiary algebra courses compared with students' prior experience?
2. How does the level of abstraction required in tertiary algebra courses compare with other mathematical topics such as analysis?
3. Should different tertiary algebra courses be offered to different clientele (e.g., future teachers and future mathematicians)?

11.3 Motivational Issues

In the discussions of the Working Group, Ed Dubinsky formulated a very general statement on motivation. He indicated his belief that the crucial elements in students' motivation to learn mathematics are the students' personal realisations: that they are able to deal with a mathematical issue (most important), that they need to deal with it, and that they have no easier alternative to deal with it. The issues discussed below fall into this general framework.

For example, Siu (2001) believes that problems in the teaching and learning of algebra have to do mainly with two aspects, abstraction and relevance (which is clearly tied to having a need for dealing with a mathematical issue). He points out that these two aspects are inter-related. A high level of abstraction may suggest to a learner that the subject is irrelevant, and thus breed feelings of indifference, anxiety, or hostility. In contrast, exhibiting the relevance of a subject can encourage the learner to put in the effort required to cope with the abstraction. The discussions of the group led to a brief prepared by Lajoie, Oktaç, Wahyuni, and Wood (as detailed in the report on the Working Group on Tertiary Algebra). This section was developed from both the discussion and the brief.

A variety of factors may affect the motivation of tertiary mathematics students, particularly those studying tertiary algebra. Some factors cause motivational problems for students. For example, there is a common perception among students that abstract algebra and number theory have no obvious practical application, so when they are offered as electives, students tend to avoid them. Can we find ways to change this common student misconception? Linear algebra and discrete mathematics do not have this problem to the same degree since their utility is accepted; they are often compulsory courses for the education of professionals using mathematics, such as computer scientists and engineers.

On the other hand, a range of other factors may motivate students positively. These include the beauty, power, and frequently the utility of the mathematics itself; all of the tertiary algebra courses touch upon great mathematics. Individual student factors (their intellectual curiosity, prior learning experiences, positive response to challenge, and, for better or worse, their desire for good grades), peer group factors, and cultural factors also play a role. Pedagogical factors (the way the tertiary algebra is taught to the students) which may assist with motivation include: the use of technology, collaborative learning, teaching for understanding and depth, interesting assessment techniques, and discovery methods.

Recent studies have addressed the reasons why some students have motivational problems while studying tertiary mathematics. For example, a group of researchers at the University of Leeds in the UK have been doing a three-year longitudinal study, *Students' Experiences of Undergraduate Mathematics* (2003), examining the affective domain with students studying university mathematics. Other research links motivation to social variables. For example, in Australia, Forgasz and Leder (1998) have researched university students' motivation. They surveyed the attitudes of students enrolled in their first year of an undergraduate mathematics course in an attempt to describe how a complex set of social factors, including gender, socio-economic status, language background, and age, influenced the students' attitudes.

The Working Group recommends that research be conducted to answer the following questions, especially in relation to tertiary students:

1. Is there a relationship between enthusiasm for a particular mathematical topic and the effective learning of the concepts behind it?
2. Why are some students motivated to learn abstract concepts and others are not?
3. What are the roles of previous and anticipated success in student learning?
4. What are the differences between teachers' and students' ideas of what is motivating?
5. What would assist pre-service teachers in seeing the relationship between tertiary algebra and the secondary algebra they will be teaching?

11.4 Pedagogy

Other discussions led to the development of briefs by Cnop, Siu, and Siu and Dubinsky. These briefs, detailed in the report of the Working Group on Tertiary Algebra, form the basis of this section, which outlines a wide variety of pedagogical approaches for teaching tertiary algebra. Some of the pedagogical approaches discussed here for courses in tertiary algebra are applications of more general approaches. Other approaches are specific to algebra or to specific algebra courses. In the first section, innovative approaches, based on theory or empirical research, are surveyed while in the second section, the potential for other approaches, as well as further important pedagogical issues, are presented.

11.4.1 Research-based approaches

Several innovative teaching approaches have been proposed which are strongly based in theory. One of these is the approach of RUMEC (Dubinsky & McDonald, 2001), which uses Action-Process-Object-Schema (APOS) Theory and the ACE Cycle to make a theoretical analysis of each topic to be learned. APOS Theory is a constructivist theory. In it, certain mental structures are specified; if a student constructs them, then he or she should be able to learn the concept in question. Instruction is designed to get the students to make the proposed mental structures and use them to learn the concept in question. The instruction is implemented following the ACE Cycle of: *Activities* (exposing students to new situations, often involving computer programming), *Classroom* (discussion and class tasks), and *Exercises* (to reinforce and extend ideas). Then, data about students' understanding are gathered and analysed in terms both of the proposed mental structures and the learning of the mathematics. The cycle is repeated as often as necessary to achieve stability and a satisfactory level of learning.

This approach has been applied to courses in abstract algebra and discrete mathematics. Research has found that students who took these courses learned significantly more than those who took traditional courses (see Asiala, Dubinsky, Mathews, Morics, & Oktaç, 1997; Brown, DeVries, Dubinsky, & Thomas, 1997; Clark, Hemenway, St. John, Tolias, & Vakil, 1999). The students also developed more positive attitudes towards the courses in general and towards abstraction in particular. The approach has been applied to develop a course in linear algebra that, at this writing, is being implemented using constructivist philosophy and involving student programming with ISETL (see Weller et al., 2002). Simultaneously research is being carried out on the effectiveness of the course.

An alternative to basing an approach on theory is to base an approach primarily on empirical research and observation of students. Several mathematics education researchers have studied the teaching and learning of linear algebra, identified the causes of the problems, and then suggested pedagogical reforms. For example, Dorier and Sierpinska (2001) identified three possible reasons why many students find linear algebra so difficult to learn. First, students may consider the axiomatic approach to linear algebra to be superfluous and meaningless. Second, students find linear algebra to be difficult due to the complex interactions between systems of representation. Third, linear algebra is cognitively difficult because it requires the ability to deal with algebraic, geometric, and abstract registers and demands flexibility in switching between them (e.g., using vectors, systems of equations, etc.). Dorier and Sierpinska also discussed the implications of the research for reform in teaching. These reforms include teachers giving students time to reflect on the underlying processes associated with linear algebra concepts, providing 'engaging' examples, and employing well-structured, well-delivered lessons. Others have conducted research with a similar focus (see Dorier, Robert, Robinet, & Rogalski, 2000; Harel, 1989; Hillel, 2000; Sierpinska, 2000; and other works by the same authors).

As mentioned above, topics in tertiary algebra often have both geometric and algebraic aspects. Many educators believe that each student will have a consistent preference for one or the other of visual and analytic learning. However, mathematicians often combine visual and analytic elements in their mathematical developments, and so it does seem important that students' abilities in both directions should be strengthened. Zazkis, Dautermann, and Dubinsky (1996) have developed a model for how individuals combine both strategies, and applied it to analyse student thinking in abstract algebra. They suggest that if a student appears to prefer to think visually about a particular problem, then he or she should be given tasks that encourage analytic thinking, and vice versa, since both offer important perspectives on the underlying mathematics.

11.4.2 Other pedagogical issues

11.4.2.1 Too much or too little in tertiary algebra courses, and with what emphases?

There are many themes from other areas of mathematics education that are relevant to the teaching of tertiary algebra courses. While the quantity of knowledge in algebra and the breadth of applications of that knowledge are continuing to grow rapidly, the rate at which students can learn seems unlikely to be able to grow as fast. The only hope for improvement in students' rate of learning is the development of more efficient teaching techniques. Some instructors complain that there is too much material for their students to learn effectively. How can we study the relationship between the quality of student learning and the quantity of material covered in a course? How can we study the speed at which students can effectively learn material in these courses? If students have learned fewer algebraic concepts but learned them really well, will they be then able to learn additional concepts and applications more easily as they are needed?

Incorporating opportunities for experimentation and discovery learning has been identified as a key issue for learning effectively, although it has implications for the amount of material that can be covered. Mathematics is not just a collection of axioms, definitions, theorems, and proofs, all displayed in a polished form. It also comprises exploration and discovery through intuitive or heuristic means. Students should be given the opportunity to experience both aspects so that they can develop an intuitive sense, but at the same time come to realise why a formal treatment is also needed, and, in some cases, actually facilitates thinking about the topic. Courses in tertiary algebra can involve a considerable amount of experimentation by students, in which they may be asked to discover various mathematical relationships. The instructor can have students try to discover certain ideas, but limit the time and give the solution to the whole class. A learner may gain almost as much from trying to discover an idea without succeeding, as from hearing the idea from someone else.

There is general agreement that drill and practice to reinforce knowledge are essential to learning. However, it is not clear how much is necessary, and research needs to be done on this question. An important point is that since drill and practice tend to "cast in stone" whatever understanding an individual has of the concepts behind the procedures being practised, care should be taken to minimise drill and practice until there is some reason to believe that the students' understanding has approached a reasonable level.

11.4.2.2 The use of CAS in tertiary algebra

After discussion in the Working Group, Cnop wrote a brief (listed in the report on Working Group on Tertiary Algebra) about symbolic mathematics packages (here referred to as CAS) and this section is based on this brief. The use of Computer Algebra Systems (CAS) is treated more extensively in Chapter 7 in this book. The syntax used by modern CAS systems is becoming closer to written mathematics and the look and feel of the interfaces have improved. They now have advanced formatting possibilities, using palettes, and a choice of formats and translators. It has become easier for the student to produce syntactically correct input, especially since documentation has also improved. Experience has shown that the students are more likely to start successfully in CAS activity if they are given templates for meaningful activities that allow experimentation, modification, and application in specific domains of interest. Such active documents can be offered in classes or laboratory sessions and can stay accessible on-line for students to access on demand.

However, there is no consensus yet over the appropriate amount or types of technology, and the appropriate pedagogies for instruction in its use. For example, in the introduction to the well-established "European Society for Engineering Education (SEFI) Core Curriculum 2002" for engineering mathematics, no specific recommendations are made concerning which and how much technology should be included in the standard curriculum (SEFI, 2002). This core curriculum gives recommendations for the amount of linear algebra, discrete structures, logic, and proofs that are to be included in standard engineering curricula. However, the integration of abstract material and computer technology is left to the individual teacher. Research is needed for all tertiary courses, but in particular for tertiary algebra courses, to assist instructors to make decisions about the amount and types of technology to be used, and the pedagogy to be employed in its use. This research is made more difficult because the technology itself is changing so rapidly.

11.4.2.3 Other approaches to teaching algebra

Distance learning is increasingly being used for tertiary algebra courses, made possible through recent advances in technology. Web-based, on-line, and distance education offer students different opportunities to interact with other students and the material and thus construct knowledge. Oktaç (2001) reports her study into how groups of students in Mexico studied linear algebra at university level through distance learning. She alerts us to the necessity of careful planning to capitalise on the benefits (e.g., motivation, communication with other students) and to avoid some of the difficulties (e.g., problematic teaching methods) that may arise in the new teaching and learning environment.

Teachers may make their teaching more illuminating by integrating historical material in a judicious and well-designed way, adding an extra dimension to the student's enthusiasm to learn. Even though in most cases the actual path taken in history was much too tortuous to be recounted in full to pedagogical advantage (see Chapter 8 for more detail), studying the historical development of a topic can aid the teacher in identifying the crucial steps and the difficulties and obstacles to learning, and in building up a reservoir of examples and problems. The teacher should be careful to give an overview of a topic or even of the whole course at the beginning, so that students know where they are headed and how it relates to previous knowledge. In the case of abstract algebra, for example, an account of the path from solutions of polynomial equations to group theory can be given. A successful course on abstract algebra, reported by Kleiner (1998), was built around a few classical problems in algebra, number theory, and geometry. Katz (2001) discusses how the history of algebra highlights the need for using many concrete examples as a basis and motivation for abstraction.

Finally, there is a huge educational literature on the use of cooperative learning, but until recently little of it focused on the tertiary level. In the past few years, however, there has been an effort by a number of mathematicians and mathematics educators to use cooperative learning in tertiary courses. At least three books have been published to support and describe this work (Dubinsky, Mathews, & Reynolds, 1997; Hagelgans et al., 1995; Rogers, Reynolds, Davidson, & Thomas, 2001).

11.4.2.4 Assessment

Clearly assessment affects how our students study and how they learn. Wood and Smith (2001) believe that assessment in university mathematics will continue to be individual and examination-based. However, they believe that the pervasive use of technology in teaching and learning mathematics will impact on the way students are examined. They suggest that the emphasis will move from testing manipulative skills to testing conceptual understanding and other cognitive skills. Alternative methods of assessment are also available. For example, requiring students to critically evaluate a given mathematical explanation (instead of asking for the explanation) can be particularly useful in tertiary algebra because it can emphasise conceptual understanding.

11.5 Towards Reform

11.5.1 A document on the teaching of tertiary algebra

There are potential parallels between the current situation in tertiary algebra and the changes that have taken place in recent years in calculus. It is generally acknowledged that the Calculus Reform Movement in the United States has been successful in bringing about changes in the ways some teachers teach calculus, and has been accompanied by greater student understanding of underlying concepts. Associated with this teaching reform movement in the American mathematical community has been the production of several pamphlets detailing particular calculus reform activities (including careful scientific assessments of their effectiveness relative to student learning, and evidence of the spread of such reform activities). The Working Group proposes that the current status of teaching and learning tertiary algebra be carefully documented now. This would provide a basis for recommending future changes in pedagogy, and would allow careful scientific evaluation of any changes that are implemented.

There is now a body of evidence including recent research by Lajoie (2001) into group theory, that learning in tertiary algebra courses is often weak and fragmented. This evidence comes from several sources: homework, examinations, and student interviews during and after the university courses. The research revealed that many students were not clear about fundamental definitions. Some made statements like "isomorphic groups are similar groups" and "linearly independent vectors are not multiples of each other". It also showed that some students were unable to think meaningfully about the relationships between concepts, and that they used computational techniques with little regard to the conceptual entities they were working with. The drawing together of all of such evidence into a single document may make tertiary algebra instructors more aware of how poor their students' learning may be and of possible solutions to teaching problems. This may assist in encouraging tertiary teachers to be more open to consider new pedagogies.

Reform pedagogies have already been proposed in aspects of tertiary algebra and are already being used in some universities. For example, as discussed above, reforms in linear algebra have been proposed by Dorier and Sierpinska (2001);

Dorier, Robert, Robinet, and Rogalski (1994); Dubinsky and McDonald (2001); Harel (1991); Sierpinska, Trgalova, Hillel, and Dreyfus (1999); and others. Textbooks have also been developed using reform pedagogies in linear algebra and other areas:

- Linear algebra (e.g., Weller et al., 2002)
- Discrete mathematics (e.g., Fenton & Dubinsky, 1996)
- Abstract algebra (e.g., Leron & Dubinsky, 1995)

However, it is worth noting that some recent textbooks about teaching particular topics do not focus on these reforms. These include:

- Linear algebra (Carlson et al., 1997; Carlson, Johnson, Lay, & Porter, 2002; and Dorier, 2000)
- Number theory (Campbell & Zazkis, 2001)
- Abstract algebra (Hibbard & Maycock, 2002)

One issue that should be addressed in reform publications is the relationship between the content of tertiary algebra and that of secondary algebra. Unlike a course in analysis that can be viewed as a continuation of calculus, students often perceive courses in tertiary algebra (especially abstract algebra) as something completely unrelated to their previous mathematical experience. This is particularly true of algebra, where the tertiary emphasis on structure can seem removed from earlier encounters with solving equations and graphing. There are two important reasons why it is important to develop and highlight points of contact between tertiary algebra and students' previous experience. The first is to help motivate all students in the study of tertiary algebra. Students need to see an integrated picture of mathematics as a whole. Second, some of the tertiary students studying algebra will become secondary school teachers. It is important to raise their awareness of ties between tertiary and school algebra so that they will feel more inspired and confident in handling the teaching of algebra in their teaching careers. In addition, each country could consider ways of developing ongoing communication on the teaching of algebra between secondary and tertiary teachers.

11.5.2 Bringing evidence of curriculum reform efforts to the mathematical community

Whenever reform attempts are made it is important to examine their effectiveness and disseminate these results and implications for the benefit of those involved in the teaching of mathematics. Several examples of long-standing efforts to bring evidence of reform in algebra teaching to the wider mathematical community have already occurred. For example, in 1990 a small group of mathematicians in the U.S.A. formed the Linear Algebra Curriculum Study Group. This group held a workshop which produced a report (Carlson, Johnson, Lay, & Porter, 1993) recommending various reforms, produced two volumes in the American Mathematical Association Notes Series on the teaching of linear algebra (Carlson et al., 1997; Carlson, Johnson, Lay, & Porter, 2002), and was involved with others for ten years in organising sessions on the teaching of linear algebra at the annual US Joint Mathematics Meetings. These sessions involved many different speakers and always had attendances in the hundreds.

In the mid-1990s the International Linear Algebra Society initiated an Education Committee. At meetings of the society, the committee has organised sessions on teaching, presented well-known speakers from mathematics education, and held "Education Days" with activities for local secondary teachers. In addition to these group activities, individuals have spoken at a variety of mathematical meetings, organised workshops on, for example, using computing in teaching linear algebra, and written short articles for various professional journals. At the least, this activity seems to have affected the content of textbooks in linear algebra. The extent to which this activity has actually affected the teaching of the subject could be studied. In France, Dorier and his colleagues have been active in communicating mathematical education work to the mathematical community. Efforts of this sort should be considered in, for example, abstract algebra and discrete mathematics.

It is apparent that more tertiary algebra instructors need to become involved in mathematical education research. There are articles (e.g., Selden & Selden, 2001) in which individual mathematics instructors can read about ways to begin mathematics education research. It is probably far easier for such individuals to join an existing group involved with tertiary mathematics education research than to begin work on their own. One such group is associated with RUMEC (www.maa.org/sigma/arume) whose work has been described above.

11.6 Conclusions

This chapter has made recommendations for both practice and for further research. The relationship between research and practice is not straightforward. Artigue (2001), in outlining and acknowledging the value of research into the learning and teaching process at university level, attributes some of the difficulties in application to the difficulty of synthesising results and to the way in which the relevance of results can be limited in time and space by the different cultural and social environments for teaching around the world. However, along with Artigue, the Working Group is convinced that by making efforts to link practice and research and by making research results and theoretical perspectives available to a large audience of practitioners, the teaching and learning of tertiary algebra can be improved.

11.7 References

Artigue, M. (2001). What can we learn from educational research at the university level? In D. Holton (Ed.), *The teaching and learning of mathematics at university level: An ICMI study* (pp. 207-220). Dordrecht, The Netherlands: Kluwer Academic.

Asiala, A., Dubinsky, E., Mathews, D., Morics, S., & Oktaç, A. (1997). Development of student understanding of cosets, normality, and quotient groups. *Journal of Mathematical Behavior, 16*, 241-309.

Brown, A., DeVries, D., Dubinsky, E., & Thomas, K. (1997). Learning binary operations, groups, and subgroups. *Journal of Mathematical Behavior, 16*, 187-239.

Campbell, S., & Zazkis, R. (Eds.). (2001). *The learning and teaching of number theory*. New York: Ablex Publishers.

Carlson, D., Johnson, C., Lay, D., & Porter, A. D. (1993). The linear algebra curriculum study group recommendations for the first course in linear algebra. *College Mathematics Journal, 24*, 41-46.

Carlson, D., Johnson, C., Lay, D., & Porter, A. D. (Eds.) (2002). *Linear algebra gems: Assets for undergraduate mathematics*. Washington, D.C.: Mathematical Association of America.

Carlson, D., Johnson, C., Lay, D., Porter, A. D., Watkins, A., & Watkins, W. (Eds.). (1997). *Resources for teaching linear algebra*. (MAA Notes Vol. 42). Washington, D.C.: Mathematical Association of America.

Clark, J., Hemenway, C., St. John, D., Tolias, G., & Vakil, R. (1999). Student attitudes toward abstract algebra. *Primus, 9*, 76-96.

Dorier, J.-L. (Ed.). (2000). *On the teaching of linear algebra*. Dordrecht, The Netherlands: Kluwer Academic.

Dorier, J.-L., Robert, A., Robinet, J., & Rogalski, M. (1994). The teaching of linear algebra in first year of French science university. In J. P. da Ponte & J. F. Matos (Eds.), *Proceedings of the 18th annual conference of the International Group for the Psychology of Mathematics Education* (Vol. 4, pp. 137-144). Lisbon, Portugal: Program Committee.

Dorier, J.-L., Robert, A., Robinet, J., & Rogalski, M. (2000). The obstacle of formalism in linear algebra. In J.-L. Dorier (Ed.), *On the teaching of linear algebra* (pp. 85-124). Dordrecht, The Netherlands: Kluwer Academic.

Dorier, J.-L., & Sierpinska, A. (2001). Research into the teaching and learning of linear algebra. In D. Holton (Ed.), *The teaching and learning of mathematics at university level: An ICMI study* (pp. 255-273). Dordrecht, The Netherlands: Kluwer Academic.

Dreyfus, T. (1991). Advanced mathematical thinking processes. In D. Tall (Ed.), *Advanced mathematical thinking* (pp. 25-41). Dordrecht, The Netherlands: Kluwer Academic.

Dubinsky, E. (1991). Reflective abstraction in advanced mathematical thinking. In D. Tall (Ed.), *Advanced mathematical thinking* (pp. 95-123). Dordrecht, The Netherlands: Kluwer.

Dubinsky, E. (1997). On learning quantification. *Journal of Computers in Mathematics and Science Teaching, 16*(1), 335-362.

Dubinsky, E., Elterman, F., & Gong, C. (1988). The student's construction of quantification I. *For the Learning of Mathematics, 8*(2), 44-51.

Dubinsky, E., Mathews, D., & Reynolds, B. (Eds.). (1997). *Readings in cooperative learning for undergraduate mathematics*. Washington, D.C.: Mathematical Association of America.

Dubinsky, E., & McDonald, M. (2001). APOS: A constructivist theory of learning in undergraduate mathematics education research. In D. Holton (Ed.), *The teaching and learning of mathematics at university level: An ICMI study* (pp. 275-282). Dordrecht, The Netherlands: Kluwer Academic.

Dubinsky, E., & Yiparaki, O. (2000). On student understanding of AE and EA quantification. In E. Dubinsky, A. Schoenfeld, & J. Kaput (Eds.), *Research in collegiate mathematics education IV* (pp. 293-289). Providence, RI: American Mathematical Society.

Fenton, W., & Dubinsky, E. (1996). *Introduction to discrete mathematics with ISETL*. New York: Springer.

Forgasz, H. J., & Leder, G. C. (1998). Affective dimensions and tertiary mathematics students. In A. Olivier & K. Newstead (Eds.), *Proceedings of the 22nd annual conference of the International Group for the Psychology of Mathematics Education* (Vol. 2, 296-303). Stellenbosch, South Africa: Program Committee.

Hagelgans, N., Reynolds, B., Schwingendorf, K., Vidakovic, D., Dubinsky, E., Shanin, M., & Wimbish, G. (Eds.). (1995). *A practical guide to cooperative learning in collegiate mathematics*. (MAA Notes Vol. 37). Washington, D.C.: MAA.

Harel, G. (1989). Learning and teaching linear algebra: Difficulties and an alternative approach to visualizing concepts and processes. *Focus on Learning Problems in Mathematics, 11*, 139-148.

Harel, G. (1991). Using geometric models and vector arithmetic to teach high school students basic notions in linear algebra. *International Journal for Mathematics Education in Science and Technology, 21*, 387-392.

Harel, G. (1999). Students' understanding of proofs: A historical analysis and implications for the teaching of geometry and linear algebra. *Linear Algebra and its Applications, 302-303*, 601-613.

Harel, G., & Sowder, L. (1998). Students' proof schemes: Results from exploratory studies. In A. Schoenfeld, J. Kaput, & E. Dubinsky (Eds.), *Research in collegiate mathematics education III* (pp. 234-283). Providence, RI: American Mathematical Society.

Hart, E. (1994). Analysis of the proof-writing performances of expert and novice students in elementary group theory. In J. Kaput & E. Dubinsky (Eds.), *Research issues in undergraduate mathematics learning* (MAA Notes Vol. 33, pp. 49-62). Washington, D.C.: Mathematical Association of America.

Hazzan, O. (1999). Reducing abstraction level when learning abstract algebra concepts. *Educational Studies in Mathematics, 40*(1), 71-90.

Hibbard, A., & Maycock, E. (Eds.). (2002). *Innovations in teaching abstract algebra*. Washington, D.C.: Mathematical Association of America.

Hillel, J. (2000). Modes of description and the problem of representation in linear algebra. In J.-L. Dorier (Ed.), *On the teaching of linear algebra* (pp. 191-207). Dordrecht, The Netherlands: Kluwer Academic.

Holton D. (Ed.) (2001). *The teaching and learning of mathematics at university level: An ICMI study*. Dordrecht, The Netherlands: Kluwer Academic.

Katz, V. J. (2001). Using the history of algebra in teaching algebra. In H. Chick, K. Stacey, J. Vincent, & J. Vincent (Eds.), *The future of the teaching and learning of algebra* (Proceedings of the 12th ICMI Study Conference, pp. 353-359). Melbourne, Australia: The University of Melbourne.

Kleiner, I. (1998). A historically focused course in abstract algebra. *Mathematics Magazine, 71*, 105-111.

Lajoie, C. (2001). Students' difficulties with the concepts of group, subgroup, and group isomorphism. In H. Chick, K. Stacey, J. Vincent, & J. Vincent (Eds.), *The future of the teaching and learning of algebra* (Proceedings of the 12th ICMI Study Conference, pp. 384-391). Melbourne, Australia: The University of Melbourne.

Leon, S., Herman, E., & Faulkenberry, R. (1997). *ATLAST computer exercises for linear algebra*. Upper Saddle River, NJ: Prentice-Hall.

Leron, U., & Dubinsky, E. (1995). An abstract algebra story. *American Mathematical Monthly, 102*(3), 227-242.

Leron, U., & Hazzan, O. (1997). The world according to Johnny: A coping perspective in mathematics education. *Educational Studies in Mathematics, 32*, 265-292.

Oktaç, A. (2001). The teaching and learning of linear algebra: Is it the same at a distance? In H. Chick, K. Stacey, J. Vincent, & J. Vincent (Eds.), *The future of the teaching and learning of algebra* (Proceedings of the 12th ICMI Study Conference, pp. 501-506). Melbourne, Australia: The University of Melbourne.

Rogers, E., Reynolds, B., Davidson, N., & Thomas, A. (Eds.). (2001). *Cooperative learning in undergraduate mathematics: Issues that matter and strategies that work.* Washington, D.C.: Mathematical Association of America.

SEFI Mathematics Working Group (2002). *Core curriculum 2002.* Retrieved May 16, 2003, from http://learn.lboro.ac.uk/mwg/core.html

Selden, A., & Selden, J. (1987). Errors and misconceptions in college level theorem proving. In J. *Proceedings of the second international seminar on misconceptions and educational strategies in science and mathematics* (Vol. III, pp. 457-470). New York: Cornell University.

Selden, A., & Selden, J. (2001). Research into the teaching and learning of linear algebra. In D. Holton (Ed.), *The teaching and learning of mathematics at university level: An ICMI study* (pp. 237-254). Dordrecht, The Netherlands: Kluwer Academic.

Sierpinska, A. (2000). On some aspects of students' thinking in linear algebra. In J.-L. Dorier (Ed.), *On the teaching of linear algebra* (pp. 209-246). Dordrecht, The Netherlands: Kluwer Academic.

Sierpinska, A., Nnadozie, A., & Oktaç, A. (2002). *A study of the relationship between theoretical thinking and high achievement in linear algebra.* Montreal, Canada: Concordia University. Retrieved May 16, 2003, from http://alcor.Concordia.ca/~sierp/downloadpapers.html

Sierpinska, A., Trgalova, J., Hillel, J., & Dreyfus, T. (1999). Teaching and learning linear algebra with Cabri. In Zaslavsky, O. (Ed.), *Proceedings of the 23rd annual conference of the International Group for the Psychology of Mathematics Education* (Vol. 1, pp. 119-134). Haifa, Israel: Program Committee.

Siu, M-K. (2001). Why is it difficult to teach abstract algebra? In H. Chick, K., Stacey, J. Vincent, & J. Vincent (Eds.), *The future of the teaching and learning of algebra* (Proceedings of the 12th ICMI Study Conference, pp. 541-547). Melbourne, Australia: The University of Melbourne.

Sowder, L., & Harel, G. (2003). Case studies of mathematics majors' proof understanding, production and appreciation. *Canadian Journal of Science, Mathematics and Technology Education, 3*(2), 251-267.

Students' experiences of undergraduate mathematics. (2003). Retrieved May 16, 2003, from http://education.leeds.ac.uk/devt/research/mathseducation/seum3.htm

Tall, D. O., & Vinner, S. (1981). Concept images and concept definition in mathematics, with particular reference to limit and continuity. *Educational Studies in Mathematics, 22*(2), 125-147.

Vinner, S. (1991). The role of definitions in the teaching and learning of mathematics. In D. Tall, D. (Ed.), *Advanced mathematical thinking* (pp.65-81). Dordrecht, The Netherlands: Kluwer Academic.

Weller, K., Montgomery, A., Clark, J., Cottrill, J., Trigueros, M., Arnon, I., & Dubinsky, E. (2002). *Learning linear algebra with ISETL.* Retrieved May 16, 2003 from http://www.ilstu.edu/~jfcottr/linear-alg/

Wood, L. N. (1999). Teaching definitions in undergraduate mathematics. *Talum Newsletter, No. 10*, April. May 16, 2003, http://www.bham.ac.uk/ctimath/talum/ newsletter/wood.htm

Wood, L., & Smith, G. (2001). Assessment in linear algebra. In H. Chick, K. Stacey, J. Vincent, & J. Vincent (Eds.), *The future of the teaching and learning of algebra* (Proceedings of the 12th ICMI Study Conference, pp. 663-667). Melbourne, Australia: The University of Melbourne.

Zazkis, R., Dautermann, J., & Dubinsky, E. (1996). Coordinating visual and analytic strategies: A study of students' understandings of the group D4. *Journal for Research in Mathematics Education, 27*, 435-457.

The Working Group on Goals and Content of an Algebra Curriculum for the Compulsory Years

Leader: Gard Brekke

Working Group Mcmbcrs:
Jack Abramsky, Michael Bulmer, Hugh Burkhardt, Ann Crawford, Claude Gaulin, Mollie MacGregor, Per-Eskil Persson, and Alla Routitsky.

The Working Group on Goals and Content of an Algebra Curriculum for the Compulsory Years. Front (L to R): Jack Abramsky, Hugh Burkhardt, Gard Brekke. Back (L to R): Mollie MacGregor, Claude Gaulin, Per-Eskil Persson, Michael Bulmer, Ann Crawford. Absent: Alla Routitsky.

Prior to the Conference, each member of the Working Group on Goals and Content of an Algebra Curriculum for the Compulsory Years of Schooling prepared a paper for the 12th ICMI Study Conference Proceedings. These papers reflected members' expertise and prior experiences in teaching and researching aspects of algebra. The individual authors can be contacted using their e-mail addresses listed at the back of this book. The authors (sometimes with co-authors) and the titles of their papers are listed:

- Jack Abramsky: *Designing a national mathematics curriculum* (pp. 7-12).
- Gard Brekke: *School algebra: Primarily manipulation of empty symbols on a piece of paper?* (pp. 96-102).
- Michael Bulmer: *Algebra in an age of numerical mathematics* (pp. 136-139).
- Hugh Burkhardt: *Algebra for all* (pp. 140-146).
- Ann Crawford: *Developing algebraic thinking: Past, present and future* (pp. 192-198).
- Mollie MacGregor: *Does learning algebra benefit most people?* (pp. 405-411).
- Alla Routitsky & S. Zammit: *What can we learn from TIMMS: Comparison of Australian and Russian TIMSS-R results in algebra* (pp. 523-530).

The group met together as a whole for discussions about the current curriculum in the countries represented, what changes might be desirable, and how changes might be implemented and its effects assessed. Our main focus was on the algebra content that should be achieved by all or most of the students in the compulsory years of schooling.

The hard work of all of the participants of the Working Group on Goals and Content of an Algebra Curriculum for the Compulsory Years of Schooling is gratefully acknowledged. Particular thanks are extended to Mollie MacGregor for authoring the chapter and to Gard Brekke for his leadership of the Working Group.

Chapter 12

Goals and Content of an Algebra Curriculum for the Compulsory Years of Schooling

Mollie MacGregor
The University of Melbourne

Abstract: This chapter is concerned with the significance of algebra for the broad population of students in the compulsory years of schooling and with what should constitute a basic algebra curriculum. Three broad reasons for scrutiny of the curriculum are identified: the growth of universal education, the challenges and opportunities brought about by information technology, and the concern arising from documented low student achievement. The chapter proposes that the reasons for all students to learn algebra are complex, and must go beyond simple assertions of utility. The final section gives appropriate goals for a basic algebra curriculum.

Key words: Compulsory education, algebra for all, curriculum, goals, values, motivation

12.1 Introduction

This chapter is concerned with the significance of algebra for the broad population of students in the compulsory years of schooling—why learn algebra?—and with the question of what should constitute a basic algebra curriculum—what should be learned? Should algebra, or some revised version of it, be taught to all students in the compulsory years of schooling, even those who are unlikely to study in a quantitative field at a high level, and if so, why? Is there an essential core of algebraic knowledge that all students should master? How can teachers prepare some students for advanced study whilst catering for the needs, interests and abilities of others? Are there values inherent in the teaching and learning of algebra that contribute to the achievement of the broad goals of secondary education? This chapter addresses these and related questions. First, reasons for scrutiny of the algebra curriculum are discussed. Next, there is an assessment of popular arguments for and against the inclusion of algebra in the mathematics curriculum for all students (aged up to 15 years) in the compulsory years of schooling. Consideration of these arguments indicates that the teaching of algebra must be carefully justified because counter arguments are easy to create. Finally, suggestions are made about what an algebra-literate citizen in the 21st century should know, and there is an outline of the goals for algebra that all students should have the opportunity to achieve before they leave school.

For the purposes of this chapter, Crawford's (2001) definition of algebraic competence is used:

1. Ability to think in symbolic language, to understand algebra as generalised arithmetic, and to understand algebra as the study of mathematical structures.
2. Ability to understand equality and equations of algebra and to apply these within real world problem-solving settings.
3. Ability to understand relationships of quantities through patterns, defining functions, and applying mathematical modelling. (Crawford, 2001, p. 192)

The ability to work with graphs of functions and relations on the *x-y* plane is also a necessary component of algebraic competence. Other aspects of working with graphical representations (e.g., graphing data with line graphs) are important for all students, but are not considered in this chapter.

12.2 Pressures for Changing the Algebra Curriculum

The traditional school algebra curriculum, as it seems to be remembered by many people who encountered it, emphasised practising manipulation skills such as simplifying expressions containing brackets or fractions, factoring expressions, and solving equations either as exercises in symbol manipulation or, less frequently, in order to find answers to word problems. However it has long been evident that such a curriculum does not meet the needs of students and society. In the last two decades, in many countries there has been a great deal of rethinking of the goals of school algebra and in some countries much experimentation with different approaches (see all chapters of this book but especially Chapter 5 and Chapter 13). In this section the three main pressures for change are surveyed: the growth of universal secondary education, the ongoing development and increasing availability of technology, and the need to improve learning outcomes for all students.

12.2.1 The growth of universal secondary education

Teaching algebra in a mass education system necessarily presents dilemmas about curriculum content and classroom strategies. These dilemmas are especially acute for algebra because the traditional curriculum was developed for the small group of students who had access to secondary schooling in the early part of the 20th century and who looked forward to technical or professional careers. With universal secondary education comes a responsibility to cater for the abilities and interests of a wide range of students.

Academic learning based on reading and writing, such as the traditional school algebra of symbol manipulation and word-problems, is not appropriate for students with weak literacy and numeracy skills. These students may prefer to acquire knowledge through verbal interaction and concrete activity. According to a survey six years ago of Australian students' literacy and numeracy achievement (Marks & Ainley, 1997), about 20% of 14-year-olds were in this category. Moreover, in some cases a student's social background and peer group does not encourage an interest in abstract and symbolic information (Boaler, 2000; Mellin-Olsen, 1987) or a commitment to learning (*Principles and Standards 2000*; Steinberg, 1996). Consequently there are different opinions about whether 'algebra for all' is possible, and, if it is, how this goal might be achieved.

One view of algebra sees it as the branch of mathematics that deals with general properties of numbers and relations between them. However algebra in school has a far wider scope. The *Principles and Standards* of the NCTM (2000), for example, recommends that students should learn to "understand patterns, relations, and functions; represent and analyse mathematical situations and structures using algebra symbols; use mathematical models to represent and understand quantitative relationships; and analyse change in various contexts". As Burkhardt (2001) points out, "'doing algebra' has a very wide range of meaning, from substituting numbers in a given formula or extending a simple pattern to constructing a formal proof. ... There is some good evidence that nearly all children can achieve the former kinds of performance, but the last kind is achieved by few" (p. 141). Ideally, a curriculum will provide for this wide range of achievement levels and give all students something of lasting value.

Silver (1997), discussing in the USA context proposals for a compulsory one-year algebra course for all students, asks what the benefits—or the negative consequences—are likely to be for a student who fails or barely passes the course. Being in an algebra class does not ensure acquisition of algebraic ideas. Moreover, a student who experiences repeated failure is likely to either passively withdraw from participation in learning or actively rebel. If all students are to learn algebra, we need to find ways to reach those who at present (for a variety of reasons) are disadvantaged, unmotivated or uncooperative. We need to consider what new tools for learning and what classroom strategies will be helpful for them, as well as questions about the content of a core algebra curriculum.

Competence in algebra is essential for particular tertiary courses and scientific professions. This fact, however, is not a motivating factor for students who are not mathematically inclined or who expect to follow a career pathway involving minimal mathematics. Some of these students see algebra as necessary for graduation or further education and training—and hence worth mastering—but otherwise unconnected with their lives and ambitions. Some accept algebra as something they should learn because one day it might be important for them. Some reject algebra entirely because they are convinced it will never be useful in the future and see no value in it for their present lives. For all types of students, the learning itself must be made worthwhile.

12.2.2 Technology creates a changed environment

"The changes brought about by computers and calculators are so profound as to require readjustment in the balance and approach to virtually every topic in school mathematics" (*Reshaping School Mathematics,* 1990, p. 2). New information technology has opened the door to many exciting new possibilities for teaching algebra and at the same time has devalued some algebraic skills. In this book, there are many examples, primarily in Chapter 6 that explores new possibilities for teaching within technological environments, and Chapter 7 which considers the opportunities and challenges of teaching with computer algebra systems. Computers and the new generation of calculators can reliably and quickly carry out many of the most treasured skills of the traditional symbolic manipulation curriculum. For lengthy algebra calculations in most situations outside school, paper-and-pencil methods are no longer appropriate.

There has also been an important trend in engineering and science towards replacing symbolic methods by numerical and visual techniques. As Bulmer (2001) states, in many professions that rely heavily on mathematics and statistics (e.g., engineering, sciences, economics) algebra as a manipulative skill is of little importance. It is being replaced by fast numerical computation and dynamic visualisation. Algebraic knowledge may be necessary for specifying a function but not for finding values. New directions for the algebra curriculum, whether in the compulsory years or beyond, must take these developments into account. However technology will not make all aspects of algebra easier; modelling, for example, requires hard thinking and practice with or without technological tools.

12.2.3 Generally unsatisfactory outcomes

An important reason for reconfiguring the algebra curriculum is to improve learning outcomes for students. As noted above, the trend to mass education made traditional algebra curricula untenable because of the wide range of abilities and interests of students now staying in school until their mid-teens and beyond. However, even in countries that have been making adjustments in recent years, algebra as currently taught is not achieving its purpose for various groups of students. Researchers have commented on the failure of the school algebra curriculum as implemented in some schools to achieve good learning for many students (see, e.g., Sowder, 1998; Stacey & MacGregor, 1999; Sutherland, Rojano, Bell, & Lins, 2001). In particular, it is not enabling all students to follow a path of algebraic reasoning; that is, to analyse real situations, formulate critical relationships as equations, apply techniques to solve the equations, and interpret the results. What some students learn (or partially learn) is a collection of rules to be memorised and tricks to be performed, having no logical coherence, very little connection with previously learned arithmetic, and no applications in other school subjects or in the world outside school. For example, Brekke's (2001) report of a large-scale assessment project in Norway showed that the principal outcome of school algebra for students aged 11-15 years was the learning of rules for the manipulation of symbols.

Algebra curricula around the world are currently being scrutinised to determine their effectiveness. For example, Burkhardt (2001) alerts us to the influence of curriculum design and development on students' learning outcomes. He reports on a current UK project that is monitoring the state of development of an array of assessment instruments to evaluate the effectiveness of specific curricula in specific classrooms. For example, one instrument assesses a variety of aspects of algebraic performance related to both short tasks and longer more complex tasks. Further evidence of the influence of curriculum on learning outcomes was provided to the Working Group on Goals and Content of an Algebra Curriculum for the Compulsory Years of Schooling by Routitsky and Zammit (2001) who showed how differences in the achieved algebraic curriculum between Australian and Russian students aligned with different curriculum emphases, using data from the Third International Mathematics and Science Study – Repeat (TIMSS-R) (see also Chapter 13). There are also differences in depth of curriculum treatment. For example Crawford (2001) concludes that, in the middle grades, USA curricula attempt to cover too many topics, not allowing for deep conceptual understanding that may be achieved in other countries.

12.3 Why Should All Students Learn Algebra?

It is accepted by many people in the mathematics education community that algebra:

- Is a necessary part of the general knowledge of members of an educated and democratic society;
- Is a prerequisite for further study of mathematics, certain higher education courses, and many fields of employment;
- Is a crucial component of mathematical literacy, which underpins a nation's technological future and economic progress;
- Is an efficient way to solve certain types of problems;
- Promotes the intellectual activities of generalisation, organised thinking, and deductive reasoning.

The sections immediately below show that justifications based upon these reason are complex and certainly not self-evident to those who look at algebra from the outside. Later in the article, the discussion turns to how the algebra curriculum can indeed respond to these important concerns.

12.3.1 Algebra as necessary general knowledge

Algebra, if perceived as a set of manipulation procedures and a problem-solving tool, is irrelevant to the lives of most people (MacGregor, 2001; Noddings, 1994). Whereas algebra is valued and used by mathematicians, scientists and engineers, the mathematical language of everyday living and most employment is numerical. For the vast majority of people, arithmetic computation, approximation, and commonsense reasoning are far more useful than algebraic manipulation. With the exception of formulas, spreadsheets, and graphs, algebraic representations of mathematical ideas are not used in most professions and trades. In any society there are people who do not see the algebra knowledge they acquired in school as important, useful, or interesting. There are many highly intelligent, well-informed and critical citizens who did not master school algebra, or if they did, have never thought about or used algebra since. The methods they use for mathematical tasks are likely to be based on arithmetic or to involve computer technology not closely related to algebraic procedures learned in school. Indeed, common practices in the workplace such as substituting in formulas, using spreadsheets, and reading graphs, which have an algebraic basis, may not be considered by most people as "doing algebra" because letters are not manipulated.

Knowledge of algebra is not essential for a basic understanding of the many social, environmental and political issues in today's world. For example, no knowledge of algebra is required to read, in an issue of *Scientific American* (January 2002), articles written by four leading scientists on global warming, energy resources, population, and biodiversity. The reader needs to understand measures, rates, percentages, decimals, and elementary probability. All the mathematical information in these four articles is expressed clearly without any use of algebra notation or graphs of functions.

12.3.2 Algebra as a gateway to qualifications and employment

To solve problems out of school and in the workplace very few people use the algebraic methods they learned in school (Dowling, 1990; Fitzsimons, 1997; Magajna & Monaghan, 1998; Packer, 1997). In contrast to the view that algebra is not relevant to the lives of most people, the *Principles and Standards 2000* states that "algebraic competence is important in adult life" (p. 37). It is difficult to justify this claim. Results from two large studies in UK some years ago of the mathematics used in employment found that algebra as taught in school was unlikely to be used in the workplace. As reported in *Mathematics Counts*,

> One of the more surprising results of the studies is the little explicit use which is made of algebra. Formulae, sometimes using single letters for variables but more often expressed in words or abbreviations, are widely used by technicians, craftsmen, clerical workers and some operatives but all that is usually required is the substitution of numbers in these formulae and perhaps the use of a calculator ... It is not normally necessary to transform a formula; any form which is likely to be required will be available or can be looked up. Nor is it necessary to remove brackets, simplify expressions or solve simultaneous or quadratic equations ... Solution of linear equations is required occasionally (*Mathematics Counts*, 1985, p. 22).

More recently Brown comments, "most occupations require generic skills such as basic literacy and numeracy, oral communication, IT skills, and teamwork rather than advanced subject knowledge" (Brown, 1999, p. 82). It is unlikely that Brown included algebra in "basic numeracy".

In seeming contrast, the USA report *Everybody Counts* (1985) states that "over 75 percent of all jobs require proficiency in simple algebra and geometry, either as a prerequisite to training or as part of a licensure examination" (p. 4). Commenting on this statement, Noddings (1994) points out that no claim is being made that the jobs themselves require knowledge of algebra or geometry. Assessment of proficiency is used as a convenient filtering and selection process for entry to training programs and employment (*Mathematics Counts*, 1985; Mellin-Olsen, 1987; Noddings, 1994; Wolf, 2002).

One reason, then, for teaching algebra to all is to give school-leavers a chance to succeed in the competition for entry to the workforce and job training. This is an especially important reason in societies where categories of jobs and professions are unevenly distributed over racial and socio-economic groups (see, for example, Moses, Kamii, McAllister, & Howard, 1989). Decisions about who is taught algebra at school, and who is not, are likely to affect the range of employment opportunities open to school-leavers.

12.3.3 Algebra and a nation's technological future

A high general level of mathematics education is often seen as underpinning a nation's growth and economic competitiveness. For example, Steen (1997) writes "the economic future [of USA] depends on strength in mathematics education" (p. 134). *Reshaping School Mathematics* (1990) states "when today's children enter the workforce, more jobs will require greater mathematical skills" (p. 3). To what extent do these calls relate to algebra? It is clear that new technology has created a situation where large numbers of jobs—traditional jobs as well as new ones being created—require mathematical knowledge and techniques that are different from much of the mathematics in general, and algebra in particular, that was learned in the past. Paulos (1988), writing on the consequences for society of widespread mathematical illiteracy, states that what people need is a better number sense—in particular, a grasp of the relative sizes of numbers and the relations between them—and an understanding of chance and probability.

Evidence of the economic benefits of any particular part of the school curriculum is sparse. According to Wolf's (2002) survey of education and economic growth worldwide, there is no clear evidence that high-quality schooling has made any difference to the relative economic performance of countries (see Wolf, 2002, pp. 38-46). It is obvious that a nation's economic and technological progress depends on a certain number of its citizens having high levels of mathematical expertise. However large numbers of jobs (e.g., in retailing, tourism, hospitality, music industry, advertising, entertainment, and many others) make use of very little mathematical knowledge beyond basic numeracy. There seem to be no sound, empirically supported grounds for arguing that if all, or most, people have algebraic skills then a nation's economy will benefit.

12.3.4 Algebra for problem solving and for training reasoning

One goal of an algebra curriculum is to introduce students to new aspects of reasoning, and especially the disciplines of generalising, organised thinking, and deductive reasoning. However, it is obvious that many people who have not learned algebra carry out generalising, organised thinking, and deductive reasoning. It cannot be assumed that learning algebra necessarily promotes the development of good reasoning skills. However algebra gives learners the opportunity to engage with abstract ideas and to experience the pleasure and satisfaction of using a powerful symbol system to support logical thinking. These opportunities and experiences must be an important aim of teaching.

Algebra provides powerful problem-solving methods, which are potentially valuable in everyday life and in employment. However, it was noted above that most people do not use these problem-solving methods in everyday life and work, either because they do not encounter problems that require these methods or because they do not feel confident to use them well. Algebra is only useful as a problem-solving method if it can be taught in such as way that students can identify where it might be useful and are confident to use it correctly.

In an effort to ensure that all students, including the less able, achieve some success in algebra courses, there has been a trend in some countries to avoid problems and solution methods that require reasoning with algebraic symbols. Students think they are "doing algebra" because there are letters instead of numbers in their exercises. However the need to operate on, and with, unknowns—the essence of algebra—has been removed from much of the material in mainstream textbooks (Stacey & MacGregor, 1999). Many students continue to use familiar ways of operating based on arithmetic, which serve them well enough in most of the "algebra" tasks they are required to do. They continue to solve problems by calculating with known numbers and working towards the answer, instead of constructing and using equations as statements of equivalence relating knowns and unknowns (Stacey & MacGregor, 2000). They do not make—and do not need to make—the transition from arithmetic to algebraic thinking. A well-designed curriculum will aim to strengthen students' arithmetic problem-solving methods, but will also provide opportunities for the growth of algebraic thinking. Such a curriculum needs to be carefully constructed and delivered if it is to give students a problem-solving tool that is more powerful and more reliable than the arithmetic methods they are inclined to keep using.

12.4 An Algebra Curriculum for the Compulsory Years

There is still debate about how much algebraic knowledge and what particular algebraic skills all students should master, and how this can be achieved. How can a curriculum cater for the wide range of abilities of students at a given grade level? Should an algebra course teach a set of testable skills, or should it promote the development, over time, of a way of thinking and reasoning? Should students be streamed, as in some systems, so that different groups of students learn different types of algebra or should they be taught together so that everyone has the same exposure to a potentially empowering curriculum? When discussing details, the debate in the Working Group reinforced the comments of Foreman (1997) that

> Mathematicians, it seems, cannot agree on the content and pedagogy of high school algebra. Most agree . . . that there are important ideas that can best be communicated by using the symbols of algebra, but they disagree about the usefulness of some topics in the traditional curriculum (p. 16)

On the other hand, throughout the Working Group discussions, there was strong agreement on the values of some basic principles. There was agreement that algebra curricula continue to need attention, even in countries where there has been substantial change. In 1990, *Reshaping School Mathematics* called for radical redesign of the school mathematics curriculum and the way it is taught, bearing in mind the profound changes brought about by the availability of technology in the classroom. A few years later, in an article called "What mathematics should students learn?", Davis (1994) again stressed the need for changes to school mathematics programs so that more students would become engaged in learning and would benefit from the time they spend in mathematics classes. Calls such as these, arising from the concerns outlined at the beginning of this article, have led to significant changes in algebra curriculum and teaching over many years and there are now many reports of innovative and successful algebra programs for the early years of algebra learning, including several in this book and many of the papers in the conference proceedings (Chick, Stacey, Vincent, & Vincent, 2001). Dougherty's (2001) report of Measure Up, an algebra program centred on middle school students discussing, comparing, commenting on, and writing about their solutions to rich problems is an example. Of course, it is easier to attain positive change in small experiments than lasting widespread improvement across whole school systems. To this end, collaborative research by researchers and teachers will be required to design, test, and select programs and approaches to suit students' various abilities and attitudes and to take advantage of the many opportunities offered by technology. The place of algebra in this readjustment should continue to be assessed by curriculum planners.

12.4.1 Arithmetic competence and algebra learning

The relationship between competence with numbers and with algebra needs to be carefully considered in a curriculum for all students. In some school systems (e.g., some states in USA), "algebra readiness" is regarded as a useful notion. Teachers tend to expect that students must master arithmetic before beginning algebra (Crawford, 2001). Students who do not reach an appropriate level in arithmetic may not be offered algebra courses. More recently it has been proposed that algebra be taught as a mandated course to all students at a specified grade level, probably grade 8 or 9 (13-14 years old). In other systems (e.g., states in Australia), algebra topics are introduced to younger students (11-12 years old) as units (e.g., textbook chapters) interspersed with other mathematical topics but sometimes having very little connection with them. For example, students might "do algebra" for several weeks and then move on to "measurement" or "chance and data" where no links are made with the algebra they been learning. The Working Group agreed with the recommendation of the *Principles and Standards* (p. 211) that the middle grade mathematics curriculum should integrate algebra with geometry, number, and measurement. However, the Australian experience shows that this must be more than an organisational change. All students—and their teachers—need to see that algebra is not an isolated mathematical topic but is interwoven with other strands of mathematics. Algebra and arithmetic understanding can develop in tandem, each enhancing the other. If students are to learn to speak the language of algebra, it needs to be used as a language across the mathematics curriculum.

There are many ways in which learning about algebra can support learning about arithmetic, which may be of benefit to students whose number learning has proceeded slowly. There is widespread support at present for algebra to be introduced in order to enrich experiences of number, beginning in the earliest grades of elementary school, and the reasons for this are persuasively argued in Chapter 4 of this book. As early as 2nd and 3rd grades (7-8 years old), algebraic thinking can be encouraged by simple problems about numbers that motivate conjecture and discussion, reasoning, solving, and verification. Several studies (e.g., Carpenter & Franke, 2001; Fujii & Stephens, 2001) have shown that young children are capable of making generalisations about numbers and operations, and can attempt to justify their generalisations. Early experience of recognising and explaining number properties and operations can provide a foundation for algebraic thinking as well as number sense.

Traditionally, the range of numbers used in algebra exercises and problems for the middle grades has been restricted to small whole numbers. In the past, this restriction was reasonable because of the difficulty of paper-and-pencil calculation if a task involved decimals or large numbers. Today, access to electronic calculators enables a wide range of numbers to be used. The student who knows that if $3x = 12$ then x must be 4 "because three fours are 12" needs to be confronted with $3x = 1212$, for example, and to think about the operation to use for solving it. Whereas a student may simply know the answer to the first question, to solve the second requires understanding that division will "undo" multiplication. As Zazkis (2001) found, the use of large numbers helps students focus on operations and structures, enabling them to reason about these structures and express them algebraically, thus enhancing their awareness of numerical operations. Likewise, no student with access to a calculator should look at the equation $3x = 0.069$ and not attempt it, saying "But I can't do decimals." In a teaching experiment reported by Cedillo (2001), students about 12 years old were able to explore and giving meaning to algebraic expressions and equations involving large numbers, decimals, and exponents, using calculators. As Cedillo comments, these students achieved success on activities normally given to much older students. Calculator use allowed them to explore patterns and functions on a firm basis of arithmetic, using their own ways of reasoning.

12.4.2 Appreciating the value of algebra

The early part of this chapter argued that teaching algebra to all students cannot be justified on the basis of its everyday usefulness: competence in using algebra beyond using formulas, understanding functional relations, and interpreting graphs of functions is not necessary for most people's lives and work. The example of the articles from *Scientific American* shows is not necessary for communicating qualitative information about social, environmental, and political issues. There is no evidence that a nation's productivity and overall economic growth will improve as a consequence of all its young people taking algebra courses. If algebra is to be of value to all, the learning must be made interesting and satisfying for its own sake, as well as for its utility. Its intrinsic value needs to be evident. Students should see algebra as "a powerful friend with extraordinary explanatory powers" (Malcolm, 1997) and not as a feared enemy. As a good teacher said, "Math is a game. It's fun to play. We play it for its own sake. It's more fun than applying it. Most of the math I teach is never used by anyone" (Davis & Hersh, 1981, p. 273). And as a student said about algebra, "It's so neat, so clean. You know when you are right. It feels so good when you get the answer". What sort of curriculum and teaching would achieve a more widespread appreciation of algebra for its intrinsic value?

Even if algebra is not seen as a powerful friend, enjoyed as a game that is fun to play, or valued as a neat and efficient way of reasoning, some level of understanding should be an outcome of a general education. All citizens should learn enough about algebra to appreciate why it is the language of science and to know how it is used in formulas, graphs, and modelling, even if they do not use it themselves. They should understand why the language of algebra is necessary for communicating with precision mathematically significant ideas and why it is a powerful tool for deductive reasoning. They should know how formulas are derived from sets of data. How to achieve goals such as these for all students is a major challenge for mathematics education.

12.4.3 Goals of a basic algebra curriculum

In 1990, *Reshaping School Mathematics* stated that a major goal of the secondary mathematics curriculum should be the development of symbol sense. Students should be able "to represent mathematical problems in symbolic form and to use these symbolic representations in relations, expressions, and equations" (p. 45). Algebra topics should include "general algorithms and families of functions (polynomial, trigonometric, exponential, logarithmic)" (p. 46) and will be linked to geometry and data analysis. In contrast, the suggested minimum curriculum for U.K. students up to 16 years of age outlined in *Mathematics Counts* (1985) does not include any formal algebra. Commenting on this document some years later, Cockcroft (1994), the chair of the committee that produced *Mathematics Counts,* stated that students should know how to substitute numbers in formulas but that knowledge of algebra beyond this simple application is not necessary for employment in industry and commerce. However Cockcroft goes on to state that teaching must provide for able students who will go further in mathematics learning to "begin to see how to generalise, how to deal with abstractions, how to understand what it is to prove a result" (p. 48). As has been stated earlier in this article, how to cater for all students' needs in the compulsory years is one of the greatest challenges for schools.

There was considerable interest in the Working Group in seeing how this issues raised by Cockcroft are being addressed now. The National Curriculum for England (effective from September 2000), which has separate streams for different ability groups, has as its focus strengthening the role of algebra—in particular, as the key to abstraction and generalisation—with a forward look at the post-16 mathematics curriculum (see Abramsky, 2001). It is hoped that students will be enabled to develop a much better understanding of algebra and an appreciation of its usefulness. According to Abramsky (2001) it seems likely that many students who would have great difficulty with a formal treatment of algebra can make significant progress if they use technological tools to work with real data. Supported by technology, students can deal with a greater variety of problem types and with harder problems than those they tried to solve by traditional methods. Students who previously had not learned enough algebra to approach any but the simplest routine problems can now model and solve more interesting problems, using formulas they write themselves. Spreadsheets, graphing programs, and dynamic geometry programs such as Geometer's Sketchpad offer opportunities for numerical and visual solutions. Nevertheless, technology does not make all aspects of algebra easier; modelling, for example, is hard to do with or without it.

In a basic algebra curriculum for all learners in the 21st century, the traditional content is likely to remain. Students will still learn the use of symbols, index notation, equations, formulas, functions, inequations, and graphs. However there will continue to be great changes in the order of topics, their relative importance, and the way they are presented. Furthermore, there is likely to be an emphasis on numerical methods for solving certain types of problems. All students should be equipped to tackle problems numerically using variables, formulas, and equations as a language to communicate with technology. There will be an emphasis on learning through problem solving instead of practising manipulative techniques first and then trying to apply them. The consensus of the Working Group was that most students are capable of mastering a coherent core of algebraic knowledge, given appropriate opportunities to learn. This basic knowledge of algebra will enable students to:

- Become confident in their ability to interpret information expressed in algebraic notation (e.g., formulas in science and statistics);
- Recognise mathematical structures and patterns, and understand that algebra is used to express such generalities;
- Interpret and use formulas, especially substituting numbers to find values;
- Know how formulas are related to, and derived from, sets of data;
- Understand the relationships between functions and graphs;
- Know at least qualitatively some important properties of linear and exponential functions, and the implications for managing personal financial

affairs, for understanding environmental issues, and for making judgments about policies and plans in many fields of business and government.

- Understand how notations and representations (e.g., tables, equations, spreadsheets, and graphs) can be used for modelling certain situations and solving problems;
- Understand arithmetic operations more deeply (e.g., the relationship between multiplication and division) to achieve a more secure grasp of basic mathematical ideas
- Use technological tools (such as spreadsheets to deal with numbers, formulas and graphs);
- Experience the pleasure of doing mathematical experiments—making conjectures, testing them and, at an appropriate level, convincing oneself and others that they are correct.

12.5 References

Abramsky, J. (2001). Designing a national mathematics curriculum. In H. Chick, K. Stacey, J. Vincent, & J. Vincent (Eds.), *The future of the teaching and learning of algebra* (Proceedings of the 12th ICMI Study Conference, pp. 7-12). Melbourne, Australia: The University of Melbourne.

Boaler, J. (2000). Mathematics from another world: Traditional communities and the alienation of learners. *Journal of Mathematical Behavior, 18*(4), 379-397.

Brekke, G. (2001). School algebra: Primarily manipulations of empty symbols on a piece of paper? In H. Chick, K. Stacey, J. Vincent, & J. Vincent (Eds.), *The future of the teaching and learning of algebra* (Proceedings of the 12th ICMI Study Conference, pp. 96-102). Melbourne, Australia: The University of Melbourne.

Brown, M. (1999). One mathematics for all? In C. Hoyles, C. Morgan, & G. Woodhouse (Eds.), *Rethinking the mathematics curriculum* (pp. 78-89). London: Falmer.

Bulmer, M. (2001). Algebra in an age of numerical mathematics. In H. Chick, K. Stacey, J. Vincent, & J. Vincent (Eds.), *The future of the teaching and learning of algebra* (Proceedings of the 12th ICMI Study Conference, pp. 136-139). Melbourne, Australia: The University of Melbourne.

Burkhardt, H. (2001). Algebra for all: what does it mean? How are we doing? In H. Chick, K. Stacey, J. Vincent, & J. Vincent (Eds.), *The future of the teaching and learning of algebra* (Proceedings of the 12th ICMI Study Conference, pp. 140-146). Melbourne, Australia: The University of Melbourne.

Carpenter, T., & Franke, M. (2001). Developing algebraic reasoning in the elementary school: Generalization and proof. In H. Chick, K. Stacey, J. Vincent, & J. Vincent (Eds.), *The future of the teaching and learning of algebra* (Proceedings of the 12th ICMI Study Conference, pp. 155-162). Melbourne, Australia: The University of Melbourne.

Cedillo, T. (2001). Learning algebra by using it: A promising approach to using calculators in the classroom. In H. Chick, K. Stacey, J. Vincent, & J. Vincent (Eds.), *The future of the teaching and learning of algebra* (Proceedings of the 12th ICMI Study Conference, pp. 171-178). Melbourne, Australia: The University of Melbourne.

Cockcroft, W. (1994). Can the same mathematics program be suitable for all students? A personal view from *Mathematics Counts*, not forgetting *Standards*. *Journal of Mathematical Behavior, 13*, 37-51.

Crawford, A. (2001). Developing algebraic thinking: Past, present, and future. In H. Chick, K. Stacey, J. Vincent, & J. Vincent (Eds.), *The future of the teaching and learning of algebra* (Proceedings of the 12th ICMI Study Conference, pp. 192-193). Melbourne, Australia: The University of Melbourne.

Davis, R. B. (1994). What mathematics should children learn? *Journal of Mathematical Behavior, 13*, 3-33.

Davis, P., & Hersh, R. (1981). *The mathematical experience*. Brighton, Sussex: Harvester Press.

Dougherty, B. (2001). Access to algebra: A process approach. In H. Chick, K. Stacey, J. Vincent, & J. Vincent (Eds.), *The future of the teaching and learning of algebra* (Proceedings of the 12th ICMI Study Conference, pp. 207-212). Melbourne, Australia: The University of Melbourne.

Dowling, P. (1990). The shogun's and other curriculum voices. In P. Dowling & R. Noss (Eds.), *Mathematics versus the National Curriculum* (pp. 33-64). London: Falmer.

Everybody counts (1989). Washington, D.C.: National Academy Press:

Fitzsimons, G. (1997). Mathematics in the vocational education and training sector. In F. Biddulph & K. Carr (Eds.), *People in mathematics education* (Proceedings of the 20th annual conference of the Mathematics Education Group of Australasia, pp. 163-169). Rotorua, New Zealand: MERGA.

Foreman, S. (1997). Through mathematicians' eyes. In L. Steen (Ed.), *Why numbers count: Quantitative literacy for tomorrow's America* (pp. 161-172). New York: College Entrance Examination Board.

Fujii, T., & Stephens, M. (2001). Fostering an understanding of algebraic generalisation through numerical expressions: the role of quasi-variables. In H. Chick, K. Stacey, J. Vincent, & J. Vincent (Eds.), *The future of the teaching and learning of algebra* (Proceedings of the 12th ICMI Study Conference, pp. 258-264). Melbourne, Australia: The University of Melbourne.

MacGregor, M. (2001). Does learning algebra benefit most people? In H. Chick, K. Stacey, J. Vincent, & J. Vincent (Eds.), *The future of the teaching and learning of algebra* (Proceedings of the 12th ICMI Study Conference, pp. 405-411). Melbourne, Australia: The University of Melbourne.

Magajna, Z., & Monaghan, J. (1998). Non-elementary mathematics in a work setting. In A. Olivier & K. Newstead (Eds.), *Proceedings of the 22nd annual conference of the International Group for the Psychology of Mathematics Education* (Vol. 3., pp. 231-238). Stellenbosch, South Africa: Program Committee.

Malcolm, S. (1997). Making mathematics the great equalizer. In L. Steen (Ed.), *Why numbers count: Quantitative literacy for tomorrow's America* (pp. 30-35). New York: College Entrance Examination Board.

Marks, G., & Ainley, J. (1997). *Reading comprehension and numeracy among junior secondary students in Australia*. Camberwell, Vic: Australian Council for Educational Research.

Mathematics counts (1985). London: Her Majesty's Stationery Office. ("Cockcroft Report")

Mathematics in the National Curriculum (1999). UK: Department for Education and Employment and the Qualifications and Curriculum Authority

Mellin-Olsen, S. (1987). *The politics of mathematics education*. Dordrecht, The Netherlands: D. Reidel.

Moses, R., Kamii, M., McAllister, S., & Howard, J. (1989). The algebra project: Organizing in the spirit of Ella. *Harvard Educational Review 59*(4), 423-443.

Noddings, N. (1994). Does everybody count? Reflections on reforms in school mathematics. *Journal of Mathematical Behavior, 13*, 89-104.

Packer, A. (1997). Mathematical competencies that employers expect. In L. Steen (Ed.), *Why numbers count: Quantitative literacy for tomorrow's America* (pp. 137-154). New York: College Entrance Examination Board.

Paulos, J. (1988). *Innumeracy: Mathematical illiteracy and its consequences*. New York: Hill & Wang.

Principles and standards for school mathematics (2000). Reston, VA: National Council of Teachers of Mathematics.

Reshaping school mathematics. A philosophy and framework for curriculum (1990). Washington, D.C.: National Academy Press.

Routitsky, A., & Zammit, S. (2001). What can we learn from TIMSS: Comparison of Australian and Russian TIMSS-R results in algebra. In H. Chick, K. Stacey, J. Vincent, & J. Vincent (Eds.), *The future of the teaching and learning of algebra* (Proceedings of the 12th ICMI Study Conference, pp. 523-530). Melbourne, Australia: The University of Melbourne.

Silver, E. (1997). "Algebra for all" - Increasing students' access to algebraic ideas, not just algebra courses. *Mathematics teaching in the middle school, 2*(4), 204-20.

Sowder, J. (1998). Editorial. *Journal for Research in Mathematics Education, 29*(5), 496-502.

Stacey, K., & MacGregor, M. (1999). Implications for mathematics education policy of research on algebra learning. *Australian Journal of Education, 43*(1), 58-71.

Stacey, K., & MacGregor, M. (2000). Learning the algebraic method of solving problems. *Journal of Mathematical Behavior* 18 (2), 149–167.

Stacey, K. (2002). Strengthening both algebraic and arithmetic reasoning to improve outcomes from the teaching of algebra. In J. Abramsky (Ed.), *Reasoning, explanation and proof in school mathematics and their place in the intended curriculum* (Proceedings of the QCA International Seminar, pp. 55-68). London: Qualifications and Curriculum Authority.

Steen, L. (1997). The new literacy. In L. Steen (Ed.), *Why numbers count: quantitative literacy for tomorrow's America* (pp. xv-xxviii). New York: College Entrance Examination Board.

Steinberg, L. (1996). *Beyond the classroom. Why school reform has failed and what parents need to do*. New York: Simon & Schuster.

Sutherland, R., Rojano, T., Bell, A., & Lins, R. (2001). *Perspectives on school algebra*. Dordrecht, The Netherlands: Kluwer Academic.

Wolf, A. (2002) *Does education matter? Myths about education and economic growth*. London: Penguin.

Zazkis R. (2001). From arithmetic to algebra via big numbers. In H. Chick, K. Stacey, J. Vincent, & J. Vincent (Eds.), *The future of the teaching and learning of algebra* (Proceedings of the 12th ICMI Study Conference, pp. 676-681). Melbourne, Australia: The University of Melbourne.

Chapter 13

Algebra: A World of Difference

Margaret Kendal and Kaye Stacey
The University of Melbourne, Australia

Abstract: This chapter serves to highlight that the teaching of algebra is very different in different educational jurisdictions. It shows some of the differences in school structures that impinge on who learns algebra and at what stage. It surveys briefly some of the differences in content and the nature of algebraic activity. There is no attempt here to be comprehensive, but instead to point out some of the dimensions of difference around the world. There are weak links between the structures of schooling and the nature of algebra, and between the use of technology and the nature and purpose of algebraic activity. However the scene is characterised more by different themes and multiple variations on these themes, than by clear connections. Links between the nature of the curriculum and what students can do are accumulating from international studies.

Key words: Curriculum, school structures, algebra, teaching, international differences, technology, formalism, integrated curriculum

13.1 Introduction

13.1.1 Scope of this survey

As we have seen in various chapters throughout this book, the learning and teaching of algebra has had a very long history during which time it has gradually extended across the world and has become available to a high proportion of the population in many countries (Fauvel & van Maanen, 2000). This chapter attempts to describe various aspects of the current curricula in many countries, and to explain some of the interesting variations that occur. The countries from which information has been included in this chapter include Australia, Brazil, Canada, China, Czech Republic, England, France, Germany, Hungary, Israel, Italy, Japan, the Netherlands, the Russian Federation, Singapore, and the USA. Because schooling and curriculum is not always the responsibility of national governments but sometimes varies across different geographical, social, or political boundaries, the surveys here often do not relate to whole countries, but to individual educational jurisdictions within them (e.g., the Canadian provinces of Quebec and British Columbia).

Data from the available sources does not permit us to make a definitive, fully comprehensive comparison of algebra curricula. However, it does allow us to make some interesting comparisons between algebra teaching around the world and to highlight some of the dimensions of difference. Section 13.2 gives an overview of who learns algebra and when, giving examples of different structures of curriculum and choices about differentiating curriculum offering by ability. Section 13.3 describes some of the differences in algebraic activities, looking at features such as the degree of formalism, the nature of problems, the relation to the real world, and the use of technology. These two sections broadly correspond to the content and student vertices of the didactic triangle depicted in Figure 13.1. Matters relating to teachers and teaching methods (the teacher vertex) have been omitted because they cannot be adequately surveyed in this chapter. Instead, for a rich variety of teacher-related examples, the reader is referred to the Conference Proceedings of the 12th ICMI Study (Chick, Stacey, Vincent, & Vincent, 2001) and to other chapters in this book. Section 13.4 provides two case studies, which illustrate the nature of some international differences in more detail.

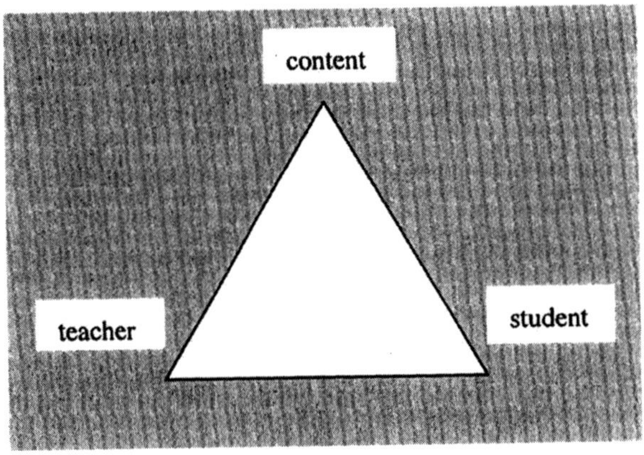

Figure 13.1. The didactic triangle.

13.1.2 Sources of information

This information in this chapter has been largely sourced from the Plenary Panel, *Algebra around the world*, organised by Romulo Lins (Brazil), which provided illustrations of the international differences in curricula for teaching algebra to the participants of the 12th ICMI Study Conference. Helen Chick (Australia) chaired the Plenary Panel and the speakers were Rosamund Sutherland (England), Toshiakira Fujii (Japan), and Jarmila Novotná (Czech Republic). They presented an overview of algebra education in their respective countries, outlined its curriculum, and provided a range of typical algebraic activities and problems. Romulo Lins (Brazil) provided the case study on Brazil in Section 13.4.2, and several other participants in the Study Conference provided material about algebra teaching in their region.

Sutherland's presentation in the Plenary Panel was based on her substantial report, *A comparative study of algebra curricula* (Sutherland, 2000), which is a major source of information for this chapter, used with permission. The report compares the algebra curricula in Australia (Victoria), Canada (British Columbia & Quebec), Europe (Hungary, France, Italy, Germany, the Netherlands, & Israel), and the Pacific Rim countries (China (Hong Kong SAR), Singapore, & Japan) with England's National Curriculum. Her comparison is also based on samples of secondary school national state examination questions and a selection of primary and secondary school textbooks.

There is considerable variation in Sutherland's sources of data. More information about curriculum, textbooks, and examination papers was available in relation to secondary education years than primary education. Curriculum information was current in 2000, but the mathematics curricula for Italy, Hungary, Hong Kong SAR, Quebec, and possibly others, were due to be updated, so may have changed. Sutherland also comments that her data does not allow reliable inferences about teachers' practices.

13.2 School Structures: Who Learns Algebra and When?

There is considerable variation in the way different educational jurisdictions structure their schools including their overall programs in all subjects and in mathematics. In consequence, many differences in decisions are made about which students are to learn algebra and at what age. Since these decisions provide an important part of the social context in which the teaching of algebra takes place, they have consequences for the goals of algebra and the nature of teaching approaches adopted. This section offers two examples of the overall variations which seem to most affect teaching.

13.2.1 Comprehensive or streamed classes

Educational jurisdictions differ considerably in the extent to which they divide students according to their achievement and aspirations, and the ages at which this happens. Some systems are *comprehensive*, catering for nearly all students in mixed ability classes well into the secondary years. For example, British Columbia, France, Italy, and Japan have comprehensive school systems until the end of third year of secondary school (and in some cases considerably beyond), whilst Quebec and the Netherlands stream a little earlier. Nearly all students go to the same types of primary and early secondary schools, and in these schools they are generally taught in mixed ability groups. In the later school years, comprehensive systems generally introduce some form of selection (either selection by the student or by the school) so that students with more similar ability or interest are in the same classes. In some Australian states, for example, students in the compulsory years of schooling (i.e., up to about age 16) are generally taught in mixed ability groups, after which students can decide for themselves how much mathematics they wish to study and at what level. As a consequence, curricula and classroom teaching in late secondary school do not cater for the whole ability range. England introduces streamed classes earlier, so that the comprehensive phase ends in the early years of secondary school. Japan provides a comprehensive education in mathematics at every level, so that all students undertake the same mathematics subjects through to the end of schooling. Japanese students are taught in mixed ability groups, although there is variation between the academic level of the schools through selection of students by competitive examinations at the end of primary school and from junior to senior high school.

Other educational jurisdictions have a policy of streaming considerably earlier. Hungary first streams all students when they are approximately 10 years old and Singapore streams the top 10% students at the same age. In Hungary, students in the higher streams at the end of primary school (approximately 10 or 11 years old) are using letters as variables and formal algebraic methods such as the balance method for solving equations. In these two systems, with early streaming, there is an early strong emphasis on symbolic manipulations and formal approaches to functions and their transformations. During the Plenary Panel, Novotná described the Czech Republic school structures and curriculum. There are three different programs from the middle primary years. She compared sections of the mathematics curriculum relevant to algebra in the basic school and the general school (two different streams of education) for grades 6 to 9. This comparison is given in Figure 13.2. For comparison, the topics undertaken in the general school are undertaken one or two years earlier than would be the case in Victoria (Australia).

From her survey of algebra curricula around the world, Sutherland (2000) concludes that educational jurisdictions tend to place less emphasis on symbolic aspects of algebra in the comprehensive phase, where classes are not streamed. It seems reasonable to propose that, in settings where all students are taught together, there has been more pressure to undertake the generational activities which aim to give meaning to the building blocks of algebra (see Kieran's Chapter 2 in this volume). Similarly, in streamed settings, higher ability classes have more emphasis on symbolic transformational aspects of algebra whilst lower ability classes may have restricted goals for algebra, or do not undertake it at all. For example, the middle secondary students in the Lycée in France and Liceo in Italy are expected to develop a much higher degree of symbolic competence than middle secondary school students in other schools. Japan is an exception. While it has a comprehensive school system, it appears to give an earlier emphasis to the use of symbolic algebra than other educational jurisdictions.

Chapter 12 in this volume addresses the vexed question of what sort of algebra would most benefit those students with lower achievement in mathematics. In this regard, it is useful to note that there is continuing debate on whether the use of streaming raises overall achievement or achievement for particular groups.

Grade	Basic School	General School
6	Consolidation of knowledge and skills (including tables, graphs, diagrams).	Core: Good mastering of numerical operations with natural and decimal numbers; drawing; word problems. Pre-algebra (expressions, variable, function and its graph, linear equations).
7	Direct and indirect proportions.	Core: Dependences, functions, orientation towards applications. Linear equations and linear functions.
8	Powers with natural exponents. Expressions. Linear equations.	Core: Good mastering of algebraic language, numerical mathematics, geometrical constructions. Calculations with polynomials
9	Polynomial functions. Solving equations with the unknown in the denominator. Systems of linear equations with two variables. Functions.	Core: Knowledge necessary for mathematical applications. Algebraic technique. Functions.

Figure 13.2. Algebra curriculum in two streams of Czech Republic schools.

13.2.2 Integrated mathematics or a layer-cake approach

Discussion of the teaching of algebra is often hindered in international arenas by a fundamentally different approach to structuring mathematics in the USA and some other educational jurisdictions. In many countries, students study mathematics across a variety of topic areas in each year, but in junior and senior high schools in the USA (middle and late secondary school), students generally study one field of mathematics per year. As a consequence, before year 9 (or year 8 for certain students) there is only minimal algebra in the curriculum, whereas year 9 typically concentrates on algebra, followed by a year of geometry before another year of algebra, earning the nickname of the *layer-cake* curriculum.

This layer-cake curriculum has been much criticised within the algebra education community. As is explained by Lins and Kaput in Chapter 4 of this volume, a desire to make a better bridge to algebra and to link it more strongly to other mathematical topics has been a major impetus behind the strong USA interest in early algebra. In addition, the "gatekeeper" role of the year of algebra has been identified as having strong social consequences, because students from educationally disadvantaged groups who are perceived as not ready for algebra are often sidetracked into courses which do not prepare them for higher education and training.

The variation between layer-cake and integrated mathematics can also make an obstacle to discussion of "what is algebra" and who is a teacher of algebra, as educators from the layer-cake countries sometimes identify "algebra" with curriculum at a particular level, whilst educators from other countries have a more diffuse interpretation. Readers of Chapter 10 in this volume need to understand these distinctions.

In summary, educational jurisdictions make decisions about how to structure the teaching of algebra in the school curriculum within the broad frameworks that they establish for the goals of schooling and in order to cater most effectively for their own students' needs and capabilities. Algebra is a relatively difficult and abstract area of mathematics and builds upon substantial prior knowledge. Thus, decisions made about the structures of schooling are highly likely to impact on its teaching.

13.3 Different Content for Algebra

In this section, we give some indications of how educational jurisdictions vary in their views of what algebra should be included in school curricula. Lins and Kaput, in Chapter 4, point out some of the difficulties in attempting to define algebra "because what one takes to be algebra depends on many cultural and other factors that vary widely across and even within communities". This view is supported by Sutherland's (2000) examination of the curriculum documents.

While it is not possible to identify precisely each country's conception of what constitutes algebra, and in any case educational jurisdictions can, and probably should, hold more than one view simultaneously (algebra is certainly multifaceted), considerable variation in these views is evident. These views of algebra are consistent with and extend the approaches to introductory algebra discussed in Chapter 5, where some examples have already been given.

Broadly expressed, algebra is conceived of as:

a) A way of expressing generality and pattern (strongly evident in British Columbia, England, Victoria, Singapore).
b) A study of symbol manipulation and equation solving (Brazil, France, Germany, Hong Kong SAR, Hungary, Israel, Italy, Russian Federation).
c) A study of functions and their transformations (France, Hungary, Israel, Japan, the Netherlands, USA).
d) A way to solve problems (usually word problems) beyond the reach of arithmetic methods (Czech Republic, France, Hungary, Italy, Japan, Hong Kong SAR, Singapore)
e) A way to interpret the world through modelling real situations, precisely or approximately (Quebec, England, Netherlands, Victoria).
f) A formal system, possibly dealing with set theory, logical operations, and operations on entities other than real numbers (Singapore, Hungary).

Since many chapters in this volume describe how some of these views are realised in introductory algebra, this chapter has been structured differently, looking at differences in how generality is treated, in the emphasis placed on symbolic manipulation and a formal approach, on use of technology and multiple representations, and how functions and real data are treated. In each section, the intention is to illustrate some of the parameters of difference, rather than to give a comprehensive account.

13.3.1 Generality and pattern

British Columbia, England, Victoria, and Singapore are examples of educational jurisdictions which emphasise algebra as a means of expressing generality and pattern, especially in the early stages. Chapter 5 gives an example of this approach in the early stages from England as well as a brief account of its rise to popularity, especially from the influential book "Routes to/Roots of Algebra" (Mason, Graham, Pimm, & Gowar, 1985). A typical introductory problem might be to find a general rule for describing the number of pavers required to border a square garden of variable side length (see Figure 13.3), first expressed in words (e.g., number of pavers is equal to 4 times side length plus 4) and later in symbols ($p = 4s + 4$).

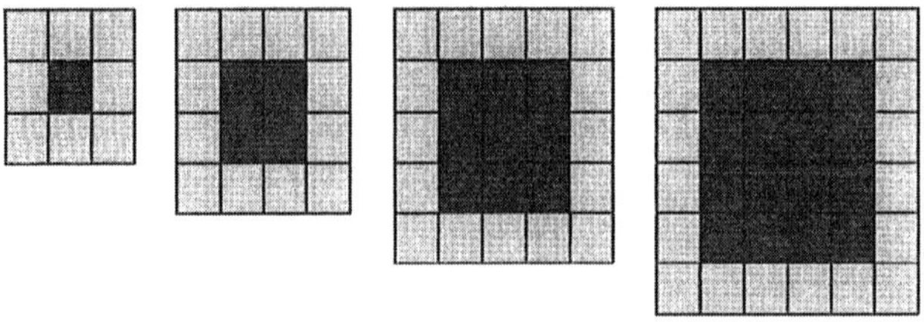

Figure 13.3. Pavers around a garden bed – a problem situation for approaching introductory algebra through generality and pattern.

This approach supplemented or replaced earlier approaches including that based simply on symbol manipulation (e.g., exercises in collecting like terms such as replacing $g+g+g+h+h$ by $3g+2h$).

Elsewhere, however, the theme of generality comes into the mathematics curriculum differently. Japanese students also work on problems exhibiting pattern. These problems are not intended as an introduction to algebra (although it is a useful preparation) but as part of a strand on mathematical relations (Ministry of Education, 1999), where students are to learn to solve problems by using "functional thinking" in three steps:

- Identify some independent variables and a dependent variable in the problem situation.
- Find a recurrent relation or functional relation between an independent variable and a dependent variable by searching systematically using a table or a graph.
- Use the pattern or relations to solve the problem.

In the Japanese primary curriculum, the intention of this work is to give students a first sense of "functional thinking" working in an arithmetic setting. This is done by encouraging students who are still working with numbers to retain and analyse the unclosed numerical forms of the answer (e.g., 4×1 + 4, 4×2 + 4, 4×3 + 4 etc. rather than 8, 12, 16, etc. in the pavers problem above). The intention is not to introduce a symbol for variables, but to lay the foundation for understanding that variables can be related (functional thinking) and for identifying mathematical relationships.

An interesting feature of the Japanese curriculum, pointed out to us by Junichi Ishida (Japan), is that this work on functional thinking is to be applied not only to problems of expressing generality, such as the pavers problem, but also to standard word problems. The pavers problem above requires the student to make an expression of generality, the relationship between side length and number of pavers, and then to use this relationship to find a number of pavers or a side length for a given number of pavers. There are clearly only two variables involved. The Japanese curriculum, however, intended that these insights also be applied to standard word problems, which are generally characterised by more than two variables linked by a series of constraints between them. This is an important difference between the perspective on functional thinking in Japan and in English-speaking countries (as represented by all of the examples of expressing generality elsewhere in this book).

The Japanese curriculum for lower secondary schools also stresses functional relationships between variables and the different representations of these relationships in the curriculum (e.g., understanding that a functional relationship can be expressed by means of a table, graph, formula). There is a strong emphasis on transforming and manipulating symbols from early primary school, and by the time the students are 15-16 years old, there is a much more demanding expectation with respect to use of symbolic algebra than most other educational jurisdictions.

13.3.2 Symbolism, formalism, and abstraction

Educational jurisdictions vary considerably in the extent to which algebra is taught as an abstract formal system, and the extent to which manipulation of symbols is practised. As noted above, in the Russian Federation, the Czech Republic, and Hungary there is an early emphasis on transforming and manipulating expressions, variables, and equations and at least some students reach high levels of proficiency at young ages. Germany introduces its students to symbolic algebra ideas of variable and simple equations in the first years of secondary school (approximately 12 years old) followed by quadratic and exponential functions.

After a relatively gentle introduction to symbolic algebra, Italian (and French) students (about 15 years old) have quite high symbolic algebra demands put upon them. For example, the Italian students are expected to be able to work with parameters and to solve inequalities. Similarly, the French students are expected to work with complex systems of equations and functions using formal set theory notation.

Japanese teachers have a carefully designed approach for moving students from arithmetic methods to formal algebra methods, which builds upon sound teacher-knowledge of intuitive methods. In order to effectively teach all of the students in their mixed ability classes at all age groups, Japanese teachers commonly base their lessons on open-ended investigations. First, students are given the problem, and time to think about how they would tackle it, with access to relevant materials. While the students work on the problem the teacher simultaneously assists them and monitors the types of solutions they produce. Next, the teacher asks particular students to present their solutions to the class; they range from very concrete solutions based on arithmetic to increasingly sophisticated solutions that involved generalising using symbolic algebra. The teacher and students together then evaluate the effectiveness of the solutions so that the power of the algebraic solution can be identified. A good example of such a Japanese lesson can be seen along with algebra lessons from Germany and the USA in the released videos from the TIMSS classroom video study (NCES, 1998).

Singapore also aims at a high level of symbolic manipulation to be achieved by relatively young students. Ng Swee Fong (2001) describes the *model method* which gives a pictorial representation of algebraic equations, to assist students to see algebra as a "meaningful study of relationships rather than a study of routine manipulations and learned algorithms" (p. 473). Ng (2001, p. 469) gives the example in Figure 13.4 to show the *model method* and the corresponding *algebra method* which is derived from it in early secondary school. This teaching is intended to provide a bridge from concrete to algebraic methods of thinking.

Problem:
Raju had three times as much money as Gopal. After Raju spent \$60 and Gopal spent \$10, they had an equal amount of money left.
How much did Raju have at first?

Using the model method

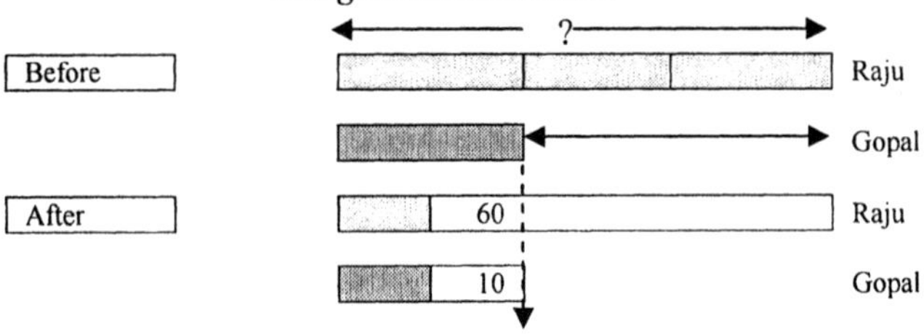

Since each box represents 1 unit, then 2 units should be equivalent to 60 – 10.
Therefore 1 unit is equivalent to 25.
If one unit is 25, then Raju should have 3 × 5.
At the beginning, Raju has \$75.

Using algebra

Let *r* represent the amount of money [number of dollars] held by Raju and *g* the amount of money [number of dollars] held by Gopal.

$$r = 3g$$
$$r - 60 = g - 10$$
$$3g - 60 = g - 10$$
$$2g = 50$$
$$g = 25$$

At the beginning, Raju has \$75. (Ng, 2001, p.469)

Figure 13.4. Solving a word problem using the Singapore *model method* and an *algebra method* (Ng, 2001).

13.3.3 Other aspects of a formal approach

There are other aspects of a formal approach to algebra, beyond symbolic manipulation. In Israel, young students (12-13 years old) study set notation including union and intersection of sets of numbers on a number line and establishing true and false sets in the solving of simple equations and this notation is also used within algebra. Solving equations, for example, in some countries (e.g., France) is requested formally as

$$\text{Find } \{ (x,y): 2x + y = 3, x - y = 0, x \in \Re, y \in \Re \}$$

with expected solution of (1,1) or {(1, 1)} whereas other countries (e.g., Australia) request the solution informally as

$$\text{Solve } 2x + y = 3 \text{ and } x - y = 0$$

with the expected solution of $x = 1$ and $y = 1$.

Other variations are evident in the emphasis given to proof and to understanding the logical basis for proof. Germany gives significant priority to these aspects of mathematics in a curriculum that emphasises generalisation, the need for proving, differences between deterministic and stochastic thinking, methods for proving, axiomatics, formalisation, generalisation versus specification, heuristic work, and variation of argumentation levels (Sutherland, 2000). Hungary has an early emphasis on using logical connectives formally in the construction of proofs where, by the end of grade 6 (approximately 11 years of age) students are expected to decide on the truth of simple statements, their negation, and correctly use logical elements of the language. In contrast, other educational jurisdictions, such as Victoria, have very little emphasis on formal proof at any level of primary or secondary school.

Varying degrees of formalism are also evident in the treatment of functions. In Israel, a formal approach to multiple representations of functions and their transformations is encouraged. Students as young as 12 or 13 years old are introduced to the concept of function as a special relation between abstract sets (domain and co-domain/range). Singapore also studies the abstract properties of functions as special types of mappings in the middle and late secondary years.

Most educational jurisdictions show very little evidence of a Bourbakian approach to mathematics. In many countries, the axiomatic approach which was a part of the "new maths" movement of the 1960s has nearly disappeared. Sutherland (2000) notes that Israel is an exception. Other than in the late secondary years in a few countries, there is little evidence of treating mathematical objects (numbers, functions, matrices, geometric transformations, etc.) as elements of abstract algebraic systems which obey specified axioms.

13.3.4 Functions and multiple representations

Although Puig and Rojano (Chapter 8) tell us that, from a historical point of view, functions are not formally a part of algebra, many educational jurisdictions currently adopt a functional approach to algebra where translations between symbolic, graphical, and numerical representations of the function are central. This is an important idea discussed in several chapters of this book. Computer and calculator technology has been important in stimulating this change in some countries.

As was noted above, a functional view of algebra can be very formal, stressing the abstract notion of function as a mapping between two abstract sets with given properties. France, Hungary, Israel, and Italy, for example, start from an emphasis on systems of equations, which tends to develop into a formal approach to functions and transformations of functions. Chapter 6 presents an approach using multiple representations and technology from Israel.

A functional approach can, however, be very informal. Quebec and the Netherlands place strong emphasis on links between different representations but do so in the context of solving reality-based problems. Both these educational jurisdictions introduce algebraic activity by emphasising relationships and connections between different representations of functions (tables, graphs, words, and formulae), paying particular attention to understanding the differences between them. Chapter 6 describes a USA function-based curriculum entitled *Core-Plus Mathematics Project* (CPMP) that promotes solving realistic problems at secondary school level through linking multiple representations of functions, again assisted by technology.

In some curricula, the real world enters primarily through traditional word problems (e.g., France, Hungary, Italy, Hong Kong SAR); in others functional relationships are sought in data (e.g., Netherlands, Victoria, England, Quebec). Questions are based on real life situations, with a strong possibility of inexact answers; technology can assist the solution of the problems. Sutherland (2000) notes that in general, where there is emphasis on solving realistic problems there tends to be less emphasis on symbolic manipulation and more emphasis on interpreting results.

In Australia, functions are often introduced from a real-world context. This can be done from situations where there is an exact model or an approximate one. For example, the cost of a taxi fare for various distances may be plotted from data supplied by the teacher, and students will note that the graph is a straight line and will go on to learn about its algebraic form, thereby introducing linear functions. Other teachers use real world data with only an approximate linear form. For example, an Australian curriculum program for teaching algebra with technology (Asp, Dowsey, Stacey, & Tynan, 1998) suggests using students' measurements of height and arm-span as shown in Figure 13.5, first in tables, then graphed and a line of best fit placed by eye (or by computer) along the data as a tool for prediction. Prediction provides the motivation for a general rule to get arm-span from height. Computer or graphics calculator technology assists these investigations, especially to deal with the large amounts of data generated by a whole class and the awkward numbers. The relative merits of approaching linear functions through exact and approximate real situations have not been explored.

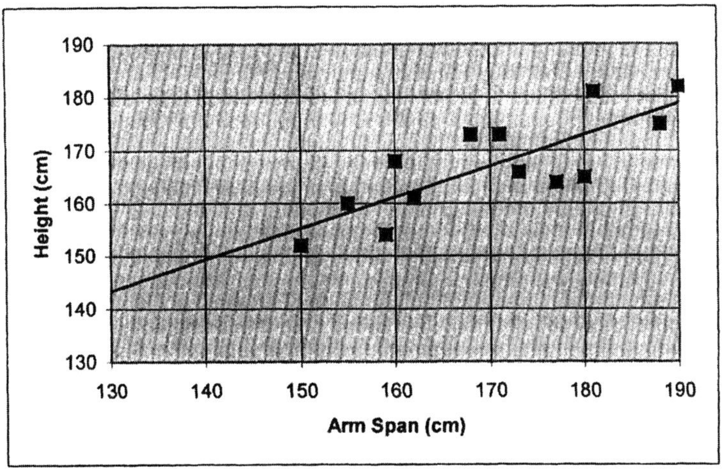

Figure 13.5. Scattergram of height against arm span in our class. Is there a linear relation?

In contrast, in the Japanese curriculum, where the lower secondary school curriculum stresses understanding functional relationships between variables and the different representations of these relationships (e.g., by means of a table, graph, formula), computer use is almost entirely restricted to teacher demonstrations and calculators of any sort are rarely used by students.

Thus, even within the educational jurisdictions that use functions and multiple representations, there are differences in the ways they are adopted. As discussed in several chapters of this book, it is believed that problem solving is enhanced through linking different representations, especially giving the symbolic representation meaning in the graphical representation. In educational jurisdictions where students have access to technology, curriculum can stress that a function models a real situation rather than describing it precisely, which also links to ideas of data handling. Elsewhere, real problems are restricted to situations where the functional relationship is precise (e.g., geometrical situations). In some curricula, the real world enters primarily through traditional word problems; in others functional relationships are sought in data. In some places, functions can be treated very formally, as abstract entities which provide a notation for many other ideas within the mathematics curriculum.

13.3.5 Using technology

In recent years, algebra education has been particularly influenced by computer and calculator technology, as described in Chapters 6 and 7. For algebra, technology is especially important because it provides easy access to numerical, graphical, and increasingly symbolic representations of mathematical functions through standard software such as spreadsheets and graphing programs as well as special purpose pedagogical software. As a consequence, many educational jurisdictions make explicit recommendations on the use of calculators and computers for algebra teaching. In some countries, technology is principally recommended for teacher demonstration for pedagogical purposes. For example, in Japan, whilst individual calculator use by students is not encouraged, the use of special computer programs by teachers to demonstrate mathematical concepts is highly regarded. Other educational jurisdictions advocate the use by students of technology to solve problems and to enhance learning. In the Netherlands, with its particularly strong emphasis on graphs, the curriculum specifies use of a simple computer program to solve problems in which the relationship between two variables plays a part. Many other countries expect students to have graphics calculators available and to learn how to use these to solve problems (e.g., USA Advanced Placement Calculus). However, these changes are far from universal, even within relatively wealthy countries. Within Australia, for example, while some educational jurisdictions encourage the use of graphics calculators in examinations, others prohibit their use.

Computer technology with symbolic algebra manipulation capabilities is becoming increasingly available in hand-held calculator technology and is endorsed by educational jurisdictions such as France, Austria, Denmark, and Victoria for use in examinations. This has led to important discussion within the mathematics education community and in this book (see Chapters 6 and 7) about the role that by-hand manipulation plays in establishing students' understanding of the underlying concepts of algebra and to how the priorities for the algebra curriculum may change.

Asian countries which have been relatively high-achieving in mathematics and also previously conservative with respect to technology use are now changing. Singapore has had a massive introduction of technology into all areas of the curriculum and Malaysia is now following, including support for the use of graphing, spreadsheet, and computer algebra technology in schools and universities. In 2003, Shanghai became the first area of China to experiment with information technology in schools.

13.4 Differences in Content Lead to Differences in Outcomes

Different views of what content of algebra is most appropriate lead to different outcomes for students. As international comparisons and cross-cultural studies become more prevalent, some of these effects are beginning to be documented. This section describes two cases.

13.4.1 Australia and Russia

The algebra section of the Third International Mathematics and Science Study (Repeat 1999) known as TIMMS-R contained the items L12 and R12 in Figure 13.6 and Figure 13.7. Routitsky and Zammit (2001) compared the success of Australian and Russian students (aged about 13 years) on the items requiring generalisation such as item L12 (in *Number patterns and simple relations*) and those relating to use of symbols such as item R12 (in *Simple algebraic expressions*). Although the success of Australia and Russian students averaged over all items was very similar, these two groups of students showed large differences in success rates and Routitsky and Zammit conclude that this is due to national differences in curriculum emphasis. On the first set of items, the Australian students achieved better than the Russian Federation students and significantly better than the international average. For example on L12 (see Figure 13.6) which tests *Number patterns and simple relations*, the Australian students achieved 70%, the Russian Federation students 59%, and the international average was 53%. This result reflects the greater emphasis on expressing generality and pattern in many of the Australian states and the Russian Federation's lesser emphasis.

L12. *In a sequence of starts and stops, an elevator travels from the first floor to the fifth and then to the second floor. From there, the elevator travels to the fourth floor and then to the third floor. If the floors are 3m apart, how far has the elevator travelled?*

A. 18m B. 27m C. 30m D. 45m

Figure 13.6. Item L12 of TIMMS-R Algebra test.

On the other hand, Russian Federation students achieved significantly better than the international and Australian students on the set of items that tested traditional algebraic topics *Simple algebraic expressions* and *Representing situations algebraically; formulas*. For example, on R12 (see Figure 13.7), which involved simple expressions, the Russian Federation students achieved 81%, Australian students achieved 34%, and the international average was 47%. This result reflects the Russian Federation's greater emphasis on understanding symbolic notation where all 12-14 year old students are expected to manipulate expressions based on formulas such as

$$(a \pm b)^2, (a \pm b)^3, a^2 - b^2, a^3 - b^3, a^4 - b^4, (\sqrt{a} + \sqrt{b})^2$$

Novotná, in her Plenary Panel presentation, demonstrated how all Russian students are expected to solve equations like $|1 - |1 - x|| = ½$ by considering the signs over different intervals (where $1 - x$ is positive or negative etc.), whilst the elite students are also expected to be able to solve such an equation graphically by building up the graph in stages from 1 – x and also by interpreting the expression as indicating that the distance of $|1 - x|$ from 1.

R12. *If k represents a negative number, which of these is a positive number?*
A. k^2 B. k^3 C. $2k$ D. $\frac{1}{2}k$

Figure 13.7. Item R12 of TIMMS-R Algebra test.

13.4.2 Algebra education in Brazil – Report from Romulo Lins

Algebra education in Brazil follows a traditional symbol-manipulation model; there are only two textbook series that try to avoid this approach. It is safe to say that well over 95% of Brazilian pupils have a traditional algebra education in schools. In primary school (mainly grade 4) there is usually some work with simple 'equations' using place holders ('boxes'); some teachers use the scale-balance metaphor. There is sometimes a little work with patterns and expression to represent them. In grades 5 and 6 pupils deal with linear equations, either using only symbols or in the application of solving simple problems. Very rarely is this work named 'algebra'.

In grade 7, work with symbol manipulation begins. It includes factoring and expanding powers of binomials, with lots of drill. There is sometime use of geometrical diagrams to 'justify', for instance, the expansion of $(a+b)^2$. Most of the year is spent on this, and in many schools this is all that is done (geometry is skipped completely). Grade 7 has always been, in Brazil, the strongest 'cut' year of school life. Many pupils fail and even those who have previously enjoyed mathematics mostly start to dislike it strongly at grade 7. "Algebra" in Brazil is almost a bad word!

Grade 8 is a continuation of the kind of work done in grade 7, with the inclusion of functions and graphs (with few applications) and quadratic equations. Sets of equations are treated either late in grade 7 or in grade 8. In high school (the next three years), students study a little more on functions in general, trigonometry, and sometimes, a more general study of polynomials (e.g., polynomial equations of degree higher than 2, polynomial division, and so on). There is very little use of technology, either computers or calculators.

Lins reports that, in his research investigations for his doctoral degree, he could (to some extent) compare English and Brazilian students. He identified a kind of 'willingness' in Brazilian students to use equations to solve verbal problems that was not apparent in English students. For example, there were instances in which grade 7 and grade 8 Brazilian students even wrote equations or sets of equations which they could not solve at all. English students used equations only very rarely, because they had little of this experience in school. The key point is that somehow the extensive drilling and the complete domination of algebra in grades 7 and 8 had almost compelled the Brazilian students to use equations in verbal problems, except when a non-algebraic solution was immediately visible.

Results from the international test of mathematical and scientific literacy, Programme for the International Student Assessment (PISA), organised by the Organisation for Economic Development and Co-operation (OECD), clearly show that Brazilian students have difficulty with solving non-standard problems that can be tackled by setting up equations and solving them. When most students leave school they know very little mathematics in general, and what is there is also quickly forgotten (based on data from university entrance exams and recent national tests).

The culture among Brazilian teachers is heavily centred on 'teaching content', even though the most recent national curricula guidelines emphasise problem solving and applications. With respect to algebra education, these guidelines clearly state that their primary objective is to help students to develop the ability to think algebraically. While the teaching of symbolic manipulation is perceived as a way to achieve this goal, it should not be the primary objective. There has always been a strong opposition to change proposed by educational authorities. Only recently, a culture of examinations external to schools began to be implemented in Brazil (again, with a huge opposition) as a way to allow educational changes to permeate through the system.

13.5 Conclusion

The conclusion of this chapter is: don't take your country's curriculum and approach to teaching algebra for granted and don't assume all other educational jurisdictions operate in a similar way—they conspicuously do not. Algebra is a large content area, too large to fit entirely within any one school curriculum, and so choices must be made. Moreover, it is a rich field with many possibilities for applications and for addressing meta-mathematical goals, such as learning about problem solving, or axiomatics, or mathematical structure, or the benefits of an organised approach. Again this means that choices can and must be made.

This chapter has clearly shown that there are worlds of differences in the way that algebra is viewed in educational jurisdictions. There are some weak links between the structures, content, teaching, and approaches. For example, in comprehensive schooling which caters for students across the ability range there tends to be less emphasis on formal symbolic manipulation and meaning for algebra is sought more to describe variation in real world situations. Much of the energy behind the search for innovative teaching methods has come from the need to make algebra meaningful to a broader range of students. The use of computer technology is also weakly linked to content and approaches to algebra, as spreadsheet and graphing facilities provide opportunities to study phenomena in new ways and in turn, these capabilities might be seen to alter priorities in learning algebra. These links between structures, content, and teaching are, however, not strong. The scene is better characterised by different themes and multiple variations on these themes, than by clear connections.

13.6 References

Asp, G., Dowsey, J., Stacey, K. & Tynan, D. (1998). *Graphic algebra: explorations with a graphing calculator*. Berkeley, USA: Key Curriculum Press.

Bednarz, N., Kieran, C., & Lee, L. (1996). *Approaches to algebra: Perspectives for research and teaching*. Dordrecht, The Netherlands: Kluwer Academic.

Chick, H., Stacey, K., Vincent, J., & Vincent, J. (Eds.). (2001). *The future of the teaching and learning of algebra* (Proceedings of the 12th ICMI Study Conference). Melbourne, Australia: The University of Melbourne.

Fauvel, J., & van Maanen, J. (Eds.). (2000). *History in mathematics education. The ICMI study.* Dordrecht, The Netherlands: Kluwer Academic.

Mason, J., Graham, A., Pimm, D., & Gowar, N. (1985). *Routes to / roots of algebra*. Milton Keynes, UK: The Open University Press

Ministry of Education. (1999). *Guidebook for course of study, Mathematics*. Tokyo: Toyokan.

National Center for Educational Statistics (NCES) (1998) *Video examples from the TIMSS videotape classroom study: Eighth grade mathematics in Germany, Japan and the United States*. Washington, USA: Author.

Ng, S.F. (2001). Secondary school students' perceptions of the relationship between the model method and algebra. In H. Chick, K. Stacey, J. Vincent, & J. Vincent (Eds.), *The future of the teaching and learning of algebra* (Proceedings of the 12th ICMI Study Conference, pp. 468-476). Melbourne, Australia: The University of Melbourne.

Routitsky, A., & Zammit, S. (2001). What can we learn from TIMSS: A comparison of Australian and Russian TIMSS-R results in algebra. In H. Chick, K. Stacey, J. Vincent, & J. Vincent (Eds.), *The future of the teaching and learning of algebra* (Proceedings of the 12th ICMI Study Conference, pp. 523-530). Melbourne, Australia: The University of Melbourne.

Sutherland, R. (2000). *A comparative study of algebra curricula*. Bristol, UK: Qualification and Curriculum Authority.

Conference Participants

Abramsky, Jack
Qualifications and Curriculum Authority, UK
abramskyj@qca.org.uk

Agudelo, Cecilia
Monash University
Melbourne, Australia
c.agudelo@education.monash.edu.au

Aczel, James
The Open University
UK
j.c.aczel@open.ac.uk

Artigue, Michèle
Université Paris
France
artigue@math.jussieu.fr

Arzarello, Ferdinando
Università di Torino
Italy
arzarello@dm.unito.it

Asp, Gary
The University of Melbourne
Australia

Atiola, Alifeleti
Tupou High School
Tonga
feleti@tupouhs.fwc.edu.to

Baker, Bernadette
Drake University
USA
bernadette.baker@drake.edu

Ball, Lynda
The University of Melbourne
Australia
lball@unimelb.edu.au

Bazzini, Luciana
Università di Torino
Italy
bazzini@dm.unito.it

Bednarz, Nadine
Université du Québec à Montréal
Canada
descamps-bednarz.nadine@uqam.ca

Belfort, Elizabeth
Universidade Federal de Rio de Janeiro, Brazil
beth@dmm.im.ufrj.br

Blanton, Maria
University of Massachusetts, Dartmouth, USA
mblanton@umassd.edu

Brekke, Gard
Telemarksforsking-Notodden
Norway
Gard.Brekke@hit.no

Britt, Murray S.
Auckland College of Education
New Zealand
m.britt@ace.ac.nz

Brizuela, Bárbara
Tufts University
USA
barbara.brizuela@tufts.edu

Brown, Laurinda
University of Bristol
UK
Laurinda.Brown@bris.ac.uk

Brown, Roger
International Baccalaureate
Organization, UK
rogerb@ibo.org

Bulmer, Michael
University of Queensland
Australia
mrb@maths.uq.edu.au

Burkhardt, Hugh
University of Nottingham
UK
Hugh.Burkhardt@nottingham.ac.uk

Campbell, Stephen R.
University of California, Irvine
USA
sencael@uci.edu

Carlson, Dave
San Diego State University
USA
jncarlson@yahoo.com

Carpenter, Tom
University of Wisconsin - Madison
USA
tpcarpen@facstaff.wisc.edu

Carraher, David
TERC
USA
david_carraher@terc.edu

Cerulli, Michele
Università di Pisa
Italy
cerulli@dm.unipi.it

Chick, Helen
The University of Melbourne
Australia
h.chick@.unimelb.edu.au

Cnop, Ivan
Vrije Universiteit Brussels
Belgium
icnop@vub.ac.be

Cooper, Tom
Queensland University of
Technology, Australia
tj.cooper@qut.edu.au

Crawford, Ann
University of North Carolina,
Wilmington, USA
Crawforda@uncwil.edu

De'Liberto, Deanna
D Squared Assessments, Inc.
USA
Ddeliberto@aol.com

Delozanne, Elisabeth
Université René Descartes
France
Elisabeth.delozanne@creteil.iufm.fr

Doerr, Helen M.
Syracuse University
USA
hmdoerr@syr.edu

Dougherty, Barbara J.
University of Hawaii
USA
bdougher@hawaii.edu

Drijvers, Paul
Freudenthal Institute, University of Utrecht, The Netherlands
P.Drijvers@fi.uu.nl

Driver, David
Brisbane School of Distance Education, Australia
DDRIVER@brisbanesde.qld.edu.au

Drouhard, Jean-Philippe
IUFM & IREM DE NICE, CREEM (CNAM), France
drouhard@math.unice.fr

Dubinsky, Ed
New York and Ohio
USA
edd@mcs.kent.edu

Mary Enderson
Middle Tennessee State University, USA
mcenders@mtsu.edu

Fearnley-Sander, Desmond
University of Tasmania
Australia
dfs@hilbert.maths.utas.edu.au

Peter Flynn
The University of Melbourne
Australia
p.flynn@pgrad.unimelb.edu.au

Franke, Megan
University of California, Los Angeles, USA
mfranke@ucla.edu

Fujii, Toshiakira
Tokyo Gakugei University
Japan
tfujii@u-gakugei.ac.jp

George Gadanidis
University of Western Ontario
Canada
ggadanid@uwo.ca

Gallardo, Aurora
CINVESTAV Instituto Politecnico Nacional, Mexico
agallard@mail.cinvestav.mx

Gaulin Claude
Laval University
Quebec, Canada
Claude.Gaulin@fse.ulaval.ca

Grugeon, Brigitte
Université Paris
France
brigitte.grugeon@amiens.iufm.fr

Haimes, David H.
Curtin University Of Technology
Australia
D.Haimes@smec.curtin.edu.au

Hatch, Gillian
Manchester Metropolitan University, UK
g.hatch@mmu.ac.uk

Heid, M. Kathleen
Pennsylvania State University
USA
ik8@psu.edu

Henry, Valerie
University of California, Irvine
USA
vhenry@telis.org

Hewitt, Dave
University of Birmingham
UK
D.P.Hewitt@bham.ac.uk

Hodgson, Bernard
Université Laval
Canada
bhodgson@mat@ulaval.ca

Horne, Marj
Australian Catholic University
Australia
m.horne@patrick.acu.edu.au

Isoda, Masami
University of Tsukuba
Japan
msisoda@human.tsukuba.ac.jp

Johnson, Sylvia
Sheffield Hallam University
UK
s.m.johnson@shu.ac.uk

Kaput, James J.
University of Massachusetts – Dartmouth, USA
jkaput@umassd.edu

Kendal, Margaret
The University of Melbourne
Australia
kendal@bacchusmarsh.net.au

Kidron, Ivy
Weizmann Institute of Science
Israel
ivy.kidron@weizmann.ac.il

Kieran, Carolyn
Université du Québec à Montréal
Canada
kieran.carolyn@uqam.ca

Kissane, Barry
Murdoch University
Australia
kissane@central.murdoch.edu.au

Kleiner, Israel
York University
Canada
kleiner@home.com

Lagrange, Jean-baptiste
IUFM Rennes
France
lagrange@univ-rennes1.fr

Lajoie, Caroline
Université du Québec à Montréal
Canada
lajoie.caroline@uqam.ca

Lee, Lesley
Pacific Resources for Education and Learning, Hawaii, USA
LeeL@prel.org

Leigh-Lancaster, David
Victorian Curriculum and
Assessment Authority, Australia
leigh-lancaster.david.d@edumail.vic.gov

Lins, Rom
UNESP
Brazil
romlins@rc.unesp.br

MacGregor, Mollie
The University of Melbourne
Australia
mollie@unimelb.edu.au

Mann, Giora
Levinsky College of Education
Israel
giorama@macam98.ac.il

Manouchehri, Azita
Central Michigan University
USA
azita@panam.edu

Marjanovic, Milosav
Mathematical Institute
Yugoslavia
djkadij@turing.mi.sanu.ac.yu

Mason, John H.
Open University
UK
j.h.mason@open.ac.uk

Menzel, Brenda
Murrayville Community School
Australia
brendam1@bigpond.com

Mesa, Vilma
University of Michigan
USA
vmesa@umich.edu

Monaghan, John
University of Leeds
UK
j.d.monaghan@education.leeds.ac.uk

Ng, Swee Fong
National Institute of Education
Singapore
sfng@nie.edu.sg

Nicaud, Jean-Francois
IRIN, University of Nantes
France
Jean-Francois.Nicaud@irin.univ-nantes.fr

Novotná, Jarmila
Charles University
Czech Republic
jarmila.novotna@pedf.cuni.cz

Oktaç, Asuman
UQAM & CINVESTAV-IPN
Canada & Mexico
oktac.asuman@uqam.ca

Olive, John
The University of Georgia
GA, USA
jolive@coe.uga.edu

Panizza, Mabel
University of Buenos Aires
Argentina
mpanizza@mail.retina.ar

Pegg, John
University of New England
Australia
jpegg@metz.une.edu.au

Persson, Per-Eskil
Sweden
persson@klippan.se

Pierce, Robyn
University of Ballarat
Australia
r.pierce@ballarat.edu.au

Puig, Luis
Universidad de Valencia
Spain
luis.puig@uv.es

Quinlan, Cyril
Australian Catholic University,
Sydney, Australia
c.quinlan@mary.acu.edu.au

Rojano, Teresa
Avenida Instituto Politécnico ,
Nacional, Mexico
mrojanoa@sep.gob.mx

Routitsky, Alla
Australian Council for
Educational Research, Australia
routitsky@acer.edu.au

Schliemann, Analúcia
Tufts University
USA
ana.schliemann@tufts.edu

Siu, Man-Keung
University of Hong Kong, China
Hong Kong, China
mathsiu@hkucc.hku.hk

Sri Wahyuni
Gadjah Mada University
Indonesia
swahyuni@indosat.net.id

Stacey, Kaye
The University of Melbourne
Australia
k.stacey@unimelb.edu.au

Stephens, Max
Victorian Curriculum and
Assessment Authority, Australia
m.stephens@unimelb.edu.au

Stump, Sheryl
Ball State University
USA
sstump@gw.bsu.edu

Sutherland, Rosamund
University of Bristol
UK
ros.sutherland@bris.ac.uk

Teppo, Anne
Bozeman, MT
USA
arteppo@theglobal.net

Thomas, Michael
University of Auckland
New Zealand
m.thomas@math.auckland.ac.nz

Trigueros, Maria
ITAM
Mexico
trigue@itam.mx

Tuyen, Nguyen Xuan
Vietnam
nxtuyen@yahoo.com

van der Kooij, Henk
Freudenthal Institute, University of Utrecht, The Netherlands
henk@fi.uu.nl

van Reeuwijk, Martin
Freudenthal Institute, University of Utrecht, The Netherlands
martinr@fi.uu.nl

Vincent, Jill L.
The University of Melbourne
Australia
jlvinc@unimelb.edu.au

Vincent, John T.
The University of Melbourne
Australia
jlv@alphalink.com.au

Warren, Elizabeth
Australian Catholic University, Queensland, Australia
e.warren@mcauley.acu.edu.au

Wijers, Monica
Freudenthal Institute, University of Utrecht, The Netherlands
m.wijers@fi.uu.nl

Williams, Anne
Queensland University of Technology, Australia
a_m_williams@bigpond.com.au

Williams, Gaye
University of Melbourne
Australia
g.williams@pgrad.unimelb.edu.au

Wood, Leigh N.
University of Technology, Sydney, Australia
leigh.wood@uts.edu.au

Yerushalmy, Michal
University of Haifa
Israel
michalyr@construct.haifa.ac.il

Zazkis, Rina
Simon Fraser University
Canada
zazkis@sfu.ca

Zbiek, Rose Mary
University of Iowa
USA
rose-zbiek@uiowa.edu

Index of Authors

Index

NEW ICMI STUDY SERIES

1.	M. Niss (ed.), Cases of Assessment in Mathematics Education. An ICMI Study. 1992 (The 6th ICMI Study)	0-7923-2089-1 HB
2.	M. Niss (ed.), Investigations into Assessment in Mathematics Education. An ICMI Study. 1992 (The 6th ICMI Study)	0-7923-2095-6 HB
3.	G. Hanna (ed.), Towards Gender Equity in Mathematics Education. An ICMI Study. 1996 (The 7th ICMI Study)	0-7923-3921-5 HB 0-7923-3922-3 PB
4.	A. Sierpinska and J. Kilpatrick (eds.), Mathematics Education as a Research Domain: A Search for Identity. An ICMI Study. 1998 (The 8th ICMI Study)	0-7923-4599-1 HB 0-7923-4600-9 PB
5.	C. Mammana and V. Villani (eds.), Perspectives on the Teaching of Geometry for the 21st Century. An ICMI Study. 1998 (The 9th ICMI Study)	0-7923-4990-3 HB 0-7923-4991-1 PB
6.	J. Fauvel and J. van Maanen (eds.), History in Mathematics Education. The ICMI Study. 2000 (The 10th ICMI Study)	0-7923-6399-X HB 1-4020-0942-9 PB
7.	D. Holton (ed.), The Teaching and Learning of Mathematics at University Level. An ICMI Study. 2001 (The 11th ICMI Study)	0-7923-7191-7 HB 1-4020-0072-3 PB
8.	K. Stacey, H. Chick and M. Kendal (eds.) The Future of the Teaching and Learning of Algebra: The 12th ICMI Study. 2004	1-4020-8130-8 HB

Information on the ICMI Study program and on the resulting publications, including Studies 1 to 5, can be obtained at www.mathunion.org/ICMI/ or by contacting the ICMI Secretary-General, whose email address is available on the ICMI website.

Printed in the United States
135202LV00002BA/5/A

9 781402 081309